ECONOMIC RISK IN
HYDROCARBON EXPLORATION

ECONOMIC RISK IN HYDROCARBON EXPLORATION

Ian Lerche
University of South Carolina
Department of Geological Sciences
Columbia, South Carolina

James A. MacKay
Texaco
E & P Technology Division
Houston, Texas

Academic Press

San Diego   London   Boston   New York   Sydney   Tokyo   Toronto

Cover photographs: Oil drill © 1993 Carlye Calvin. Dice © 1995 PhotoDisc Inc.

This book is printed on acid-free paper. ∞

Copyright © 1999 by ACADEMIC PRESS

All Rights Reserved.
No part of this publication may be reproduced or transmitted in any form or by any means, electronic or mechanical, including photocopy, recording, or any information storage and retrieval system, without permission in writing from the publisher.

Academic Press
a division of Harcourt Brace & Company
525 B Street, Suite 1900, San Diego, California 92101-4495, USA
http://www.apnet.com

Academic Press
24-28 Oval Road, London NW1 7DX, UK
http://www.hbuk.co.uk/ap/

Library of Congress Catalog Card Number: 98-86364

International Standard Book Number: 0-12-444165-3

PRINTED IN THE UNITED STATES OF AMERICA
98 99 00 01 02 03 EB 9 8 7 6 5 4 3 2 1

Contents

PREFACE XIII

1

Introduction

I. Overview 1
II. Geologic Uncertainty 3
III. Some Statistical Concerns 5
IV. Geotechnical Estimates 8
V. Economic Estimates 9
VI. Risk Estimates 10
VII. Practical Concerns 12
    A. Positive Aspects 12
    B. Negative Aspects 13

2

Risk-Adjusted Value and Working Interest

I. Introduction 17
II. General Methods of Constructing Working Interest and *RAV* Formulas 18
    A. *Cozzolino's Formula* 19
    B. *Hyperbolic Risk Aversion* 21

III. A Numerical Illustration 23
    A. Results from Cozzolino's Risk Aversion Formula 23
    B. Results for the Hyperbolic Risk Aversion Formula 26
    C. Comparison of Results 27
IV. Negative Expected Values 28
V. Conclusions 30

3

Uncertainty and Probability Estimates for Risk-Adjusted Values and Working Interest

I. Introduction 31
II. RAV Estimates with Uncertainties 33
    A. General Considerations 33
    B. Numerical Illustrations 33
III. Discussion and Conclusions 51

4

Portfolio Balancing and Risk-Adjusted Values under Constrained Budget Conditions

I. Introduction 55
II. Deterministic Portfolio Balancing 56
    A. Relative Importance 57
    B. Profitability 57
    C. Costs 57
    D. Budget Constraints 58
    E. Finding the Best Working Interest 59
III. Numerical Illustrations with Fixed Parameters: Comparison of Three Opportunities 64
    A. Budget of B = $20 MM 65
    B. Budget of B = $11 MM 65
    C. Budget of B = $4 MM 66
IV. Probabilistic Portfolio Balancing 66
V. Numerical Illustrations with Variable Parameters 68
    A. Variable Values 68
    B. Variable Values and Costs 69
    C. Variable Values, Costs, and Success Probabilities 71
    D. Very Low Budget 73
    E. Low to High Budget Comparison 75

VI. Comparison of Parabolic and Cozzolino *RAV* Results 78
   A. *Deterministic Results* 78
   B. *Statistical Results* 79
VII. Discussion and Conclusions 83
VIII. Appendix: Weighted *RAV* Optimization 85

5

SIMILARITY, DEPENDENCE, AND CORRELATION CONSIDERATIONS FOR RISK-ADJUSTED VALUES AND WORKING INTEREST

I. Introduction 87
II. General Arguments 89
   A. *Similar Opportunities* 90
   B. *Nearly Similar and Dissimilar Opportunities* 92
III. Correlated Behavior 93
   A. *Correlated Value and Cost for a Single Opportunity* 93
   B. *Geologically Correlated Opportunities* 99
IV. Numerical Illustrations 104
   A. *Correlation Effects for a Single Opportunity* 104
   B. *Correlation Effects between Opportunities* 105
V. Discussion and Conclusions 133
VI. Appendix: Cumulative Depth Considerations for a Single Opportunity 135

6

MODIFICATIONS TO RISK AVERSION IN HIGH GAIN SITUATIONS

I. Introduction 137
II. Modifications to the Cozzolino *RAV* Formula 140
   A. *Overview* 140
   B. *Practical Considerations* 141
   C. *"Crossover" from the Classical Cozzolino Formula* 143
   D. *Numerical Illustrations* 144
   E. *Weighting the Uncertainties* 154
   F. *Volatility-Weighted Estimates of Potential Opportunities* 156
   G. *Relative Importance of Each Category* 157
   H. *Other Weighting Measures* 157

III. Discussion and Conclusions   158
IV. Appendix A: Modifying Risk Formulas   159
   A. Modifications to the Cozzolino (Exponential Weighting) Formula   159
   B. An Algorithm for Determining V_{max} of Equation (6.4)   160
V. Appendix B: Modifications to the Parabolic Risk Aversion Formula   161

7

CORPORATE FUNDING REQUESTS, FIXED BUDGETS, AND COST BALANCING

I. Introduction   163
II. Specific Project Requests   164
III. General Project Requests   167
   A. Preference Indices   167
   B. Value per Dollar Spent ("Bang for the Buck")   168
   C. Analytic Example for Two Projects   170
   D. Numerical Examples for Two Projects   171
IV. Multiple Requests, Fixed Budgets, and Funding Strategies   172
   A. General Analysis   172
   B. Analytic Example   177
   C. Numerical Illustration   180
V. Appendix: Approximate Probability Behavior for $B \ll B_0$ 182

8

MAXIMIZING OIL FIELD PROFIT IN THE FACE OF UNCERTAINTY

I. Introduction   187
II. Nind's Formula for Present-Day Worth   188
   A. General Analysis   188
   B. Numerical Illustration   190
III. Probability and Relative Importance   193
IV. Numerical Example   194
   A. Probabilities and Ranges   194
   B. Sensitivity and Relative Importance   201
V. Discussion and Conclusions   208

9

THE VALUE OF ADDED INFORMATION: CATEGORIES OF WORTH

I. Introduction   211
II. Increased Expected Value and Decreased Uncertainty   213
   A. *Quantitative Analysis*   213
   B. *Numerical Illustration*   217
III. Constant or Decreased Expected Value and Decreased Uncertainty   220
   A. *Quantitative Analysis*   220
   B. *Numerical Illustration*   224
IV. Unanticipated Benefits of Data Acquisition   225

10

COUNTING SUCCESSES AND BIDDING STATISTICS ANALYSES

I. Introduction   227
II. Corporate Successes and Failures: Simpson's Paradox   228
III. Bid Analysis and Inferred Corporate Strategies for Lease Sale 147   232
   A. *Bid Distributions*   233
   B. *Summary*   241
IV. Block Statistics for Lease Sale 157 and Bid Ratios   242
   A. *Overview*   242
   B. *Bid Ratios*   242
   C. *Bid Distributions*   243
V. What Is a Bid Worth?   249
   A. *Bid Probabilities and Worth Uncertainties*   252
   B. *Numerical Illustrations*   264

11

ECONOMIC MODEL UNCERTAINTIES

I. Introduction   267
II. A Simple Production Model   268

III. A Simple Economic Model   270
   A. Revenue   270
   B. Production Costs   270
   C. Exploration Costs   271
   D. Development Costs   271
   E. Depreciation Recovery   272
   F. Net Cash Flow   272
IV. Combined Production and Economic Models   274
V. Minimum Average Selling Price   275
VI. Probable Profit Including Uncertainties   276
VII. Numerical Illustrations   278

12

BAYESIAN UPDATING OF AN OPPORTUNITY

I. Introduction   281
II. General Concepts of Bayesian Updating   282
III. Numerical Illustrations   283
   A. Testing for HIV   283
   B. Testing for Oil Fields from Bright Spot Observations   285
   C. Decision to Abandon an Opportunity   287
IV. Oil Pipeline Spills   294
   A. First Year Updating   296
   B. Second Year Updating   297
   C. Third Year Updating   297
   D. Fourth Year Updating   298
   E. Probability of a Spill   298

13

OPTIONS IN EXPLORATION RISK ANALYSIS

I. Introduction   301
II. An Example of Current Option Value and Future Decisions   303
III. Option Value Based on Available Information Only   306
   A. Maximal Gains Option, O_p 308
   B. Minimal Loss Option, O_n 310
   C. Numerical Illustrations for Maximal Gains Option   311
   D. Risk Tolerance Factors   314
   E. A Numerical Illustration   317

F. Effects of Variable Gains, Costs, Working Interest, and Success Probability   318
   G. Effects of Massive Changes in Gains, Costs, and Success Probability   325
   H. Options and Portfolio Balancing   327
   I. Portfolio Balancing with Common Working Interest   334
   J. Bids and Options   337
IV. Option Values Based on Future Models   342
   A. Two-Commodity Exchange Model   343
   B. A Numerical Illustration   344
V. Appendix A: Including Option Value in Costs   345
VI. Appendix B: Portfolio Balancing Using the Cozzolino Formula   347
VII. Appendix C: Portfolio Balancing with Conservative Optioning   348

EPILOGUE   351

APPENDIX
Numerical Methods and Spreadsheets   353

BIBLIOGRAPHY   393
INDEX   401

Preface

Exploration for hydrocarbons is a high-risk venture. On the one hand, the geologic concepts are uncertain with respect to structure, reservoir, seal, and hydrocarbon charge; on the other hand, the economic assessments of the potential profitability of a venture are uncertain with respect to total costs, probability of actually finding and producing an economically worthwhile reservoir, volume and type (oil/gas) of hydrocarbons, and the future selling price of the product. These intertwined uncertainties of the geologic and economic models make for high-risk decisions, with no guarantee of successfully striking hydrocarbons at any given drilling site.

The ability to provide quantitative measures of both geologic and economic risks is therefore paramount if the petroleum industry is to do a better job of making decisions that lead to profitable operations. In addition, companies cannot merely assess the overall scientific or economic risks; they must also be able to determine from a quantitative analysis which components are causing the largest uncertainties in the risk assessment. Once those are known, work can be prioritized in attempts to narrow the range of uncertainty by focusing on those components causing the uncertainty, without spending an inordinate amount of time and effort on less significant factors. This mode of operation is effective in terms of time and effort, and it is also a cost-effective measure for determining the worth of performing the required improvements.

Because of the high risk involved in petroleum exploration, individual corporations also need to evaluate the decisions associated with what fraction of their budget *should* be committed to a given exploration opportunity when many opportunities are available, and also what fraction of each opportunity *should* be taken. To couch such factors in terms of the tolerance to risk within a corporate framework, including the scientific and economic uncertainties both for single opportunities and for a portfolio of opportunities, is one of the major issues addressed in this book.

The motivation for addressing a large number of the individual topics developed here arose serendipitously in the middle of a short course we were teaching when the optimum working interest had to be evaluated for a prospect. Although a numerical code had been developed at Texaco to handle the problem, it seemed to both of us that there should be a way to solve the problem analytically, thereby minimizing calculational effort. Thus was started an apparently simple effort to do a small piece of work. Once the initial problem was solved, however, we wondered to what extent a large number of related problems could be reduced from requiring major computational efforts to either analytic formulas or even simpler spreadsheet calculations. Our goal was to speed calculation time in the hope that any reductions in complexity would lead to greater insight into the root causes for behaviors and patterns being computer produced.

The results presented in this volume represent our efforts to simplify problems of exploration risk assessment from a dominantly economic viewpoint. The corresponding scientific risk assessment has already been considered in some detail in a previous work (*Geologic Risk and Uncertainty in Oil Exploration,* Ian Lerche, Academic Press, 1997), and here we assume that the reader understands how scientific models are used to determine uncertainty, risk, and strategy in evaluating the potential resource assessment, its uncertainty, and the major contributors to that uncertainty. Throughout the present volume, however, we make use of some of the same or similar methods and do, perforce, evaluate some of the scientific uncertainties on the road to evaluating economic risk and associated uncertainties. In this way, it is not absolutely critical that the reader know all of the methods of all scientific risking in detail, but it is helpful to understand the basic precepts, which is why the brief descriptions here are of value.

The continued evolution of linked quantitative scientific and economic risks seems to be one of the major pathways by which hydrocarbon exploration risk can at the least be controlled, if not actively diminished, as petroleum corporations struggle to maximize gains and minimize costs and failures (dry holes). It also seems that the quantitative evaluation of such processes should help in making decisions about pursuing particular opportunities and in determining the likely costs associated with a particular exploration option.

In our opinion, about 80 to 90% of the volume should be readily comprehensible to an able geology graduate student who is involved in the arena of quantitative methods in the geosciences. Oil company scientific professionals and those whose job it is to make economic and decision analyses in a corporate setting should also find much of value. We will have succeeded in our task if others, more able than ourselves, can modify and develop the ideas presented here to raise them to a much higher level so that mysticism and personal prejudice can be replaced by objective realism, and so that economists and scientists can interact as an integrated team in exploration risk assessment.

In addition, in the Appendix, which deals with numerical procedures, we present some examples of spreadsheets and how they can be used. An electronic version of numerical procedures that can be downloaded by interested readers from the Internet is provided under the static AP Net web page at http://www.academicpress.com/pecs/download.

We are grateful to the industrial associates of the Basin Modeling Group at the University of South Carolina who have supported our efforts over the years: Unocal, Phillips Petroleum, Texaco, and Saga Petroleum. We are particularly grateful to Texaco for its assistance in permitting one of us (J.A.M.) to continue to develop the ideas expressed here. Many individuals have supplied ideas, data, supportive recommendations, and critical comments over the period of this endeavor. Most helpful have been Pete Rose (Telegraph Exploration), Elchin Bagirov, Mike Walls, Alex Kemp, John Harbaugh, Ellen-Sofie Kjensmo, Kenneth Petersen, Brett Mudford, Lee Hightower, and Tony Doré. We are also grateful to Donna Black, who, once again, transcribed illegible scrawl to rational English. Particular thanks go to the French Academy of Sciences for their award of a visiting professorship in geology, which enabled I.L. to spend a year in France, during which the bulk of this volume was written. Elf-Aquitaine (Production) was kind enough to permit I.L. to spend the year in France at their complex in Pau, and thanks are extended for all the courtesies afforded I.L. by Elf during this sojourn.

Texaco is thanked for publication permission.

As is not unusual, it seems that friends and families suffer the most as one labors over the months to bring a book to fruition. To all of you we extend our thanks.

Ian Lerche
Columbia, South Carolina

Jim A. MacKay
Houston, Texas

1

INTRODUCTION

I. OVERVIEW

Every project undertaken by a corporation has some element of risk, and such is particularly the case when one is involved in hydrocarbon exploration and/or bidding situations. There is always the chance of failure, with some total investment lost, versus the chance of success, with a gain that exceeds the total investment. When the estimated possible loss is high, to the point where a significant fraction of the capital asset value of the corporation would be at risk, most hydrocarbon exploration corporations then either (1) reject a project because it exceeds the total risk tolerance the corporation can sustain or (2) take less than 100% working interest in a project to limit potential losses to values below that which the corporation considers harmful to its fiscal health.

Evaluations of the chance of failure, of the total exploration costs, of the potential gains, and of the appropriate working interest to take are the main components that make up exploration risk analysis. Each of these factors is usually beholden, to a greater or lesser extent, to several components of intercoordinated development.

At some fundamental level, a *geologic concept* is customarily put forward, constrained by whatever data are available. This concept attempts to provide (1) estimates of the likelihood of a structural and/or stratigraphic trap being present and of the volume of hydrocarbons it might contain; (2) the likelihood of the presence of a reservoir of sufficient porosity and permeability within the trap area that hydrocarbons could be produced from the reservoir; (3) the likelihood of hydrocarbons being produced in the basin, of their migration to the reservoir, and of their being capable of charging the reservoir; and (4) the likelihood of a seal not only being present at the time of hydrocarbon charging but of persisting through to the present-day. The combination of geological, geochemical, and geophysical information, and of basin modeling capabilities, is used at this stage to attempt to estimate likely volumes of hydrocarbons in place at a particular site today.

Another area of development allows for uncertainties in the basin model assumptions, model parameters, and in the quality, quantity, and sampling distributions of the available data used as constraints. Consideration of these uncertainties provides probable ranges for the volumetric estimates of in-place hydrocarbons. This stage is usually referred to as *resource assessment*, and it yields cumulative probabilities of the volumetric distribution of potential hydrocarbons.

Most often the volumetric distribution values at 10, 50, and 90% cumulative probability are used as representative measures of the range of uncertainty. However, this range of uncertainty is no better than the influence of the uncertainties in model assumptions, the parameters, and in the data, all of which went into creating the resource assessment.

At the same time that the resource assessment is being made based on geological concept results, a scientific *risk assessment* analysis is being performed, also based on the geologic concept. This risk assessment procedure evaluates the probability of two factors: the probability of finding hydrocarbons in the reservoir and the probability of being able to produce any hydrocarbons through evaluation of permeability and stratigraphic variations in the reservoir, and the types of hydrocarbons that are estimated to occupy the reservoir—heavy/light oil, gas, condensate—as well as estimates of each fraction likely to be in the reservoir. Naturally, each of these factors also has its own range of uncertainty, which has to be evaluated based on both available data and on models.

Up to this point, the exploration risk assessment is purely scientific with no economic factors incorporated, and it provides some idea of the probability of finding hydrocarbons, their amount and type, their producibility, and also ranges of uncertainty for each of the quantities.

Financial aspects of risk are now added to the scientific risk aspects in three stages. First an *engineering design* study is usually done so that some estimates are available of production rate and its profile with time, of the timing of production in relation to exploration and reservoir delineation studies, to recovery efficiency, to facilities required, and to total operating cost. Because this pre-project engineering design study is done ahead of drilling, all of the factors involved in that study are also uncertain and, once again, the uncertainties need to be quantified.

Once the engineering design phase is done an *economic analysis* is normally performed based on anticipated contract terms, cash flow, economic measures of worth, anticipated inflation and escalation factors, and the expected internal corporate discount rate and product prices. From a combination of the risk assessment study, the engineering design study, and the economic analysis, one then has an idea of the likely total costs of the project, the likely total gains, and the probability of successfully finding hydrocarbons, together with uncertainty estimates for each.

Decision analysis then encompasses the study of problems such as the optimum working interest (i.e., fraction) of the project that a corporation should take; the value of improving knowlege about the project by paying for more information (e.g., 3D seismic) prior to deciding to make more significant investments; the cost of the project in relation to the risk tolerance of the corporation; the worth of the project in relation to a portfolio of other opportunities with a fixed corporate budget; the sensitivity of the project outcome in relation to the uncertainties; and strategies for involvement in, or later disengagement from, a project.

If the decision is made not to drill, then the total pre-drill costs to the corporation have to be absorbed by the corporation as part of its business costs and are usually, but not always, included in the corporate cost of capital or discount factor. If the decision is made to drill, then two results are possible: Either the drilling is successful with economically attractive hydrocarbons being found, or the drilling is unsuccessful with either a completed evaluation that indicates subcommercial hydrocarbons were found or a true dry-hole (no hydrocarbons) or, worse, an incomplete evaluation results, which does not resolve the uncertainty at all (a junked hole). A post-drill review is then performed to see if estimates of parameters prior to drilling matched those found when drilling was successful and, if not, the reasons why not. If the drilling is a failure economically, the review is performed to attempt to uncover the reasons for failure and to learn from that failure. In either event, the actual drilling information is then used to update the geologic concept, the reserve assessment, and the scientific risk assessment, thereby providing a Bayesian update using the value of the information to aid in further decisions associated with this project or future projects.

Figure 1.1 provides a flow diagram indicating how these various components fit together.

II. GEOLOGIC UNCERTAINTY

At the early stages of an exploration project when data are limited (most often to surface geology, gravity, aeromagnetic and seismic surveys), even the use of quantitative basin analysis models is fraught with difficulties in providing accurate measures of geologic confidence. Most situations analyzed end up yielding one of about five subjective confidence statements along the broad lines of "virtually certain," "reasonably confident," "some uncertainty," "probably not too good a chance," and "really pretty poor

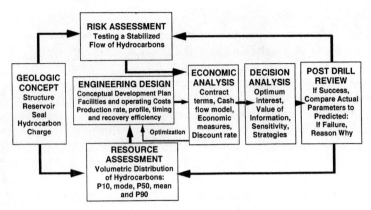

FIGURE 1.1   Flow diagram indicating the various components that go into an exploration risk analysis assessment and some of the feedback links for updating information. (From R. M. Otis and N. Schneidermann, © 1997, reprinted by permission of the American Association of Petroleum Geologists.)

geologically." Such subjective measures have to be quantified if they are to be used at the resource assessment or risk assessment stages. An empirical range of conversions is provided in Figure 1.2 from such subjective expressions of confidence to more appropriate objective probability ranges.

In terms of a rapid, but inaccurate, first estimate of the probability of finding hydrocarbons, such geologic expressions are often used for the probabilities of existence of a structure, S, of a reservoir, R, of the trap being sealed, T, and of a hydrocarbon charge being present, HC, which,

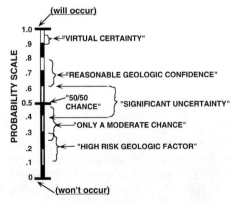

FIGURE 1.2   One scale for converting qualitative geological chance statements to approximate quantitative ranges of probability. (Modified and used with permission of P. R. Rose.)

	PROBABILITY
RESERVOIR	0.5
	x
STRUCTURE	0.8
	x
HC-CHARGE	0.9
	x
SEALED TRAP	0.7
"CHANCE OF SUCCESS"	**= 0.25**

FIGURE 1.3 Sketch indicating how geologic chance factors are multiplied to produce a success probability for finding hydrocarbons. (Modified and used with permission of P. R. Rose.)

when all multiplied together give the finding probability. Figure 1.3 gives an example of this rough-and-ready approach. Notice that "success" is meant in a geologic sense rather than an economic sense because there is, as yet, no estimate provided of the economic worth of the potential hydrocarbons. Also note that no range of uncertainty has so far been given to each of the probability estimates in Figure 1.3, so that one does not yet know the range of uncertainty on the estimated success probability.

III. SOME STATISTICAL CONCERNS

Part of the problem in estimating success uncertainties has to do with the types of distribution from which are drawn each of the uncertainties; part of the problem has to do with the manner in which mean values are calculated; and part of the problem has to do with the functional way in which a quantity of interest is calculated (multiplicatively in the present simple illustration for success probability).

The underlying distribution types are often not well known for an exploration project so it is not unusual to see different model distributions used; five of the more common distributions are triangular, normal (bell-shaped or Gaussian), log-normal, uniform, and exponential as sketched in Figure 1.4. Care must be taken to describe which distribution is being used so that one ascertains the effect on output uncertainties of each input uncertainty (a practical example of this influence on uncertainty is provided in a later chapter).

At the same time it is important to be aware of the way in which mean values are calculated; usually three different estimates are considered: arithmetic, geometric, and harmonic. For N values $x_1, x_2 \ldots, x_N$, the arithmetic mean is given by

$$\langle x \rangle_{am} = N^{-1} \sum_{i=1}^{N} x_j;$$

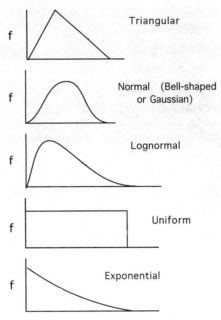

FIGURE 1.4 Some different types of probability distributions that are used to evaluate uncertainties. (Used with permission of E. Capen and P. R. Rose.)

the geometric mean by

$$\langle x \rangle_{\text{gm}} = \left(\prod_{i=1}^{N} x_i \right)^{1/N};$$

and the harmonic mean by

$$\langle x \rangle_{\text{hm}} = N \left(\sum_{i=1}^{N} 1/x_i \right)^{-1}.$$

These three estimates of the mean differ, sometimes substantially. For instance, a trip at 60 mph and a return trip at 30 mph have an average speed of 45 mph (arithmetic), 42.43 mph (geometric), and 40 mph (harmonic), for a total dynamic range of difference of 5 mph (about 12% variation). Clearly, the mean value has an uncertainty depending on the functional method used for its calculation, which is over and above the statistical uncertainty arising from the choice of the type of distribution.

In a large number of problems it happens that either sums of variables or products of variables arise. In such situations two common mathematical formalisms are often employed: (1) the sum of independent, normally distributed, random variables is itself normally distributed; (2) the product of independent, log-normally distributed, random variables is itself log-normally distributed. Thus, in such cases, the underlying distribution types are often taken as normally or log-normally distributed to ease calculations.

It is also an empirical fact that even if the underlying distributions are not normally or log-normally distributed, the sums and products are often very closely approximated by normal and log-normal distributions, respectively, often to a much closer degree than geologic parameter ranges can be constrained, making the empirical approximation of great practical value. For a log-normal distribution (closely approximating oil and gas field statistics) the main characteristics are (1) that there are large numbers of volumetrically small fields and a few very large fields, but the volume of the large fields far exceeds the combined volumes of the small fields; and (2) on log-normal probability paper the cumulative distribution is represented by a straight line. For other types of distribution, the cumulative distribution is often accurately approximated by a straight line between about 5 to 95% cumulative probability on log-normal probability paper. For instance, as shown in Figure 1.5, the observed cumulative gas production distribution per reservoir for the WX-4 sand of the Wilcox Formation is close to log-

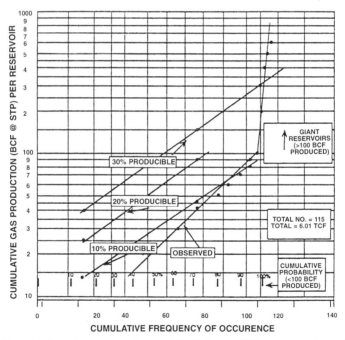

FIGURE 1.5 Cumulative gas production (BCF) through 1989 per reservoir as a function of cumulative frequency of occurrence, based on 115 reservoirs. Note the "break" at around 100 BCF into "giant" and "small" reservoirs, each of which is separately log-normally distributed. The three lines labeled 30%, 20%, and 10% producible, respectively, are the assessed producibility curves for the small fields only.

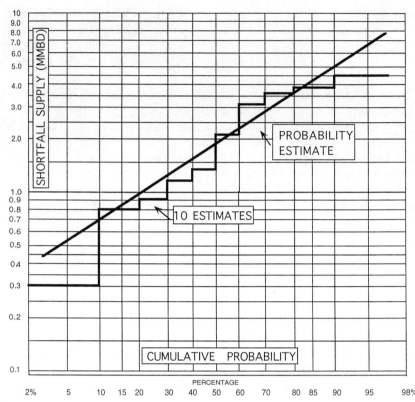

FIGURE 1.6 Cumulative distribution of Stanford Energy Forum 1991 estimates of U.S. shortfall (bbl/day) by year 2000.

normal, while Figure 1.6 shows the cumulative distribution of estimates made in 1991 at the Stanford Energy Forum for U.S. shortfall (bbl/day) by the year 2000, again showing the approximately log-normal nature of the cumulative distribution.

IV. GEOTECHNICAL ESTIMATES

Part of the determination of the resource assessment and scientific risk analysis components of a complete exploration risk analysis is predicated on the ability to provide technical estimates on (1) dimensional aspects of a potential hydrocarbon target, such as thickness, area, volume, depth, tilt, closure; (2) reservoir characteristics, such as porosity, permeability, gas-to-oil ratio (GOR), hydrocarbon-recovery efficiency; (3) geochemical factors, such as source rock richness and depth, maturity, total organic carbon (TOC), and kerogen types; (4) migration factors of hydrocarbons to the potential reservoir, such as transmission effectiveness (migration loss), the paths taken, dispersal, solution versus free-phase transport; (5) trap integrity with time, such as leakage, recharge, flushing, tilting; (6) the relative and absolute timing of hydrocarbon migration versus structural and stratigraphic dynamic events both in the reservoir and in the migration routes; (7) and other factors, such as biodegradation and percent inert gases. These components are then used to provide the geologic chance factors and resource uncertainty. But it is clear that each and every model used to assess these components must be uncertain, else one would be able to give 100% accuracy, which, given the rank wildcat worldwide exploration drilling success average of around 1 well in 10, is not the case.

It is important that one be willing to demonstrate how uncertain each geotechnical prediction is and what the probability is that the model is incorrect. One must also provide some conversion of descriptive knowledge to a quantitative numerical description and attempt to separate preconceived ideas of what one would like to have occur from the objective reality of what could occur, even with the limited data one has to constrain the models used.

Geotechnical estimates also play a role in the engineering design phase because estimates of depth to the potential reservoir, of the stratigraphic sequences present, and of overpressure all indicate the likely drilling conditions to be encountered, while reservoir area estimates provide an idea of the acreage that needs to be developed, and reservoir characterization estimates are used to yield an indication of the completion conditions. Thus pre-project engineering design costs are also influenced by the quantitative estimates of conditions anticipated to be encountered while drilling from the geologic concept.

V. ECONOMIC ESTIMATES

In addition to the engineering design cost, estimates of several other major costs have to be evaluated: the cost to acquire a lease, the cost of inflation, the cost of unanticipated events (e.g., a stuck drill-string), the cost of capital, the costs of royalties and taxes, etc. Once all such costs are modeled—and each corporation uses variously diverse models each of which is again subject to model assumptions, model parameters and limited data constraints—one then has some idea of the costs of the exploration opportunity together with their uncertainty.

On the other hand, there is the problem of estimating potential gains from the project. Here the problem is dominated by attempts to estimate the future well-head price of product not only at the time a project comes on-line (which can be up to a decade or more after an initial exploration investment) but also throughout the life of the field. The difficulty is apparent: In order to provide a cash flow estimate on a yearly basis, from which one then estimates total lifetime gains, a rate of production model is required for product. This model must include the type of product (gas, oil, and condensate), the primary, secondary, and tertiary enhanced oil recovery (EOR) estimates, and when they occur, together with estimates for field size, possible satellite fields, and so on. And the future selling price of product is a major unknown in such estimates.

In short, economic models suffer from the same drawbacks as scientific models, being dependent on model assumptions, model parameters, and lack of data. The anticipated gains, if success occurs, are then also uncertain. And such uncertainty must also be allowed for when assessing a project for economic worth.

VI. RISK ESTIMATES

On the supposition that all success probability estimates have been made, that all cost and gain estimates have been made and reduced to fixed-year money (thereby minimizing inflation factors), that no further information is available, and that all uncertainties for all factors have been properly evaluated quantitatively, it is then necessary to evaluate the risk to the corporation of becoming involved in a project. The point here is that the potential for a project *not* to be successful will involve the corporation in financial loss if that eventuality occurs. If the loss is large, it could cause substantial fiscal damage to the corporation, not excluding bankruptcy. Accordingly, most corporations limit their potential for catastrophic loss by setting a risk tolerance amount. This amount is used to limit potential losses by placing more value on taking a fraction smaller than 100% working interest in a project; the riskier the project the lower the fraction (i.e., the higher the cost relative to potential gains and the lower the estimated success probability so the higher is the risk of a loss). Corporations are highly risk averse when the risk tolerance amount is set low and approach risk neutral when it is high. There are also dangers associated with using unrealistic values of parameters in the exploration risk analysis process, such as setting an unrealistically high discount factor, being too pessimistic on well-head price forecasting, overestimating taxes or drilling costs, and arranging for unnecessary further seismic programs. Such factors can both increase expected costs and lower expected gains, and so can drive the worth of a project so low that it falls into the uneconomic category. Corporations that are overly risk averse usually find that they tend to focus on only small reserve ventures, that they ignore the leverage factor of numerous projects, that they automatically opt out of taking any working interest in any unorthodox high gain, but high risk, potential projects, and tend to be so excessively cautious that they cannot move effectively to take a position for a potential opportunity within a time-limited framework.

But risking a single project is not usually done in isolation. Two other conditions are customarily in force: a suite of projects has to be evaluated simultaneously, and ranked in some way, to maximize the total potential gains to the corporation relative to total costs and to costs per project; and most corporations have a fixed budget cycle so that either a portfolio of potential opportunities is risk-evaluated relative to a fixed budget or, as the budget is steadily being spent, each project is risk-evaluated relative to the residual budget. Such portfolio balancing considerations form part of the elements of a complete risk analysis of a single project. Figure 1.7

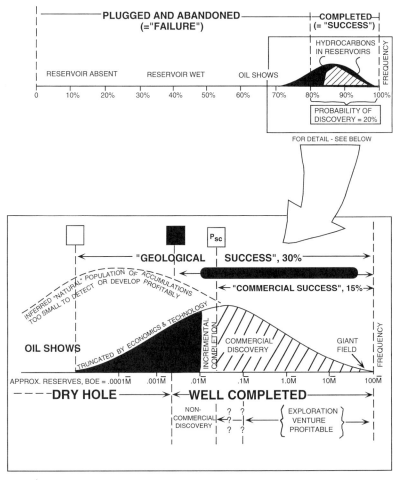

FIGURE 1.7 Sketch illustrating how the interaction of geological chance factors and economic risk together limits the ability to be commercially successful in producing hydrocarbons. (Modified and used with permission of P. R. Rose.)

sketches how the interaction of geologic and economic concerns influences the commercial viability of a project.

Even when the corporate risk tolerance is low it is still possible to take some working interest in a risky project. Clearly, zero working interest is the lower limit and 100% working interest is the high limit. Somewhere between these two extremes there must be an optimum working interest to take which depends on both the parameters of the project (success probability, estimated gains, and estimated costs) and on the corporate risk tolerance. This problem is addressed in the next chapter.

VII. PRACTICAL CONCERNS

A. POSITIVE ASPECTS

A hydrocarbon exploration corporation has to have as its principal priority a concern for its owners, the shareholders. Thus the motivation of staff and management has to be to increase corporate assets and to provide dividends.

Within that quintessential framework the allocation of capital for exploration opportunities should be based on objective, reproducible determinations of the merits of each project. In addition, the *same* quantitative risk analysis procedure should be used on a corporate-wide basis to risk and rank projects, else one cannot draw a fair comparison between the worth of different projects.

It is also manifest that business decisions be made based on a good understanding of the scientific risk analysis, the economic analysis, and the uncertainties inherent in both sets of model estimates. Failure to do so will lead to personal bias overriding the objective measures in place, thus making the business of exploration risk analysis more one of subjective prejudice than one of rational objectivity.

In addition, ongoing review and up-dating of the tools and methods used in both scientific and economic areas, and the continual updating of the worth of prospects and their risks and ranking based on later-acquired information (i.e., a Bayesian updating), are absolutely required if one is to maintain a dynamic equilibrium that shuffles and reevaluates projects to maximize total likely gains to a corporation in relation to likely costs, in light of ever-increasing amounts of information.

To encourage the corporate-wide application of quantitative exploration risk analysis, it is almost an absolute requirement that there be education and training of staff in integrated, multidisciplinary groups. There is little point in keeping, say, the economists and planners isolated from the geoscientists because then neither has an appreciation or understanding for the work of the other group. There is also little point in constructing or using highly sophisticated deterministic models of economic or geoscientific processes in light of the uncertainties, nor is there much point in constructing incredibly detailed distributions of parameter values; by and large, neither the models nor the available data warrant such minutae. Instead one should attempt to be constrained by the level of uncertainty and so deal with the broad trends provided by cumulative probability behaviors, which are usually more than adequate at the exploration stage, as is demonstrated later.

It usually transpires that within a multidisciplinary group there are a few individuals with a flair for helping other staff implement the total exploration risk analysis package. Such experts should, of course, be encouraged to provide assistance and training to both operational groups and to senior management.

A corollary to having multidisciplinary groups in place is that *constructive* reviews of emerging projects can be done, so that as little as possible is overlooked, with the advantage of delineating more sharply the uncertainties and, possibly, of narrowing the ranges of uncertainty. A second advantage is that economic and geotechnical estimates are both less likely to be biased by individual preference when they are subjected to the scrutiny of one or more multidisciplinary groups. And one can always put into effect a penalty/reward system that penalizes consistent production of biased estimates but rewards consistent production of unbiased estimates.

Perhaps four of the more definitive aspects of improving an exploration risk analysis procedure are these:

1. To insist upon objective appraisal of comparisons between predictions versus results, both scientific and economic, so that one learns from the mismatches where the weak areas are in the process and figures out how to improve them
2. To carry out thorough, nonconfrontational, post-project reviews of both successful wells *and* unsuccessful wells with the findings documented, so that others can learn from the results of the "post-mortems"
3. To ensure that responsibility *and* accountability occur at all levels of the corporation, so that personal bias is minimized in favor of a rational approach
4. To make sure that the geoscientific personnel understand the economic aspects of a project and, more importantly, to mandate that economic and planning personnel understand the geoscientific aspects!

B. NEGATIVE ASPECTS

As is not unusual in a multidisciplinary hydrocarbon exploration company, factors exist that can easily lead to the failure of risk analysis procedures. Thus the inability of management to keep abreast of developments in quantitative risk analysis procedures for both the scientific and economic aspects is almost sure to cause problems with implementation of a corporate-wide risk analysis program. Equally, insistence on staff using risk analysis methods without the corresponding proper training and education merely courts disaster.

Perhaps of similar, or even greater, concern are the difficulties attendant on different risk analysis procedures being used by different parts of a corporation, so that no fair and objective comparison is possible between different projects. Another concern of comparable negative impact is the ultraconservative decision to order yet another seismic survey or yet more slim-line drillings or yet more gravity measurements, or allied collections of information at some cost, when it is already clear that the project economic uncertainty cannot, or will not, be resolved by such information. In such cases, the decision maker has driven up project costs just to ensure that all bases are covered without any hope of narrowing uncertainty.

On the risk analysis computer program side, one must always guard against the danger that just because someone has available a program, it will be used to make the risk analysis component more important than the prospect generation component. There is also the difficulty that simple-to-use risk analysis software will be mistaken for competent software. Pretty inputs and pretty outputs without underlying good science and economics are merely pretty. The same argument is valid if one treats software for handling risk analysis procedures as black boxes. Lack of understanding of what a program does is no excuse for poor or incorrect decisions. And one should be especially suspect of claims that particular risk analysis software can solve all problems. There are just too many considerations for any one piece of software to have included them all.

A problem that arises quite often is the inability of individuals, both on the geoscientific end *and* on the economic end, to adjust estimated parameter values and ranges based on better model capabilities, more data, or historical precedent. All too often a rigid, cast-in-stone attitude is taken that defeats the whole purpose of risking prospects.

One further negative aspect, and one which triggered a good amount of the work reported in the body of this text, is the problem of computer turnaround time. Any risk analysis program that takes many hours to run through once is not likely to be used because it limits the ability to run through many scenarios to determine ranges of probable outcomes. This sort of time problem occurs not only at the economics end of the game but also with the scientific portion.

We provide an appendix at the end of the volume with some spreadsheet examples worked through in detail, and with the relevant spreadsheet numerics spelled out. More detailed spreadsheet information in electronic form is available by downloading from the Internet under APNet web page http://www.academicpress.com/pecs/download.

It is hoped that such procedures will enable any competent geoscientist to evaluate simply the relevant economic risk analysis components of their work without needing to be beholden to others, who may appreciate the scientific work less, for the corresponding exploration risk analysis.

2

RISK-ADJUSTED VALUE AND WORKING INTEREST

I. INTRODUCTION

In the case of the oil industry, individual corporations tend to estimate the expected value of a project and then multiply the result by a risk factor to assess the prioritized worth versus risk to the corporation.

Nearly twenty years ago Cozzolino (1977a, 1978) showed how to allow for risk tolerance factors in estimating a risk-adjusted value for a project. Cozzolino (1977b, 1978) used an exponential model of fractional working interest in a project that should be followed in order to ascertain the maximum likely return on a project subject to the constraints of risk tolerance avoidance.

The requirements of such a Cozzolino-type model force an exponential risk aversion factor to be employed by any corporation wishing to use the procedure. However, not all corporations choose exponential risk aversion; some have a hyperbolic tangent type of risk weighting, others have empirical models based on prior evaluations of projects and the anticipated value to the corporate assets, and so on.

It would be useful if a method could be designed along the lines of Cozzolino's basic precepts, but that allows arbitrary functional forms (as employed by different corporations) to be used to assess the relevant working interest (W) and risk-adjusted value (RAV) of a project. In this way one could then determine quickly the sensitivity, precision, uniqueness, and resolution of W and RAV not only to variations in expected gains and/ or losses in relation to chances of success/failure and to variations in risk tolerance (RT), but also to the functional form of risk aversion formula being used to make the estimate. This ability would then go a long way toward allowing evaluation of projects under a variety of settings.

The purposes of this chapter are to show precisely how such a general methodology can be set up and to illustrate the practical use of such a framework by direct example.

II. GENERAL METHODS OF CONSTRUCTING WORKING INTEREST AND RAV FORMULAS

Cozzolino (1977a, 1978) constructed his RAV formula based on utility theory applied to a chance node decision tree diagram, such as that given in Figure 2.1a. The notation of Figure 2.1a is as follows: V = value (positive); p_s = probability of success; C = expected cost of nonsuccess (positive); $p_f (\equiv 1 - p_s)$ = probability of the project not succeeding. At the chance node point of Figure 2.1a, the expected value, or weighted average of the two possible outcomes of the project, is

$$E_1 = p_s V - p_f C, \qquad (2.1)$$

which is positive provided $p_s V > p_f C$. The second moment of the project value is $E_2 = p_s V^2 + p_f C^2$, so that a measure of the uncertainty in the outcome is provided by the variance

$$\sigma^2 = E_2 - E_1^2 \equiv (V + C)^2 (p_s p_f). \qquad (2.2)$$

A measure of risk is often assigned by the volatility, ν, defined by

$$\nu = \sigma/E_1 \equiv (V + C)(p_s p_f)^{1/2}/(p_s V - p_f C), \qquad (2.3)$$

which evaluates the stability of the estimated mean value, E_1, relative to the fluctuations about the mean. A small volatility ($\nu \ll 1$) implies that there is little uncertainty in the expected value, while a large volatility ($\nu \gg 1$) implies a considerable uncertainty in the expected value.

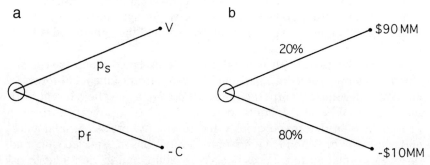

FIGURE 2.1 (a) Sketch of the chance node diagram indicating a value of V with a probability of p_s, together with a loss of $-C$ at a probability of p_f ($\equiv 1 - p_s$). (b) As for part (a), but with V = \$90 MM, p_s = 20%, p_f = 80%, C = \$10 MM, as used in the numerical illustration in text.

While the expected value, E_1, of a project may be high, and the volatility small, nevertheless it can be the case that if failure does occur then the total project costs, C, may be so large as to bankrupt, or cause serious financial damage to, the corporation. Under such conditions it makes corporate sense to take less than 100% working interest in the project. A smaller fraction of the project will cut potential gains but will also cut catastrophic potential losses. Thus, with a working interest fraction, W, the expected value to the corporation at the chance node is

$$E_1(W) = p_s(WV) - p_f(WC) \tag{2.4}$$

on the assumption that fractional working interest does not change the probabilities of success or failure of the project.

The effective corporate value is reduced from V to WV, while the potential losses are reduced from C to WC.

A. COZZOLINO'S FORMULA

Cozzolino's (1977b, 1978) determination of risk-adjusted value (RAV), in relation to the risk tolerance ($\equiv$ risk threshold, RT) of a corporation can be given simply using an analogy from geochemistry.

Imagine two energy states existing with activation energies $E_1(\equiv WV)$ and $E_2(\equiv -WC)$. At a given temperature, T, the rate at which a compound can be lost by decay to the state E_1 is given by the Arrhenius formula:

$$p_s \exp(-E_1/RT), \tag{2.5a}$$

where R is the gas constant and p_s is the probability of the reaction pathway being along the path determined by energy state E_1. Decay along the path of E_2 is then proportional to

$$p_f \exp(-E_2/RT), \tag{2.5b}$$

and, because only two paths exist, $p_s + p_f = 1$.

Thus the total decay is then proportional to

$$p_s \exp(-E_1/RT) + p_f \exp(-E_2/RT). \tag{2.5c}$$

If one were to represent the total decay of Eq. (2.5c) by an equivalent activation energy, RAV, through a single equivalent pathway, then the equivalence demands of an Arrhenius formula are

$$\exp(-RAV/RT) = p_s \exp(-E_1/RT) + p_f \exp(-E_2/RT). \tag{2.6}$$

Hence,

$$RAV = -RT \ln\{p_s \exp(-WV/RT) + p_f \exp(WC/RT)\}. \tag{2.7}$$

The intrinsic assumption here is that the activation energy rates of conversion are controlled by the exponential Arrhenius formula. With RT understood as risk tolerance, Eq. (2.7) is Cozzolino's formula relating estimated risk aversion value, RAV, to risk tolerance, RT, working interest, W, and the value, V, cost, C, and probabilities of success/failure, with $p_s + p_f = 1$.

1. Maximum Working Interest

Note that for a given value of RT, Eq. (2.7) has a maximum value of RAV with respect to working interest, W, when W takes on the value

$$W_{max} = \frac{RT}{(C+V)} \ln\left(\frac{p_s V}{p_f C}\right) > 0 \text{ if } E_1 > 0, \quad (2.8)$$

and at this value of W_{max}, the maximum value of RAV is

$$RAV(max) = -RT \ln\left\{p_s\left(\frac{p_f C}{p_s V}\right)^{V/(V+C)} + p_f\left(\frac{p_s V}{p_f C}\right)^{C/(C+V)}\right\} \quad (2.9)$$
$$\equiv VW_{max} - RT \ln\{p_s(1 + V/C)\}.$$

Note that the requirement $0 \le W \le 1$ (no less than 0% or more than 100% working interest can be taken in a project) then implies, from Eq. (2.8), that

$$p_f C < p_s V < p_f C \exp\{(C+V)/RT\}. \quad (2.10)$$

If inequality (2.10) is *not* satisfied then RAV does not have a maximum in the range $0 \le W \le 1$, so that RAV is then either monotonically increasing or decreasing as W increases from zero to unity.

As W tends to zero then, from Eq. (2.7), one has $RAV(W = 0) = 0$ and

$$\left.\frac{dRAV}{dW}\right|_{W=0} = p_s V - p_f C > 0. \quad (2.11)$$

Thus RAV is positive increasing at small W provided $p_s V > p_f C$, that is, the expected value at the chance node of Figure 2.1a is positive. In such a case, because there is no maximum in the range $0 \le W \le 1$, the largest positive value of RAV occurs at the maximum $W = 1$, indicating that 100% interest in the project should be taken. Equally, if $p_s V < p_f C$, then RAV is negative decreasing throughout $0 \le W \le 1$, indicating that the project should not be invested in at all.

When inequality (2.10) is satisfied, then there is a range of values of W around W_{max} where RAV can be positive, so that some positive risk-adjusted return is likely even if a working interest is taken other than that which maximizes the RAV.

2. Apparent Risk Tolerance

By rearrangement, Eq. (2.8) can be used in a different manner in the form

$$RT = W_{max}(C+V)/\ln(p_s V/p_f C). \quad (2.12)$$

If an arbitrary working interest value, W, is used in Eq. (2.12) to replace the optimum value $W_{\max}$, the apparent risk tolerance, RT_A, is then given by the left side of Eq. (2.12). This apparent risk tolerance expresses the ability to see to what extent a corporate mandate of risk tolerance has over-risked or under-risked a particular project, or the extent to which a particular working interest choice permits the apparent risk tolerance to be in reasonable accord with the corporate-mandated value. It is particularly useful in determining what the corporate attitude toward risk tolerance is based on prior working interest decisions.

3. "Break-Even" Working Interest

The "break-even" value of RAV is conventionally set to zero, which, for the Cozzolino formula (2.7), occurs at a working interest W_0, determined from

$$p_f \exp\{W_0(C + V)/RT\} + p_s - \exp(W_0 V/RT) = 0, \quad (2.13a)$$

provided $0 \leq W_0 \leq 1$.

For $V \gg C$, Eq. (2.13a) has the approximate solution

$$W_0 \cong 2RTV^{-1} \ln\{1 + (3Cp_f)/Vp_s\}, \quad (2.13b)$$

so that $W_0 < 1$ in $RT \leq \frac{1}{2} V [\ln\{1 + (3Cp_f/Vp_s)\}]^{-1}$.

4. Maximum Risk Tolerance

Occasionally, a corporation requests that a particular fixed working interest be taken. The question then is "How does the RAV relate to the risk tolerance?" In this situation RAV has a maximum, $RAV(\max)$, with respect to RT at a value of RT_m given through

$$RT_m = \frac{W\{-p_s V \exp(-WV/RT_m) + p_f C \exp(WC/RT_m)\}}{\ln\{p_s \exp(-WV/RT_m) + p_f \exp(WC/RT_m)\}} \quad (2.14)$$

with

$$RAV(\max) = W\{p_s V \exp(-WV/RT_m) - p_f C \exp(WC/RT_m)\}. \quad (2.15)$$

The nonlinearity of Eq. (2.14) with respect to RT_m precludes an analytic expression being available expressing RT_m in terms of W, V, C, p_s, and p_f, but simple hand-calculator values for RAV versus RT (for fixed values of the remaining parameters) can be used to estimate quickly whether the risk tolerance for a required working interest is less than the corporation limit. An example of this point is given later.

B. HYPERBOLIC RISK AVERSION

Not all corporations model risk attitudes with a risk aversion that is exponentially weighted; an exponential weight corresponds to a risk aversion factor, $RAF(V;RT)$, proportional to $\exp(-E_1/RT)$. Alternative formulas for risk aversion are as many and as varied as the individual corporations involved. Here the hyperbolic rule is investigated corresponding to a risk aversion factor proportional to $1 - \tanh(E_1/RT)$. The idea behind a hyperbolic rule is that there is greater stability in the management of high-loss scenarios than there is with the exponential rule.

In this case, the equivalent to the Cozzolino formula is to suppose that each state is weighted with the hyperbolic tangent so that one would write a total weighting as

$$p_s\{1 - \tanh(WV/RT)\} + p_f\{1 - \tanh(-WC/RT)\}$$
$$\equiv p_s + p_f + p_f \tanh(WC/RT) - p_s \tanh(WV/RT) \quad (2.16)$$
$$= 1 + p_f \tanh(WC/RT) - p_s \tanh(WV/RT).$$

If the total weighting, Eq. (2.16), is again taken to correspond to an exponential equivalent in the form

$$\exp(-RAV/RT) \equiv 1 + p_f \tanh(WC/RT) - p_s \tanh(WV/RT),$$

then the risk-adjusted value, RAV, is given through

$$RAV = -RT \ln\{1 + p_f \tanh(WC/RT) - p_s \tanh(WV/RT)\}, \quad (2.17)$$

which is equivalent to Cozzolino's formula, but for hyperbolic tangent weighting of risk aversion rather than exponential.

For any functional form it is then clear how to proceed. For instance, with a weighting proportional to $1 - G(E_1/RT)$, where G is an arbitrary, but specified function, then one weights each state in the same manner to obtain the general equivalent formula

$$RAV = -RT \ln\{1 - p_f G(-WC/RT) - p_s G(WV/RT)\}. \quad (2.18)$$

Thus all risk aversion factors can be brought, by a general reduction, to corresponding risk aversion values related to the risk tolerance, RT, and the working interest, W.

For the hyperbolic RAV formula, Eq. (2.17), a maximum of RAV, RAV (max), with respect to working interest exists at $W = W_m$, where W_m is given through

$$\cosh(W_m C/RT) = (p_f C/p_s V)^{1/2} \cosh(W_m V/RT) \quad (2.19a)$$

with approximate solution

$$W_m \approx \frac{1}{2}(RT/V) \ln(4p_s V/p_f C), \quad (2.19b)$$

provided $0 \leq W_m \leq 1$, while the break-even value, $RAV = 0$, occurs at $W = W_0$ given by

$$p_f \tanh(W_0 C/RT) = p_s \tanh(W_0 V/RT) \quad (2.20a)$$

with approximate solution

$$W_0 \cong \frac{RT}{C} \ln \frac{p_s}{p_f}, \qquad (2.20b)$$

provided $0 \leq W_0 \leq 1$, that is, provided $RT \leq C \ln(p_f/p_s)$.

Similar arguments for fixed working interest, but variable values of risk tolerance, can be gone through as for the Cozzolino formula, but it is easier to visualize patterns of behavior with a numerical example.

III. A NUMERICAL ILLUSTRATION

To illustrate the similarities and differences between the Cozzolino formula, Eq. (2.7), and the hyperbolic formula, Eq. (2.17), for risk aversion values consider the situation of Figure 2.1b in which the probability of success is 20% ($p_s = 0.2$) and so the probability of failure is 80%, in which the value of the success path is $V = \$90$ MM; while the cost of failure is $C = \$10$ MM. It is requested that the best working interest in the project be evaluated for risk tolerances ranging from $RT = \$10$ MM through to $RT = \$100$ MM.

Note from the information supplied that $E_1 = p_s V - p_f C = \$10$ MM while the volatility is $\nu = 4$, so that there is a large degree of uncertainty in the expected value of the project at the chance node.

A. RESULTS FROM COZZOLINO'S RISK AVERSION FORMULA

For the parameters given above the Cozzolino formula can be written in the parametric form

$$RAV = -10W[x^{-1} \ln\{0.2 \exp(-9x) + 0.8 \exp(x)\}] \qquad (2.21a)$$

with $10W/RT = x$.

For a fixed value of RT it follows that

$$W = RT\,x/10 \qquad (2.21b)$$

so that W increases as x increases, while for a fixed working interest one has

$$RT = 10W/x \qquad (2.21c)$$

so that RT increases as x decreases.

Thus, as x is varied in the range $0 \leq x \leq \infty$, all possibilities are covered, RAV can then be plotted either as a function of W for fixed RT, or as a function of RT for fixed W.

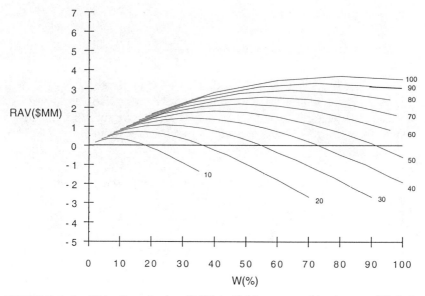

FIGURE 2.2 Risk-adjusted value (RAV) in $MM versus working interest, $W(\%)$, for various risk tolerance values (labeled on each curve in $MM). Note that the RAV has positive and negative values until RT crosses about $55 MM, when any working interest up to 100% will be profitable. The Cozzolino risk adjustment formula was used with the data of Figure 2.1b.

1. Fixed Risk Tolerances

Plotted on Figure 2.2 are the estimated RAV values (in $MM) as the working interest, W, varies from 0 to 100% for different risk tolerance values from $RT = \$10$ MM to $100 MM. We can see from Figure 2.2 that as RT increases, a greater degree of risk is considered accpetable to the corporation; then the range of working interest broadens over which a positive RAV is obtained. This broadening represents the point at which, as the risk tolerance increases, the RAV tends asymptotically to $W(p_s V - p_s C) > 0$. Thus the range of working interest over which a positive value for RAV can be found increases with increasing RT.

The break-even value ($RAV = 0$) then provides a range of working interest, W, versus risk tolerance where $RAV > 0$. Plotted on Figure 2.3 are the ranges of values of W versus RT at which $RAV = 0$. The implication from Figure 2.3 is that the larger the risk tolerance value, the greater the working interest that should be taken in the project.

2. Fixed Working Interest Values

Plotted on Figure 2.4 are the curves of RAV versus RT for different working interests. Note from Figure 2.4 that as the risk tolerance increases beyond the maximum estimated value of the project, then RAV is closer

A NUMERICAL ILLUSTRATION

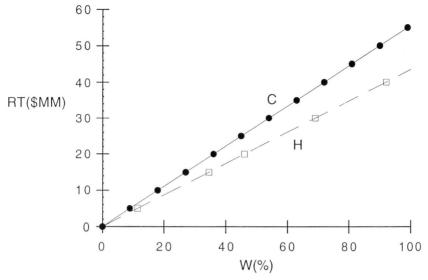

FIGURE 2.3 Plot of the "break-even" conditions (such that $RAV = 0$) for risk tolerance RT ($MM) versus working interest, W (%), for both the Cozzolino risk adjustment formula (curve C) and the hyperbolic risk adjustment formula (curve H) using the data of Figure 2.1b. RT values higher than the limits at $W = 100\%$ imply that 100% working interest should be taken in the project.

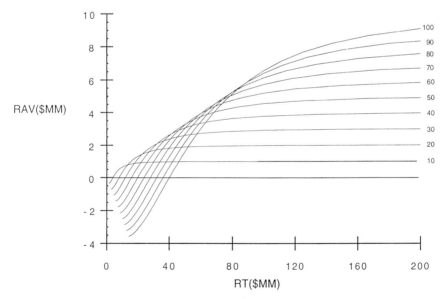

FIGURE 2.4 Risk-adjusted value (RAV) in $MM plotted against risk tolerance (RT) in $MM, for various percentage working interests, as labeled on the curves at intervals of 10%. The Cozzolino risk adjustment formula was used with the data of Figure 2.1b.

to $E_1 = p_sV - p_fC = \$10$ MM. Thus one can read from Figure 2.4 what a particular RT corresponds to in terms of a positive or negative RAV. For instance, if the risk tolerance is $RT = \$20$ MM then it makes sense to accept a working interest of greater than about 25% in order to have a positive RAV.

B. RESULTS FOR THE HYPERBOLIC RISK AVERSION FORMULA

For the parameters given for the numerical illustration of Figure 2.1b, the hyperbolic RAV formula, Eq. (2.17), can be written parametrically through

$$RAV = -10W \{x^{-1} \ln(1 + 0.8 \tanh x - 0.2 \tanh 9x)\} \quad (2.22a)$$

with, as before,

$$RT = 10W/x. \quad (2.22b)$$

1. Fixed Risk Tolerances

Figure 2.5 gives the estimated RAV values (in $MM) versus the working interest, W, in percent, for different risk tolerances ranging from $RT = \$10$ MM to $RT = \$100$ MM. Again note that a greater positive range of RAV values is obtained as RT increases. Perhaps of interest here is to compare

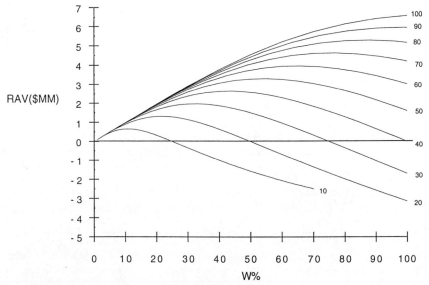

FIGURE 2.5 As for Figure 2.2 but with the hyperbolic RAV formula. Note that RAV is positive at all $0 \le W \le 100\%$ once RT crosses about $42 MM, a smaller value than for the Cozzolino formula exhibited in Figure 2.2.

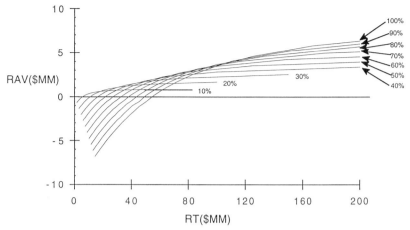

FIGURE 2.6 As for Figure 2.4, but with the hyperbolic *RAV* formula.

the range of positive *RAV* values for the risk aversion model formula of Cozzolino with the results of the hyperbolic model.

Plotted in Figure 2.3 on the ordinate is the range of *RT* values against *W* on the abscissa which produce positive *RAV* values (for the same *W* and *RT*) from the Cozzolino formula (labeled C) and the hyperbolic formula (labeled H), respectively.

2. Fixed Working Interests

Figure 2.6 provides the estimated *RAV* values (in $MM) versus the risk tolerance *RT* (in $MM) as the working interest increases from 10 to 100%.

Note that, just as for the Cozzolino exponential risk aversion formula, the hyperbolic formula provides that the *RAV* tends to E_1W as *RT* increases. The range of *W* and *RT* values yielding positive *RAV* values is plotted in Figure 2.3 versus the equivalent ranges from the Cozzolino formula.

C. COMPARISON OF RESULTS

1. At a fixed working interest, comparison of Figure 2.4 with Figure 2.6 shows that the Cozzolino risk aversion formula is basically more conservative in its estimates of *RAV* as risk tolerance increases. It takes a higher *RT* to yield an *RAV* comparable to that from the hyperbolic formula. The difference for the example given is roughly a factor of 2. The reason for this behavior is that $\tanh(x)$ at large x varies as $1 - 2\exp(-2x)$, so that the x variation is almost twice as rapid as compared to the variation $\exp(-x)$ occurring in the Cozzolino formula.

2. At a given risk threshold, comparison of Figure 2.2 with Figure 2.5 indicates that the Cozzolino formula will require a higher *RT* to obtain the

same range of working interest that yields a positive RAV as compared to the hyperbolic risk aversion formula. The Cozzolino result accepts only a narrower range of working interest at a fixed risk tolerance, with the corollary that the expected positive RAV values are systematically lower than those obtained using the hyperbolic formula.

3. The "break-even" ranges for RT and W which yield $RAV \geq 0$ are compared in Figure 2.3 where, for a given working interest, the Cozzolino formula requires a greater risk tolerance to produce a positive RAV; while for a given risk tolerance, the Cozzolino formula would advise a lower working interest be taken to produce a positive RAV than would the hyperbolic risk aversion formula.

The dominant reason for the systematic offset of results between the Cozzolino, exponential dominated, risk aversion formula and the hyperbolic formula is the functional dependence. Crudely speaking the hyperbolic tangents vary but slowly in $0 \leq \tanh x \leq 1$ over all x, while the exponentials vary without bound. This gentler variation of the hyperbolic contribution implies that the risk aversion estimates are not as sharply varying as parameter values change—leading to broader ranges of working interest and lower estimates of risk tolerance to produce positive domains of high RAV values.

IV. NEGATIVE EXPECTED VALUES

The RAV formulas for different modeled behaviors can also be used to assess corporate strategy and philosophy even when there is a negative expected return.

For example, consider a well costing $50 MM that is *obligatory* so that $W = 1$. There is a 20% chance that the well will make $90 MM, otherwise it will be a $50 MM loss. In this case the expected value at the chance node of Figure 2.1a is $E_1 = -\$22$ MM. It is conventionally taken that the expected value is the risk neutral cash amount that one should be willing to pay to buy out of the obligation, otherwise it is more appropriate to drill and take the chance (80%) of a $50 MM loss. However, note that the magnitude of the volatility, v, is 2.5, representing a considerable uncertainty on the estimated mean value, in the sense that within one standard error, $\sigma \equiv vE_1$, one has a range of expected return of $E_1 \pm \sigma$, that is, $E_1(1 \pm v)$, which stretches from $-\$66$ MM through to $+\$33$ MM. Thus, the uncertainty on the mean value should be included in assessing whether the risk neutral cash amount to buy out of the obligation is really at E_1 or within some tolerance factor around E_1.

The RAV formulas can be used with such problems as follows. Suppose the corporation is risk averse with a risk tolerance of $RT = \$50$ MM; then the RAV using the Cozzolino formula is $-\$39.6$ MM, while the RAV from the hyperbolic tangent formula is $-\$17.5$ MM. Thus the Cozzolino formula tends to emphasize the *negative* fractional uncertainty around the mean value, while the hyperbolic tangent procedure tends to emphasize the *positive* fractional uncertainty. In a sense, the Cozzolino formula is pushing more toward the chance of a greater likelihood of encountering the catastrophic loss of $50 MM, and so suggesting that it is better to pay about $39.6 MM to buy out of the obligation to be absolutely sure of avoiding the loss.

The hyperbolic tangent RAV recognizes that there is a chance that the uncertainty on the expected value could permit a positive expected value (to within one standard error), and so suggests that one should allow for this chance in assessing RAV and risk neutral cash settlements of the obligation.

To illustrate the range of behaviors, three cases were run in which the obligation requires $W = 1$, the value $V = \$90$ MM, $RT = \$50$ MM, and $p_s = 0.2$, but with cost, C, at the three values $50, $40, and $30 MM, respectively, as shown in Table 2.1.

Note that as the negative expected value increases toward zero, the magnitude of the volatility increases to the point where the uncertainty on the expected value is providing a greater chance that there is likely to be a positive *return* (to within one standard error). The hyperbolic tangent formula for RAV uses this chance to suggest that the cost to buy out of the obligation should be tempered by the likelihood of a positive return, whereas the Cozzolino formula for RAV tries to minimize total potential catastrophic damage by suggesting that it is better to buy out of the obligation at a high price, relative to the expected mean value, rather than to gamble on the chance of absolute failure.

So the Cozzolino formula tends to be pessimistic while the hyperbolic tangent formula is more optimistic. Essentially, then, the use of individual corporate RAV formulas to assess risk is tied to the corporate philosophy on risk: conservative, neutral, or aggressive. The Cozzolino formula and the hyperbolic tangent formula for RAV reflect this difference in corporate attitude to risk, which was the point of this illustration.

TABLE 2.1 Risk-Adjusted Values for Different Cost Estimates

Cost ($MM)	E_1 ($MM)	|Volatility|	Cozzolino RAV ($MM)	Hyperbolic tangent RAV ($MM)
50	−22	2.5	−39.6	−17.5
40	−14	3.7	−29.8	−14.7
30	−6	8.0	−20.0	−10.8

V. CONCLUSIONS

It is important to assess quantitatively corporate aversion to risky projects within the framework of a given corporate risk threshold. On the one hand, there is the need to ensure that the corporation does not end up bankrupt if the project fails, but on the other hand there is the need to figure out what working interest should be taken in the project to maximize potential gains should the project prove profitable.

Clearly, the competing demands of these two corporate positions must be evaluated in order to assess likely corporate involvement in a project.

Until recently, Cozzolino's (1977a, 1978) procedure was the only method readily available for enabling such evaluations to be made. The danger, then, is that the evaluatioins made of risk-adjusted value and working interest are particularly beholden to the model procedure, without the ability to evaluate the uniqueness, resolution, precision, and accuracy of the results in relation to variations in intrinsic assumptions of the model-dependent RAV.

The advantage to having available alternative functional forms for RAV assessment is that one can then compare and contrast the sensitivity of model results under different assumptions. In this way one can determine those factors which fall within broadly similar ranges for the majority of models versus those factors which are especially sensitive to the intrinsic assumptions of a particular model and so are less likely to reflect accurately corporate risk philosophy.

The other advantage is that different corporations use risk aversion factors that are not necessarily of the exponential form, as is required by the Cozzolino (1977a, 1978) risk aversion formula. Therefore, the ability to put forward a general formula for risk aversion values, as done here, means that each corporation can now use its own weighting formula, and carry through the corresponding assessments of RAV in relation to risk tolerance and working interest within a corporate required framework.

The particular numerical illustrations presented here address the points of model dependence of results, and show the magnitude and distortion of systematic offsets in behaviors for risk and working interest. Any set of model formulas for RAV can now be evaluated for their relative contrasts with the general method given here; this fact is the main point of the chapter.

3

Uncertainty and Probability Estimates for Risk-Adjusted Values and Working Interest

I. INTRODUCTION

At a chance node for a hydrocarbon exploration project, such as that given in Figure 3.1, there is a value V, a cost C, and a probability of success of p_s (and a probability of failure $p_f \equiv 1 - p_s$). One of the standard concerns is to estimate the working interest, W, and the risk-adjusted value, RAV, for the project given a corporate risk tolerance (RT). As noted in the previous chapter, such matters have been researched and developed since the application of utility theory to hydrocarbon exploration opportunities as illustrated by the Cozzolino (1977a, 1978) risk adjustment formula, which exponentially weights the success and failure branches of the chance node diagram (Figure 3.1) with respect to WV and $-WC$, where W is the fractional working interest. Generalizations to allow for other than exponential weighting, and to provide analytic algebraic results for maximum working interest, maximum RAV, risk neutral RAV, and so on, have recently been developed and are presented in Chapter 2.

However, the results are often based on the position that there is no uncertainty in V, C, p_s, or W. In reality, there is uncertainty in all four of these quantities for a variety of reasons: (1) The total value, V, depends on economic models of projected future selling price of hydrocarbons, inflation and escalation costs, and allied fiscal factors; while at the same time the quantitative basin model results, which provide an assessment

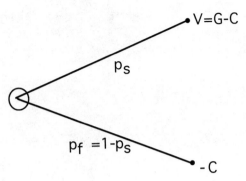

FIGURE 3.1 Chance node diagram illustrating value, V, cost, C, probability of success, p_s, and probability of failure p_f ($\equiv 1 - p_s$). The expected value at the chance node is $E_1 = p_s V - p_f C$, at a working interest, W, of 100%; and is WE_1 at $W < 100\%$.

of hydrocarbon charge to a putative reservoir, are themselves uncertain (Thomsen and Lerche, 1997). Thus the conversion of hydrocarbon charge assessment to a cash value, V, already contains an uncertainty due to the geologic model estimates. (ii) The probability of success, p_s, is related not only to exploration drilling and production concerns but also to the probability of uncertainty of hydrocarbons being in a reservoir at the present day, that is, to the probability that the basin model estimates are correct (Thomsen and Lerche, 1997); the estimated project costs, C, at the time the project is being evaluated depend on assessments of future drilling costs, including uncertainty due to unanticipated drilling conditions, site preparation or possible regulations; the estimated fractional working interest, W, is uncertain because there is usually flexibility in the values of W which can be optimized to produce a positive RAV for a given risk tolerance and, as the value, cost, and success probability change, the range of working interest yielding positive RAV also changes. The range also changes as the risk tolerance varies, and also because, even without any uncertainty in the values of p_s, V, and C, there is still an uncertainty, measured by the variance $\sigma^2 \equiv (p_s p_f)(V + C)^2$, at the chance node of the decision tree of Figure 3.1, due to the wide difference in value of a success or failure. Thus, each of V, C, p_s, and W has some degree of uncertainty. Two questions then become of major concern:

1. How does one quantify the effect of uncertainties on the RAV and on the likelihood of a positive return from a project?
2. Which of the uncertainty factors is most important in influencing RAV, so that one has some idea of where to concentrate effort in order to minimize the uncertainty?

The purpose of this chapter is to address these two concerns.

II. RAV ESTIMATES WITH UNCERTAINTIES

A. GENERAL CONSIDERATIONS

The Cozzolino (1977b, 1978) formula (exponential weighting) for RAV takes the form

$$RAV = -RT \ln\{p_s \exp(-WV/RT) + p_f \exp(WC/RT)\}, \quad (3.1)$$

while the similar hyperbolic tangent weighting formula (MacKay and Lerche, 1996c) for RAV takes the form

$$RAV = -RT \ln\{1 - p_s \tanh(WV/RT) + p_f \tanh(WC/RT)\}. \quad (3.2)$$

As noted in Chapter 2, a general, risk-weighted formula for a function $G(x)$, with a risk aversion factor proportional to $1 - G(WV/RT)$ $\{1 - G(-WC/RT)\}$ for the success (failure) branch of the chance node decision tree diagram of Figure 3.1, can be shown (MacKay and Lerche, 1996c) to yield

$$RAV = -RT \ln\{1 - p_s G(WV/RT) - (1 - p_s) G(-WC/RT)\}. \quad (3.3)$$

B. NUMERICAL ILLUSTRATIONS

1. General Observations

Because both the Cozzolino and the hyperbolic tangent RAV formulas are simple algebraic expressions, it is relatively easy to evaluate numerically the expressions for RAV for any given values of V, C, p_s, W, and RT.

This simplicity then lends itself to direct Monte Carlo simulations in which individual values of V, C, p_s, W, and RT are selected from predefined distributions (triangular, uniform, normal, log-normal) with ranges specified, and the RAV calculated for each set of values. A Monte Carlo random number selection procedure can then be invoked to choose many input values, and the distribution of RAV values computed for the selected input distributions and their ranges. In addition, one obtains the effect of each individual factor contributing to the variance in RAV, and so the ability to compute the relative importance of each contribution to the uncertainty in RAV.

The nub of the problem is to have a set of procedures for determining the individual components V, C, p_s, W, and RT, and their likely ranges of uncertainty from other considerations.

In the case of the value, V, of a project, the dominant point is to have some method for providing an estimate of volume of hydrocarbon charge to a potential reservoir, and also a measure of the expected uncertainty in the charge estimate and hydrocarbon mix (gas-to-oil ratio, GOR). Once a reservoir hydrocarbon charge is estimated, then economic considerations convert this value to a cash amount, V, together with increased ranges of uncertainty brought about by the uncertainties in the economic parameters and assumptions. For example, consider an estimated likely hydrocarbon charge of 1 MMSTB of oil with a 10% uncertainty in volume (i.e., $1^{+0.1}_{-0.1}$ MMSTB) (STB $\equiv$ stock tank barrels).

If any oil found were to sell at the fixed price of \$20/STB then the value V (in \$MM) of the estimated hydrocarbons in place would lie between[1]

$$V = \$20^{+2}_{-2} \text{ (oil)}. \tag{3.4}$$

If the selling price of oil is itself uncertain, then an even greater degree of uncertainty is introduced to V.

But the basic problem is still to assess the hydrocarbon charge and its uncertainty. Fortunately, precisely that problem is the bailiwick of basin analysis models, which estimate the influences of hydrocarbon generation, expulsion, secondary migration, and drainage area in determining likely hydrocarbon charge to a reservoir together with uncertainties, and also the probabilities of being correct in the assessment of one potential reservoir in relation to others (Thomsen and Lerche, 1997). Thus, the estimated value, V, can then be provided, as can the success probability, p_s, and their likely ranges of uncertainty.

The dominant factors controlling cost estimates and their uncertainties are usually bid prices, seismic survey costs, rig costs, subsurface drilling conditions (e.g., overpressure development), and also factors having to do with future inflation estimates, taxes, royalties, etc. While not usually as variable or uncertain as estimates of hydrocarbon volume, nevertheless project cost estimates are often uncertain by up to 50% historically, and occasionally much more.

The variation in risk tolerance that a corporation is prepared to accept is itself a matter, quite often, of historical precedent within a corporation, or often guided by risk thresholds limiting damage to a corporation should a project prove to be unsuccessful (MacKay and Lerche, 1996a). The point here is that cash equivalent values can be provided for V, C, and RT, and for their ranges of uncertainty, based on scientific and economic projections of attainable goals.

In respect to working interest, the situation is slightly different. For each set of parameters for V, C, p_s, and RT, it is possible (MacKay and Lerche, 1996a) to determine analytically (exactly or approximately) the optimum working interest (OWI) to maximize RAV, and also the minimum and maximum working interest values which return a risk neutral RAV of zero, so that any working interest in the minimum to maximum range will yield a positive RAV (see Chapter 2). However, it can also happen that some range of working interest is a prescribed requirement in order to participate

---

[1] This estimate of value also has to have the costs subtracted.

in a project (e.g., an interest greater than 20%). In such cases there is a prescribed minimum W, obviously also a maximum of $W \leq 100\%$. Within that range the working interest can vary. Clearly, as the parameters V, C, p_s, and RT vary, the RAV will take on different values for different working interest prescriptions. Hence, there exists the possibility for allowing W to be uncertain within a prescribed range.

In addition, some care has to be exercised in assigning the probability functional behavior for each of the intrinsic variables V, C, p_s, W, and RT. The reasons are that both p_s and W are restricted to the maximum allowable ranges of $0 \leq p_s \leq 1$, $0 \leq W \leq 1$; while C and RT are each restricted to lie in $\{0, \infty\}$ although value, V, can range in $\{-C, \infty\}$. Thus a normal distribution for any of the variables cannot be allowed (without truncation) because a normal distribution admits negative values; equally p_s and W cannot be log-normally distributed (without truncation) because a log-normal distribution covers the range $\{0, \infty\}$, which exceeds the $\{0, 1\}$ range of p_s and W. However, within the framework of these minor restrictions, the variation of V, C, p_s, W, and RT can be as broad as required.

Here the behavior of both the Cozzolino (exponential weighting) and hyperbolic tangent formulas for RAV is investigated under a variety of ranges of parameters to illustrate dominance and relative importance of individual factors contributing to the uncertainty in RAV.

2. Uniform Probability Distributions

Consider the situation where hydrocarbon charge (HC) estimates to a reservoir have been made using basin modeling techniques, perhaps in the manner of Thomsen and Lerche (1997), yielding ranges of $3 <$ HC (MMSTB) < 9, and with a central estimate of 6 MMSTB. For the purposes of this example, it is given that hydrocarbons can be sold at a fixed price of $15/STB. At the same time, basin modeling estimates are also made of the likelihood of the reservoir being undisturbed after hydrocarbon charging, suggesting that there is only a combined probability of between 10 and 30% of a seal being in place and of seal integrity being maintained through to the present day.

Costs of licenses, royalties, taxes, and drilling are set at $5 MM if no overpressure is found, but may be as high as $15 MM if high overpressure occurs at shallow depths. The corporate risk tolerance for similar wells under similar conditions has ranged in previous years from $50 MM to $150 MM, depending on the capital asset and yearly cash flow picture of the corporation. Thus, on the basis of the above information one would set the ranges

$$\$45 \text{ MM} \leq V \leq \$135 \text{ MM}, \tag{3.5a}$$

$$\$5 \text{ MM} \leq C \leq \$15 \text{ MM}, \tag{3.5b}$$

$$0.1 \leq p_s \leq 0.3, \tag{3.5c}$$

$$\$50 \text{ MM} \leq RT \leq \$150 \text{ MM}. \tag{3.5d}$$

Note that the minimum RT ($50 MM) is already large compared to the maximum estimated cost ($15 MM), so that risk threshold should not be a major factor in influencing probabilistic RAV values. For each of the variables, V, C, p_s, and RT, Monte Carlo runs were done to provide probabilistic results, with the random values for each variable being selected from uniform populations, bracketed at their respective minima and maxima by the ranges given above. The working interest was taken to be fixed at 100%.

Figure 3.2a exhibits the relative importance (in %) of the uncertainties in p_s, C, V, and RT in contributing to the uncertainty for the Cozzolino RAV formula, while Figure 3.2b exhibits the same information for the hyperbolic tangent RAV formula. As anticipated, in both cases RT has the smallest influence on RAV, while both p_s and C dominate, contributing a total of about 80% of the uncertainty, with p_s being about 50% of the total uncertainty.

Thus, in terms of relative importance to the uncertainty in the RAV, it is not so much the volume of hydrocarbons that dominates in this case, but rather the sealing conditions of the reservoir after hydrocarbon charging which influence the uncertainty on p_s, the probability of success.

In addition, while the range of costs is only a small fraction of the value, the problem here is that the value, V, has to be multiplied by the success probability (≈ 0.2), so that $p_s V$ is comparable to $(1 - p_s)C$, indicating an expected return (at the center values of each variable) of $p_s V - (1 - p_s)C \approx \$10$ MM, which is comparable to the cost values. Hence, the relative importance of cost on the RAV in both cases arises because the ratio $[p_s V / (1 - p_s)C]$ is of order unity.

It is not so much that cutting costs would be the ideal improvement arena, but rather some way to assess whether the range of p_s is really at 0.2 ± 0.1 or whether p_s is larger or smaller, because the larger the central value of p_s the smaller is the influence of cost, while the smaller the central value of p_s the less likely is the opportunity to be profitable. If geological information indicates that there is no further improvement possible in the determination of p_s and its range of uncertainty, then either cost uncertainty should be addressed (possibly through turn-key bids) or value estimates reinvestigated, or a hedging position taken to prevent catastrophic loss. But in any event, the point is that the relative importance of each factor provides an objective way of determining where effort should be placed to minimize the uncertainty in the RAV.

Apart from the *relative* importance of individual factors contributing to the uncertainty in the RAV, there is also the *absolute* value of the variance to consider. The point here is that relative importance informs on *which* factors need to be addressed for improvement, while absolute uncertainty informs on *when* factors need to be addressed for improvement. If the absolute uncertainty on the RAV is very much less than the mean value,

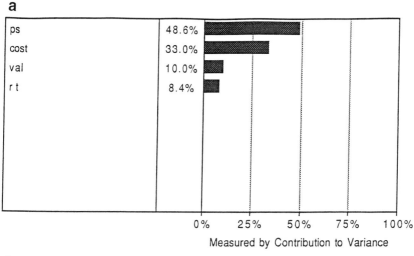

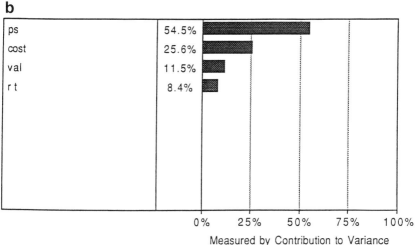

FIGURE 3.2 Relative importance (%) of factors contributing to the uncertainty in RAV. (a) The Cozzolino formula; (b) the hyperbolic tangent formula, for uniform probability distributions of p_s, V, C, and RT, with working interest set to $W = 100\%$.

then there is little uncertainty anyway, even when the relative importance indicates which factors are dominating the variance.

Thus, Figures 3.3a and 3.3b, for the Cozzolino formula and the hyperbolic tangent formula for RAV, respectively, show the cumulative probability of obtaining an RAV less than particular amounts. Both curves are approximately log-normally distributed due to the central limit theorem, as can be observed by viewing the figures as histograms.

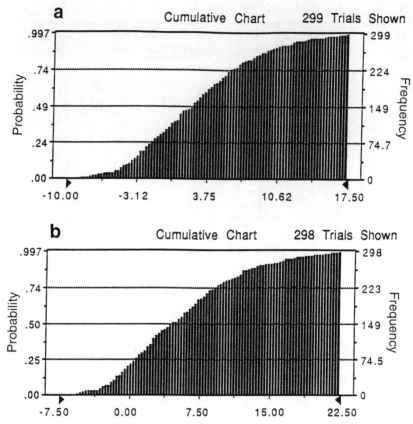

FIGURE 3.3 Cumulative probability plots of RAV for the uniform probability distributions of p_s, V, C, and RT, with $W = 100\%$; (a) the Cozzolino formula; (b) the hyperbolic tangent formula.

Two values are of importance: (1) the cumulative probability, p_0, of an $RAV \geq 0$; (2) the values RAV_{10}, RAV_{50}, and RAV_{90}, which occur at 10, 50, and 90% cumulative probability values, respectively, and which measure the range of likely positive returns to be expected given the intrinsic uncertainties.

Reading from Figures 3.3a and 3.3b one has

$$p_0 \cong 35\% \text{ (Cozzolino formula)} \qquad (3.6a)$$

and

$$p_0 \cong 20\% \text{ (hyperbolic tangent formula)} \qquad (3.6b)$$

so that there is a 65 to 80% chance of a positive RAV no matter which risk aversion formula is used.

Again reading off from Figures 3.3a and 3.3b one has:

	Cozzolino ($MM)	Hyperbolic tangent ($MM)
RAV_{10}	−3.9	−1.8
RAV_{50}	2.4	4.6
RAV_{90}	10.3	13.8

The volatility, v, of the 50% RAV is given by

$$v = (RAV_{90} - RAV_{10})/RAV_{50}, \qquad (3.7)$$

which takes on the values $v = 5.9$ (Cozzolino) and $v = 3.4$ (hyperbolic tangent), both of which are significantly larger than unity. The volatility provides a quantitative measure of uncertainty on the RAV, because if v is small compared to unity then there is little uncertainty on RAV_{50}, while a large value of v ($\gg 1$) provides an indication that there is significant uncertainty on RAV_{50}. Thus, while the estimate RAV_{50} is fairly uncertain, nevertheless there is between a 2:1 to 4:1 odds-on chance that the project should be undertaken, as measured by the value p_0 where $RAV = 0$.

In addition to relative importance and probabilities for RAV, two other factors are of relevance: the expected value (not risk weighted) and the optimum working interest that should be taken—rather than assigning a 100% working interest as has been done so far.

Referring to Figure 3.1, the expected value, E_1, at the chance node (at $W = 100\%$) is

$$E_1 = p_s V - (1 - p_s)C \qquad (3.8a)$$

and the variance σ^2 is given through

$$\sigma^2 = E_2 - E_1^2 = p_s(1 - p_s)(V + C)^2. \qquad (3.8b)$$

Thus, even for fixed values of V and C, because there is only a probability, p_s, of success, the expected value has an uncertainty of $\pm\sigma$. Clearly, when V, C, and p_s also have uncertainties then such add to the total uncertainty on the expected value. Note for reference that if each of p_s, V, and C were to be set at their midpoint values of 0.2, $90 MM, and $10 MM, respectively, then $E_1 = \$10$ MM, $\sigma = \pm \$4$ MM.

Figure 3.4a presents the cumulative probability plot for the Monte Carlo runs indicating a mean value of $E_1 = \$9.5$ MM (within $0.5 MM of the analytic calculation) and a standard error of $\pm \$8$ MM, nearly double the value of σ due to the variations in p_s, V, and C around their central values. Note also from Figure 3.4a that the probability of obtaining a positive value for E_1 is about 90% so that, even without risk weighting the project, the suggestion is that it is likely to be worthwhile financially.

Figure 3.4b presents the relative importance (in %) of the uncertainty in each of p_s, V, and C in contributing to the variance in the expected value. (RT has no importance because E_1 does not depend on RT.) In this

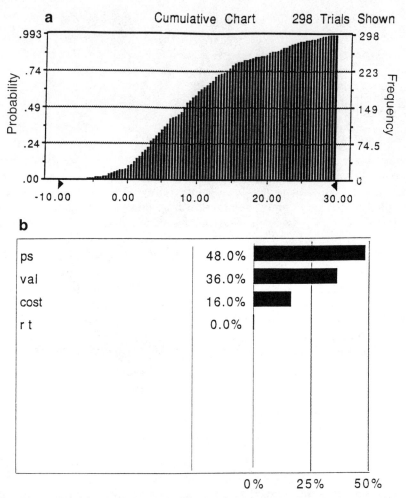

FIGURE 3.4 (a) Cumulative probability plot of expected value, E_1, for the uniform probability distributions of p_s, V, C, and RT, with $W = 1$; (b) relative importance of factors contributing to the uncertainty in E_1. Note the absence of any effect of RT (because E_1 is independent of RT).

case note that the uncertainty in p_s is contributing almost 50% of the uncertainty, with uncertainty in the value, V, contributing about 36%, and cost uncertainties only about 18%. Again, then, the strong indication is that it is the success probability, p_s, which needs to be addressed in order to improve matters further, but now with a secondary emphasis which suggests that determining a better value for V is more important (by a factor 2) than doing a better job on narrowing the cost range of uncertainty.

The optimum working interest, OWI, that one should take in the project is also of interest. In the previous chapter we showed for the Cozzolino RAV formula that OWI was precisely given by

$$OWI = \min[\{RT/(C + V)\} \ln[p_s V/\{(1 - p_s)C\}]; 1] \qquad (3.9)$$

in $p_s V > (1 - p_s)C$, else $OWI = 0$; and at this value of OWI the maximum RAV is

$$RAV_{max} = V \times OWI - RT \ln\{p_s(1 + V/C)\}. \qquad (3.10)$$

For the range of values of RT, p_s, V, and C, the Monte Carlo simulations were again used to provide a cumulative probability plot of OWI, displayed in Figure 3.5a.

Reading from Figure 3.5a, the mean value of OWI is about 62%, while the values at 10, 50, and 90% cumulative probability are $OWI_{10} = 5\%$, $OWI_{50} = 69\%$, and $OWI_{90} = 100\%$ respectively, for a volatility of $(OWI_{90} - OWI_{10})/OWI_{50} = 1.4$.

Thus, about 60–70% optimum working interest is likely the best to take in the project given the uncertainties, but a conservative risk corporation could take as low as 5%, whereas a more aggressive risk corporation should aim for 100% interest. Either choice would, of course, change the apparent risk tolerance. Figure 3.5a indicates that there is a 30% probability that an OWI of 100% is appropriate, with an 80% probability that an OWI greater than 25% should be taken.

The relative importance of contributions to the variance in OWI is shown in Figure 3.5b, indicating that success probability and cost are the dominant contributors, with risk tolerance of lesser importance, and with very little dependence on value. The reason for this surprisingly small dependence on value is that the ratio of V/C is larger than 3 for the ranges given so that the major sensitivity of Eq. (3.9) is to $\ln\{p_s/(1 - p_s)\}$, which varies rapidly as p_s varies in $0.1 \leq p_s \leq 0.3$. Thus the dominance of uncertainties in p_s overrides all other contributions to the uncertainty in OWI. In addition, cost, C, becomes of major importance due to the fact that the ratio V/C, while large compared to unity, still varies considerably, so that OWI takes on the value unity about 30% of the time. Clearly, the dominant message here is that the costs are sufficiently low, the risk tolerance sufficiently high, and the value sufficiently large that a significant fraction (~30%) of the Monte Carlo runs opt for 100% OWI, thereby narrowing the probability range to 0–70% over which OWI values are sensitive to p_s, C, V, or RT.

In general, the risk-adjusted values place much less emphasis on the value, V, than does the expected value; while the OWI (being roughly linear with respect to cumulative probability) places little emphasis on the value but considerable emphasis on success probability and cost.

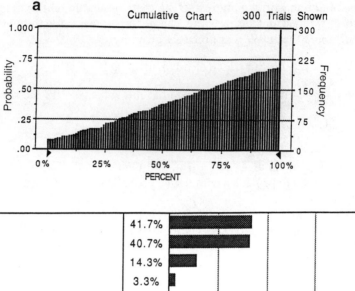

FIGURE 3.5 Optimum working interest (*OWI*) for the Cozzolino *RAV* formula, which maximizes *RAV*, for uniform probability distributions of p_s, *V*, *C*, and *RT*, with *W* = 100%. (a) Cumulative probability plot of *OWI*; (b) relative importance (%) of factors contributing to the uncertainty in *OWI*.

Figure 3.6 displays a trend chart illustrating the differences between the estimates for expected value, *RAV* (Cozzolino), and *RAV* (hyperbolic tangent) based on cumulative probabilities at different confidence intervals, indicating that the expected value is both more positive and has a wider variance relative to the two *RAV* plots, with the hyperbolic tangent *RAV* having a slightly larger variance than the *RAV* from the Cozzolino method.

While it might be thought that the *RAV* curves should have a greater variance than the expected value because they involve the extra uncertain variability of *RT*, the point is that their dependences on *RT*, and also on

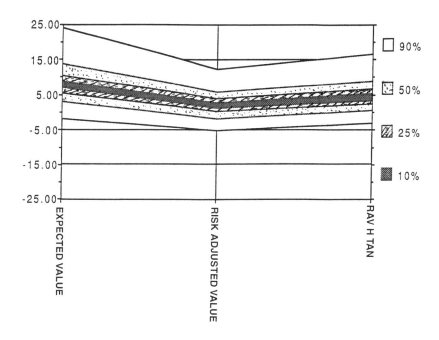

FIGURE 3.6 Trend chart showing the cumulative percentage bands comparison between expected value, *RAV* (Cozzolino), and *RAV* (hyperbolic tangent) with centering on the median values, for uniform probability distributions of p_s, V, C, and RT, with $W = 100\%$.

p_s, V, and C, is highly nonlinear so that there is more suppression of the variance than might otherwise have been anticipated.

3. Working Interest Uncertainties

The previous numerical example was predicated on the assumption that a corporation takes 100% working interest in the project. However, in reality, it is often the case that a group of corporations will band together, with each taking less than 100% working interest, such as in Gulf of Mexico Lease Sale 147 (Lerche, 1994).

In a general sense it is often the case that a corporation is offered the choice of some range of fractional participation (greater than zero but less than 100%); the corporation then has to decide what working interest to take.

To illustrate the influence of uncertainty in working interest on RAV, we again consider the situation of uniform probability distributions for V, C, p_s, RT, and also W, where the range of each variable is:

$$0.1 \leq p_s \leq 0.3, \tag{3.11a}$$

$$\$5\,\text{MM} \leq C \leq \$15\,\text{MM}, \tag{3.11b}$$

$$\$45\,\text{MM} \leq V \leq \$135\,\text{MM}, \tag{3.11c}$$

$$\$50\,\text{MM} \leq RT \leq \$150\,\text{MM}, \tag{3.11d}$$

$$0 \leq W \leq 50\%. \tag{3.11e}$$

Again, Monte Carlo runs were done for both the Cozzolino and hyperbolic tangent RAV formulas. Note from Figures 3.7a and 3.7b, which show the relative importance of individual contributions to the uncertainty in RAV for the two RAV formulas, that the uncertainty in p_s still dominates, but that uncertainties in V and W now jointly contribute about 50–60% to the uncertainty in RAV, with risk tolerance, RT, and cost, C, providing only small contributions. The Cozzolino RAV formula is more sensitive to uncertainty in value than in working interest (a 1.5:1 ratio), while the hyperbolic tangent RAV formula is nearly equally uncertain to both value and working interest uncertainties. The expected value $E_1 \equiv W\{p_s V - (1 - p_s)C\}$ is directly proportional to the working interest and, as shown in Figure 3.7c, this proportionality is reflected in the relative importance contributions, with working interest, success probability, and value all approximately equally important in contributing to the uncertainty in expected value, with contributions from cost uncertainty small and no dependence on RT, of course.

For the cumulative probabilities for RAV from the Cozzolino and hyperbolic tangent formulas, and also for the expected value, Figures 3.8a–c show, respectively, the variations; note the influence of working interest uncertainties on the shapes of the curves relative to those of Section II,B,2. Reading from Figure 3.8 the behavior can again be summarized:

	Cozzolino	Hyperbolic tangent	Expected value
RAV_{10} ($MM)	−0.14	0.03	E_{10} ($MM) = 0.04
RAV_{50} ($MM)	1.34	1.64	E_{50} ($MM) = 1.71
RAV_{90} ($MM)	4.67	6.13	E_{90} ($MM) = 6.31
$p_0(\%)$ ($RAV = 0$)	15%	9%	$p_0(E = 0) = 9\%$
|Volatility|	3.6	3.8	3.7

In this case all three measures indicate comparable volatilities but all three also indicate a probability of between 85 and 91% of a positive RAV, while

FIGURE 3.7 Relative importance (%) of factors contributing to the variance in RAV and expected value, when uniform probability distributions are assigned for p_s, V, C, and RT, together with uncertainty in the working interest, W, which is also given a uniform probability distribution. (a) RAV (Cozzolino); (b) RAV (hyperbolic tangent); (c) expected value.

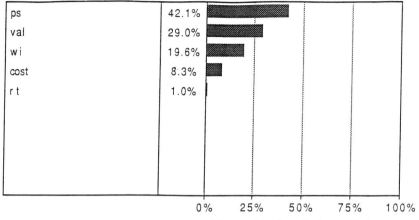

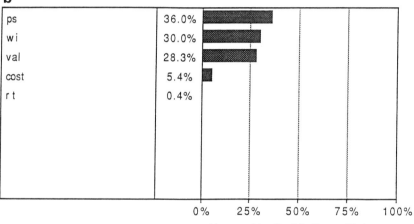

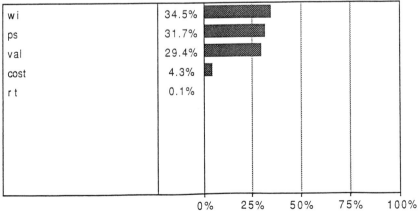

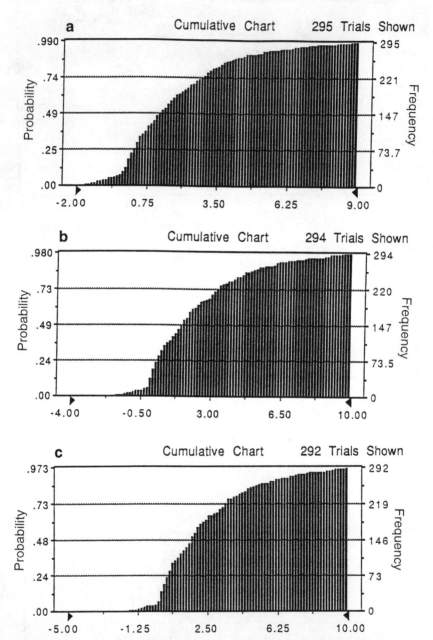

FIGURE 3.8 Cumulative probability plots using the uniform probability distributions for p_s, V, C, RT, and W. (a) *RAV* (Cozzolino); (b) *RAV* (hyperbolic tangent); (c) expected value.

the 50% probability *RAV* value is between $1.3 MM and 1.7 MM. Thus, the addition of a working interest lower than 100% cuts the potential for catastrophic losses, while reinforcing the positive aspects of the project.

The question arises as to how the range of working interest allowed ($0 \leq W \leq 50\%$) stands in relation to the *OWI* formula, which is independent of the input *W*. Shown in Figure 3.9a is the cumulative probability of *OWI* as p_s, *V*, *C*, and *RT* vary; Figure 3.9b shows that uncertainty in p_s and cost dominate (80%) the uncertainty in *OWI*, while *RT* and *V* uncertainties only jointly contribute about 20% to the uncertainty. Reading from Figure 3.9a, there is a 65% chance that a working interest *less than* 100% should be

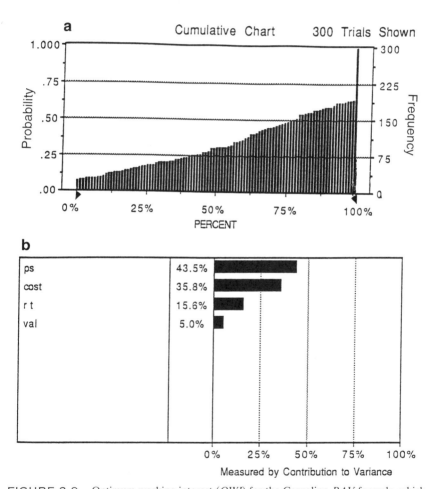

FIGURE 3.9 Optimum working interest (*OWI*) for the Cozzolino *RAV* formula, which maximizes *RAV*, using the uniform probability distributions for p_s, *V*, *C*, *RT*, and *W*. (a) Cumulative probability plot of *OWI*; (b) relative importance (%) of factors contributing to the uncertainty in *OWI*.

taken in the project, but an 80% chance that a working interest *greater than* 30% should be taken.

The upper end of the input range $0 \leq W \leq 50\%$ is met at a cumulative probability of 30%, suggesting that there is a 70% chance that a greater than 50% working interest should be taken. Thus, while the opportunity to invest in the project provides a high chance of a positive *RAV*, the offer of a working interest fraction less than 50%, while itself a good deal, is not the best deal that could have been accepted if offered; a higher working interest would have been preferred. Note that the volatility on the *OWI* is only 1.2 while $OWI_{50} = 77\%$, so that the estimate is ruggedly stable, and can be used to bargain for a higher working interest in the project, depending on the conservative, neutral, or aggressive corporate philosophy on risk taking.

4. Negative Expected Values

As the costs of a project rise relative to value, both the Cozzolino and the hyperbolic tangent *RAV* formulas indicate that, if possible, less working interest should be taken. However, often a positive working interest commitment in a project is mandated in order to maintain an exploration position in an active area of a country. In such cases the *RAV* formulas provide an estimate of the amount that should be spent to buy out of such an obligation rather than face the massive costs that could arise if no success occurs.

For instance, it can (and does) happen that a govenment agency mandates a company take a working interest of, say, around 50% as long as the project develops with mounting costs and success is uncertain. The fractional working interest allowed to a foreign corporation reduces to 25% if the costs drop, with the project more profitable. The government retains the right to increase the stake of its own national oil corporation and so decrees a lesser working interest (including zero) for the foreign corporation (often predetermined in the contract). Governments recognize that investments with essentially no chance of success are bad for both the foreign company and for the national oil company, so governments are often agreeable to cash payment in lieu of further investment.

Thus with an uncertainty on costs, and an uncertainty on mandated working interest, the problem is to figure out the buyout amount that should release the corporation from the obligation.

To illustrate the way in which such problems can be addressed, consider that p_s, V, W, and RT all have the same ranges of uncertainty as in the previous example, but now increase the costs by a factor 10, to lie in the range of $50 MM $\leq C \leq$ $150 MM. All parameters are again drawn from uniform probability distributions.

A set of Monte Carlo calculations was again made for *RAV*s from the Cozzolino and hyperbolic tangent formulas, and also for expected value, as shown by the cumulative probability plots of Figures 3.10a–c, respectively.

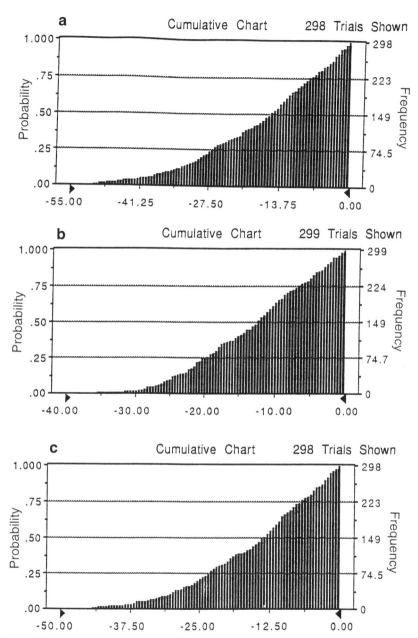

FIGURE 3.10 Cumulative probability plots of RAV and expected value for the buyout situation of high cost, when uniform probability distributions are assigned for p_s, V, C, RT, and W. (a) RAV (Cozzolino); (b) RAV (hyperbolic tangent); (c) expected value.

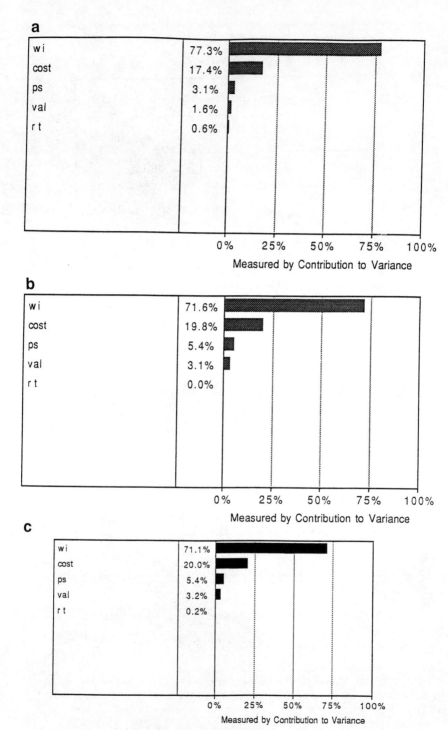

FIGURE 3.11 Relative importance (%) of factors contributing to the uncertainty in buyout amount using the uniform probability distributions for p_s, V, C, RT, and W. (a) RAV (Cozzolino); (b) RAV (hyperbolic tangent); (c) expected value.

In this case the interpretation of Figure 3.10 is as follows: From Figure 3.10a, which illustrates the Cozzolino RAV formula results, there is a 90% chance that one should pay at least $3 MM to buy out of the obligation, and there is a 50% chance that one should pay at least $15 MM, but only a 10% chance that one should pay more than $36 MM to buy out of the obligation. Likewise, for the hyperbolic tangent RAV (Figure 3.10b), the 90% chance indicates a buyout price of at least $3 MM, a 50% chance one should buy out at more than $12 MM, but only a 10% chance that one should buy out at more than $25 MM. The expected value probability curve (Figure 3.10c) indicates a 90% chance one should buy out at more than $3 MM, a 50% probability of a buyout at more than $14 MM, and only a 10% probability that one should buy out at more than $31 MM. If the government buyout amount is set higher than these ranges then it is preferable to complete the project. Depending on corporate strategy, some value in the range between $3 MM and $30 MM will satisfy the corporate decision makers.

The question of which components of uncertainty are relatively important in contributing to the buyout ranges of RAV values is illustrated in Figures 3.11a–c for Cozzolino RAV, hyperbolic tangent RAV, and expected value, respectively. As anticipated, the working interest mandate dominates at over 70% of the uncertainty in RAV, with the uncertainty in cost picking up about 20%. The uncertainties in value, success probability, and risk tolerance have but small roles to play. Clearly the aim here is to lower the mandated working interest by negotiation and to narrow the range of its uncertainty; in addition, costs need to be kept to the bare bones. But considerably more effort (3.5 times as much) should be spent on negotiation with the government to arrange a different working interest mandated contribution, in order to lower the anticipated buyout value to somewhere less than the $3–30 MM range (90 to 10% probability) indicated by the mandates in place.

III. DISCUSSION AND CONCLUSIONS

The importance has long been recognized of providing quantitative measures of risk aversion in relation to value, cost, success probability, corporate risk tolerance, and of estimating working interest fractions for hydrocarbon exploration projects. The Cozzolino (1977a, 1978) risk-adjusted value formula (which uses exponential weighting) was the predominant tool used in the oil exploration theater to assess risk until recently, when generalizations to allow for any functional form of weighting were developed (MacKay and Lerche, 1996a–d). Thus the intercomparison of different corporate model results for RAV can now be made, so that one is no longer so beholden to the specifics of a single available formula.

The use of RAV formulas is often predicated on precise values being available for the input parameters p_s, V, C, W, and RT. Different values can, of course, be inserted into the RAV formulas so that some idea of stability, sensitivity, and accuracy of results to changes in the input parameters is available.

However, the fact that the estimated values of p_s, V, C, W, and RT all carry some degrees of uncertainty, from both geologic and economic conditions, and the fact that the probability distributions for the uncertainties are not too well known, means that the RAV assessments from any and all models should also incorporate these uncertainties. In this way probabilistic ranges of uncertainty in RAV estimates, and the relative importance of factors contributing to the uncertainty, can be investigated to determine where and when to focus effort to improve matters.

What has been done here is to emphasize how one can bring to bear statistical considerations in assessing uncertainty in RAV estimates, and the importance of determining probabilities for positive RAV, and of probable ranges of buyout values for a high-cost project.

The numerical illustrations indicate how the RAV probabilities depend on the model functions (Cozzolino, hyperbolic tangent) used to provide RAV estimates. In addition, a mandated range of working interest can be addressed as an extra variable contributing to the probabilistic range of RAV; while negative RAV values for a high-cost project can be used to assess the probable buyout amount one should be prepared to pay depending on corporate risk philosophy.

The integrated use of modeling methods, geochemistry, geology, seismic processing, economic costs, future selling price, and fiscal risk can all impact the determination of economic reserves and so of risk-adjusted value. The methods described here show how, and when, the individual subdisciplines have to be addressed to improve matters. Perhaps one of the more interesting facets of such assessments is to have objective, reproducible methods expressed in simple form to demonstrate *why* the improvements are needed, and the level of increased resolution that needs to be obtained. At the same time the exigencies of reality must be kept firmly in mind in terms of work cost, personnel, and reporting deadlines. In addition, in order for a composite group of oil corporation personnel to interact together positively, it is almost mandatory to have some form of objective measures of need for improvement of individual factors influencing reserve and profit estimates; the alternative is subjective measures, with the attendant inferences for selective bias by individuals in what they believe needs improving rather than what can be shown needs to be improved.

Effectively, what is being risked here is the ability to protect the corporation from massive financial ruin. If the expected value is not too negative, then there is always a chance of not losing money and so one recommends that drilling be carried on; but as the expected value becomes even more negative the prospect of making any money becomes so remote that it is better to pay the buyout amount—unless it, too, is an extremely large amount when it is a less expensive loss to carry on drilling. But if there is any chance of bankruptcy occurring then it is always the case that the buyout will be paid, or the working interest adjusted to a smaller value to minimize the bankruptcy chance. And that is, indeed, what risk tolerance is all about in the negative expected value situation.

4

PORTFOLIO BALANCING AND RISK-ADJUSTED VALUES UNDER CONSTRAINED BUDGET CONDITIONS

I. INTRODUCTION

Because the exploration for hydrocarbon reserves is a high-risk game in terms of the probability of finding success, procedures for limiting corporate economic exposure are always involved in the decision to pursue particular available opportunities and in decisions concerning the working interest fraction, W, that a particular corporation would prefer to take in a given opportunity. Cozzolino's (1977a, 1978) formula is the most widely used in terms of assessing the risk-adjusted value (RAV) of an opportunity for a given risk tolerance, RT, of a corporation, when probability of success, p_s, value, V (present-day currency), and costs, C (present-day currency), have been evaluated. Cozzolino's formula permits a determination of the optimal working interest (OWI) that maximizes RAV. All references to costs, C, are considered to be positive, with adjustment of formulas to include the appropriate negative signs for all calculations involving costs.

In addition to Cozzolino's procedure (which requires that each corporation use exponential weighting as a risk strategy), MacKay and Lerche (1996c) have recently shown how to incorporate any corporate mandated weighting within the framework of different RAV model behaviors. A comparison of exponential weighting with hyperbolic tangent weighting was given in MacKay and Lerche (1996c) to illustrate numerically the differences and similarities in results from RAV model assumptions. MacKay and Lerche (1996d) have also shown recently that when RT, p_s, V, C, and even W are themselves uncertain, then it is possible to provide a probabilistic range of RAV yielding profit ($RAV > 0$), and also to provide quantitative measures of the relative importance (RI) of the contributions of each uncertainty in the input variables RT, p_s, V, C, and W to the uncertainty in RAV. In this way the need to put further effort into narrowing down uncertainty on the input variables can be prioritized in terms of relative importance to the RAV probability range.

Thus, from an individual opportunity perspective, the ability to categorize RAV within the requirements of any particular corporate setting would seem to be well under control.

However, while each opportunity presented to, or available to, a corporation is capable of being evaluated in isolation, the difficulty is that most corporations do not have an unlimited budget so that they cannot take the optimal working interest for each opportunity; the reason is corporate liability, in the sense that if each project were to fail with probability p_{fi} for the ith project, with a total cost of C_i, and if OWI_i were to be taken in each opportunity, then for N such projects the corporate liability on cost *exposure* is $CO = \sum_{i=1}^{n} C_i \, OWI_i$. In the unlikely event that a corporation is operating on a continuing, steady, cash flow basis, it might choose to limit corporate liability to cost *expenditure*, given as $CE = \sum_{i=1}^{n} C_i \, p_{fi} \, OWI_i$. This situation is rare because of the long delays (3 or more years usually) before a project can yield a positive cash flow. Most corporations insist that working interests W_i be taken such that $CO(W)$ ($\equiv \sum_{i=1}^{N} C_i \, W_i$) be less than or equal to the available budget, B. In such cases one cannot take the optimal working interest in each opportunity. The problem then is to find some procedure to optimize the total RAV from the N opportunities while staying within the mandated corporate budget B, that is, the portfolio of opportunities is balanced. The classic method of resolving the various optimum working interests is with a linear optimization scheme.

The primary purpose of this chapter is to provide an analytic technique, which can be solved with a small calculator, for determining portfolio balancing with a fixed total cost exposure (or cost expenditure) limit for predetermined input parameter values for each opportunity. In addition, if the individual input variables for each opportunity are themselves uncertain then, as will be shown, it is possible to generalize the procedure to yield a probabilistic interpretation, which illustrates how to determine the probability for optimizing the total RAV, and the probable range of working interest for each opportunity.

II. DETERMINISTIC PORTFOLIO BALANCING

Consider a sequence of opportunities, $i = 1, 2, \ldots, N$, in which risk aversion values, $RAV(i; W_i)$ have been computed for each opportunity as a function of working interest, W_i for a corporate risk tolerance, RT. The

RAV values can be computed from any of the risk aversion formulas given in MacKay and Lerche (1996c). The optimum working interest, $OWI(i)$, for each opportunity is also calculated so that the maximum RAV, $RAV_{max}(i, OWI)$, is also known. In making these estimates it is assumed that value V_i, costs C_i, success probability p_{si}, and risk tolerance RT are all *precisely* known. The expected worth of an opportunity is then $E_i = p_{si}V_i - p_{fi}C_i$ (with $p_{fi} = 1 - p_{si}$), while the standard error, σ_i, of the expected worth is $\sigma_i = |(V_i + C_i)| (p_{si}p_{fi})^{1/2}$ so that the volatility v_i of the expected worth is $v_i = \sigma_i/|E_i|$.

A. RELATIVE IMPORTANCE

The relative importance, RI_i, of the ith opportunity in relation to all others is usually defined either as

$$RI_i(W_i) = RAV(i; W_i) / \sum_{j=1}^{N} RAV(j; W_i) \quad \text{(unweighted);} \quad (4.1a)$$

or, if weighted inversely with respect to volatility, as

$$RI_i(W_i) = \frac{\{RAV(i; W_i)/v_i\}}{\sum_{j=1}^{N} \{RAV(j; W_j)/v_j\}}. \quad (4.1b)$$

Here $RAV(i, W_i)$ is positive or zero only. In either event the relative importance of each opportunity is available.

B. PROFITABILITY

The contribution of each opportunity to the total profitability, P, is then usually defined through

$$P_i = RAV(i, W_i) \quad \text{(unweighted)} \quad (4.2a)$$

or

$$P_i = \{RAV(i, W_i)/v_i\} / \sum_{j=1}^{N} 1/v_j \quad \text{(weighted),} \quad (4.2b)$$

with

$$P = \sum_{j=1}^{N} P_i. \quad (4.2c)$$

C. COSTS

Maximum costs of each opportunity, for a fractional working interest W_i, are C_iW_i, and have to be borne by the corporation at the buy-in time.

If the opportunity does not succeed, then no future revenues are ever available against which to replenish the corporate coffers, so a net cost of C_iW_i is the fiscal drain against the corporation. If the opportunity succeeds, then future revenues allow later replenishment of the corporate outlay. However at the time of buy-in to the project, the corporation only has a budget B to use with the estimated costs that it knows it must bear. Two limiting cases are available:

1. *Cost exposure*, defined as total costs

$$C_O = \sum_{i=1}^{N} C_iW_i, \qquad (4.3a)$$

measures the authorization for expenditure and participation in each and every opportunity.

2. *Cost expenditure*, defined as total probable costs

$$C_E = \sum_{i=1}^{N} C_iW_ip_{fi}, \qquad (4.3b)$$

measures the likely amount of cost that one would have to bear on a long-term *continuing cash flow* basis once the failure probability of each opportunity is allowed for. In most situations, it is the cost exposure limit with which a corporation has to deal against budget, and portfolio balancing will be carried through here on a cost exposure basis. Cost expenditure is often used for strategic planning over the 5- to 10-year scale, whereas cost exposure is more usually appropriate for strategic planning on the 1- to 3-year scale.

D. BUDGET CONSTRAINTS

A total budget B is available. The problem is to figure out what fraction of the budget should go to each opportunity in order to maximize the total portfolio of profitability from each. Two cases are available.

1. High Budget

Suppose first that the budget is sufficiently high that participation in each opportunity could be at the OWI_i for each opportunity. Each project is then maximized in terms of its RAV. The total cost exposure is

$$C_O(\max) = \sum_{i=1}^{N} C_i\,OWI_i, \qquad (4.4a)$$

so the budget should be at

$$B \geq C_O(\max), \qquad (4.4b)$$

DETERMINISTIC PORTFOLIO BALANCING

which, for parabolic risk aversion (see Eq. 4.7), is

$$C_O(\max) = RT \sum_{i=1}^{N} \{C_i/(v_i^2 E_i)\}. \tag{4.4c}$$

If budget funds remain $\{B - C_O(\max)\}$ they are returned to the corporation for other projects.

2. Low Budget

If the total budget, B, is less than total cost exposure, $C_O(\max)$, then optimal working interest cannot be taken in each opportunity. One has to settle for less than optimal profitability. The question is to figure out a procedure for balancing the portfolio of opportunities so as to maximize profitability, recognizing that the profitability will be less than optimal. It is this question that is now addressed.

E. FINDING THE BEST WORKING INTEREST

To show how the procedure works at maximizing portfolio balancing, in the next section of the chapter several numerical illustrations of increasing complexity are considered. The Cozzolino (1977a, 1978) risk-adjusted value formula can be used, with

$$RAV = -RT \ln\{p_s \exp(-WV/RT) + p_f \exp(WC/RT)\}, \tag{4.5}$$

which has a maximum at the optimum working interest

$$OWI = RT(V + C)^{-1} \ln\{(p_s V)/(p_f C)\} \tag{4.6a}$$

and, at $W = OWI$,

$$RAV_{\max} = V\,OWI - RT \ln\{p_s(1 + V/C)\}. \tag{4.6b}$$

A useful approximation to the Cozzolino RAV formula is the parabolic RAV formula:

$$RAV = WE\left\{1 - \frac{1}{2}Wv^2 (E/RT)\right\}, \tag{4.7}$$

where the mean value is $E = p_s V - p_f C$ (taken to be positive), and the volatility, v, is given through

$$v^2 = \{(p_s V^2 + p_f C^2)/E^2\} - 1. \tag{4.8}$$

The parabolic RAV formula has a maximum of $RAV_{\max} = \frac{1}{2} RT/v^2$ occurring on $V_{\max} = v^{-2} RT/E$ provided $W_{\max} < 1$, and $RAV = 0$ on $W = 0$ and on $W = 2W_{\max}$ (by definition of a parabola). As can be seen from the above formulas, RAV_{max} and W_{max} are simpler to compute with the parabolic formula, yielding essentially the same results as does the more complex exponential formula.

Here the unweighted procedure will be used together with the parabolic RAV formula for illustrative purposes only, because analytically exact expressions can be written down for the working interests and the total RAV. We then show how to generalize the procedure for any functional form of RAV.

The parabolic RAV formula is more risk averse at high working interest values than is the Cozzolino (1977a, 1978) RAV formula, but is almost identical at low working interests. This property of the parabolic RAV formula parallels real decisions where managers are often quick to farm out an opportunity to around 50% involvement, even when the risk tolerance is high, but are much less willing to take small working interests in an opportunity. If attempts to model this diversification preference are attempted by lowering the risk tolerance in the Cozzolino formula, then the entire range of working interest involvement is affected. However, if one uses the parabolic RAV formula at the same risk tolerance, only the high end of the range of working interests is influenced (by reduction)—precisely what is required to correct for this managerial preference.

1. Parabolic RAV Formula

For each opportunity $i = 1, \ldots, N$, there is some best W_i. The task is to obtain an explicit formula describing W_i. The procedure for so doing is as follows. The total RAV of N projects is

$$RAV = \sum_{i=1}^{N} RAV_i(W_i), \qquad (4.9)$$

where it is understood that $W_i \geq 0$ and $W_i \leq 2W_{max,i}$. It is also understood that the total budget is constrained, $B < RT \sum_{i=1}^{N} \{C_i/(E_i v_i^2)\}$, so that optimum working interest in each and every opportunity *cannot* be taken.

There are N values, $W_1, \ldots, W_N$, to obtain from Eq. (4.9). If no constraints are in place, then each W_i would be at its optimum of OWI_i. But there is the single constraint of

$$B = \sum_{k=1}^{N} C_k W_k, \qquad (4.10)$$

which can be used to write the jth opportunity working interest as

$$W_j = C_j^{-1} \left(B - \sum_{k=1}^{N}{}' C_k W_k \right), \qquad (4.11)$$

where the prime on the summation means that the term $k = j$ is to be omitted. Then rewrite Eq. (4.9) as

$$RAV = \sum_{I=1}^{N}{}'' RAV_I(W_I) + RAV_i(W_i) \qquad (4.12)$$
$$+ RAV_j \left\{ C_j^{-1} \left(B - \sum_{k=1}^{N}{}' C_k W_k \right) \right\},$$

where the double prime means omit the terms at $I = i, I = j$.

Inspection of Eq. (4.12) shows that only the two end terms involve W_i. Then the maximum of RAV with respect to W_i occurs when

$$C_i^{-1} \frac{\partial RAV_i(W_i)}{\partial W_i} - \frac{\partial RAV_j(x)}{\partial x} C_j^{-1} = 0, \qquad (4.13)$$

where $x = C_j^{-1} (B - \sum_{k=1}^{N'} C_k W_k)$. For the parabolic RAV formula one has

$$\frac{\partial RAV_i(W_i)}{\partial W_i} = E_i(1 - W_i v_i^2 E_i/RT). \qquad (4.14)$$

Hence, Eq. (4.13) becomes

$$E_i C_i^{-1} (1 - W_i v_i^2 E_i/RT) - E_j C_j^{-1} \left\{ 1 - (v_j^2 E_j/RT) C_j^{-1} \right. \qquad (4.15a)$$
$$\left. \times \left(B - \sum_{k=1}^{N}{}' C_k W_k \right) \right\} = 0,$$

that is,

$$E_i C_i^{-1} (1 - W_i v_i^2 E_i/RT) = E_j C_j^{-1} (1 - W_j v_j^2 E_j/RT). \qquad (4.15b)$$

But i and j are arbitrary choices, hence each side of Eq. (4.15b) must be a constant, independent of i or j. Let this constant be H so that

$$W_i = (1 - C_i H E_i^{-1}) RT/(v_i^2 E_i), \qquad i = 1, \ldots, N. \qquad (4.16)$$

Then use the fact that

$$\sum_{i=1}^{N} C_i W_i = B \qquad (4.17)$$

to determine H, leading to

$$W_i = RT(E_i v_i^2)^{-1} \left[\frac{1 - (C_i/E_i) \left[\sum_{j=1}^{N} \{C_j/(v_j^2 E_j)\} - B/RT \right]}{\sum_{j=1}^{N} \{C_j^2/(v_j^2 E_j^2)\}} \right], \qquad (4.18)$$

provided $0 \leq W_i \leq \min\{2OWI_i, 1\}$, and under the low budget constraint. Inspection of Eq. (4.18) shows that the budget constraint is just the requirement that

$$\sum_{j=1}^{N} (C_j/v_j^2 E_j) > B/RT \qquad (4.19)$$

and, when this inequality is in force, then each $W_i \leq OWI_i$.

The requirement that $W_i \geq 0$ is that

$$\alpha_i \equiv \sum_{j=1}^{N} \{C_j/(v_j^2 E_j)\} - \left\{\sum_{j=1}^{N} C^2/(v_j^2 E_j^2)\right\} E_i/C_i \leq B/RT. \qquad (4.20)$$

Hence, order the summations in Eqs. (4.18), (4.19), and (4.20) with respect to α_i, with $\alpha_1 < \alpha_2 < \alpha_3 < \cdots < \alpha_N$.

Then as the budget, B, is systematically decreased, $W_N = 0$ when $B = RT\alpha_N$, and the Nth opportunity is removed from consideration. As B systematically decreases, in turn W_{N-1}, W_{N-2}, etc., reach zero and these opportunities are discarded. Thus at any given budget it is relatively simple to determine which opportunities should be invested in, and also the working interest that should be taken.

The analytical exact formula for determining W_i, as given through Eq. (4.18), then maximizes the total RAV of all of the N opportunities under the fixed budget constraint, so that optimum working interest cannot be taken in each and every opportunity.

To draw a close parallel with the single parametric representation occurring with the Cozzolino RAV formula for obtaining best working interests to balance the portfolio of opportunities, one can also write Eq. (4.16) in the form

$$W_i = OWI_i (1 - C_i H/E_i),$$

with

$$H = \frac{\left(\sum_{i=1}^{N} C_i OWI_i - B\right)}{\left(\sum_{i=1}^{N} C_i^2 OWI_i/E_i\right)} \geq 0.$$

Then note that $W_i = 0$ on $H = E_i/C_i$. Thus, if the summations are organized in order $E_1/C_1 > E_2/C_2 > \ldots > E_N/C_N$ then, as B decreases, so H increases. Then, as H crosses E_N/C_N, the Nth opportunity has $W_N = 0$ and so is discarded. The remaining $N - 1$ opportunities are then used to recalculate the summations from $i = 1$ to $N - 1$, and H increased (B decreased) until $W_{N-1} = 0$; the process is then repeated. The last opportunity to be discarded is when $H = E_1/C_1$, when $B = 0$. Thus, as the budget decreases from the maximum of $B_{\max} = \sum_{i=1}^{N} C_i OWI_i$ (at which optimal working interest can be taken in each and every opportunity) to $B = 0$, the various opportunities are steadily discarded in order of their values of E/C; this parameter is similar to the economic evaluation tool of risked present worth investment (PWI).

2. Cozzolino RAV Formula

The procedure used with the parabolic RAV formula to obtain an analytic expression for each W_i in the portfolio can also be used with the Cozzolino RAV formula, although in this case a closed-form analytic solution for W_i is not possible, but a single parametric representation does work. Proceeding through the differentiations of the previous section, the difference arises at the point where the explicit functional form of the RAV is used to compute $\partial RAV(W)/\partial W$. Instead of the parabolic $RAV(W)$ formula of Eq. (4.7), if one uses the Cozzolino RAV formula of Eq. (4.5), a little bit of arithmetic then yields the following optimization:

$$W_j = RT(V_j + C_j)^{-1} \ln\{p_{sj} p_{fj}^{-1} (V_j/C_j - H)/(1 + H)\}, \qquad (4.21)$$

where the single parameter H is to be determined from

$$B = RT \sum_{j=1}^{N} C_j (V_j + C_j)^{-1} \ln\{p_{sj} p_{fj}^{-1} (V_j/C_j - H)/(1 + H)\}. \qquad (4.22)$$

Note that the Cozzolino RAV formula has

$$OWI_i = RT(V_i + C_i)^{-1} \ln\{p_{si} p_{fi}^{-1} (V_i/C_i)\}, \qquad (4.23)$$

which occurs on $H = 0$.

Also note from Eq. (4.21) that

$$W_j = 0 \quad \text{on} \quad H = E_j/C_j \equiv (p_{sj} V_j - p_{fj} C_j)/C_j. \qquad (4.24a)$$

Thus, if the terms in the summation in Eq. (4.22) are ordered in respect of decreasing values of E_j/C_j, with $E_1/C_1 > E_2/C_2 > \cdots > E_N/C_N$, then as H increases from zero, W_N will first go to zero and, thereafter, at higher H, the Nth opportunity is not considered. As H increases further, W_{N-1} will next go to zero, and so on. If one proceeds in this manner, the final term to drop is W_1, at which point $B = 0$. Thus, increasing $H \geq 0$ corresponds to a steadily decreasing budget, with $H = 0$ corresponding to a budget so that OWI can be taken in each and every opportunity. A practical strategy for determining H given the budget B is as follows. Rewrite Eq. (4.22) in the form

$$\ln(1 + H) = (a - B/RT) b^{-1} + b^{-1} \sum_{j=1}^{N} \frac{C_j}{(C_j + V_j)} \ln(V_j/C_j - H), \qquad (4.24b)$$

where

$$a = \sum_{j=1}^{N} (C_j + V_j)^{-1} C_j \ln(p_{sj}/p_{fj}) \qquad (4.25a)$$

and

$$b = \sum_{j=1}^{N} C_j/(C_j + V_j)^{-1} \quad (4.25b)$$

and under the conditions that only those terms in the respective sums are to be included for which $H < E_j/C_j$.

Then, for a given budget B, Eq. (4.24) can be iterated to provide H, with the $(n + 1)$th iteration given through

$$\ln(1 + H_{n+1}) = (a - B/RT)\, b^{-1} \\ + b^{-1} \sum_{j=1}^{N} C_j(C_j + V_j)^{-1} \ln\{(V_j/C_j) - H_n\}, \quad (4.26)$$

starting from $H_0 = 0$ on the zeroth iteration, $n = 0$. Thus, at each iteration one checks to see if $W_j \geq 0$ (from Eq. 4.21) with the respective H_n. If $W_j < 0$ then that term in the summations of Eqs. (4.25a), (4.25b), and (4.26) is ignored, and the iteration repeated. Pragmatically, convergence is extremely rapid, with no more than 5–10 iterations ever needed. In this way H is determined for a fixed budget, B.

The procedure is clear: For a given budget, steadily increase H until the appropriate solution to Eq. (4.22) is obtained *including* dropping terms as H crosses each E_i/C_i. Then, with H thus obtained, merely insert H into Eq. (4.21) to obtain the working interests that optimize the portfolio of opportunities.

Thus, portfolio optimization for the Cozzolino formula is reduced to a single parametric determination for a given budget. Clearly for *any* functional form chosen for $RAV(W)$ the logic procedure is applied as illustrated here.

This general argument, although explicitly developed here for unweighted RAV formulas, is also appropriate if any weighting is done. The logic proceeds as above, *mutatis mutandis* (see the appendix to this chapter).

III. NUMERICAL ILLUSTRATIONS WITH FIXED PARAMETERS: COMPARISON OF THREE OPPORTUNITIES

Consider three opportunities A, B, and C with the following characteristic parameters:

Opportunity A: $V = \$110$ MM; $p_s = 0.5 = p_f$; $C = \$10$ MM; $RT = \$30$ MM.
Opportunity B: $V = \$200$ MM; $p_s = 0.5$; $p_f = 0.5$; $C = \$100$ MM; $RT = \$30$ MM.
Opportunity C: $V = \$300$ MM; $p_s = 0.4$; $p_f = 0.6$; $C = \$120$ MM; $RT = \$30$ MM.

Opportunities A and B each have a mean value of $E = \$50$ MM while opportunity C has $E = \$48$ MM. Using the parabolic formula, opportunity A has $RAV(A)_{max} = \$10.417$ MM at an $OWI(A) = 0.417$; opportunity B has $RAV(B)_{max} = \$1.667$ MM at $OWI(B) = 0.067$; and opportunity C has $RAV(C)_{max} = \$0.816$ MM at $OWI(C) = 0.034$, for a total possible maximum RAV of \$12.9 MM. Note that $E(A)/C(A) = 5$; $E(B)/C(B) = 0.5$; and $E(C)/C(C) = 0.3$ so that, as the budget decreases, eventually opportunity C is expected to be least worthwhile, followed by opportunity B.

Optimization of total RAV at different budgets has been carried out under two conditions: (1) using the analytic formula for parabolic RAV given in the preceding section and (2) comparing the results obtained from the analytic formula with those arising from a Simplex solution solver, which has been the standard corporate workhorse until the current advent of analytic representations. Both procedures yielded identical results under all conditions addressed, but the new analytic method is much faster numerically than the Simplex search method.

A. BUDGET OF $B = \$20$ MM

The value of $C \times OWI$ is \$4.16 MM for opportunity A, \$6.667 MM for opportunity B, and \$4.082 MM for opportunity C, for a total cost of $\Sigma C_i OWI_i = \$14.915$ MM.

Thus, the budget of \$20 MM exceeds the total costs at the optimum working interest for each opportunity, so that OWI is taken in each at a cost of \$14.915 MM; a budget return of \$5.085 MM to the corporate coffers is then made.

B. BUDGET OF $B = \$11$ MM

In this case the budget is less than the total needed to invest in each opportunity at its OWI, but is large enough so that no single opportunity should be discounted. Both the parabolic analytic formula and the Simplex method return the optimum values of:

$W(A) = 0.4033$; $RAV(A) = \$10.406$ MM; $C(A)W(A) = \$4.0328$ MM
$W(B) = 0.0452$; $RAV(B) = \$1.4946$ MM; $C(B)W(B) = \$4.5248$ MM
$W(C) = 0.0204$; $RAV(C) = \$0.6847$ MM; $C(C)W(C) = \$2.4424$ MM

for a total RAV of \$12.585 MM. In this case note that the optimization of total RAV has kept the working interest in opportunity A very close to its OWI, and has kept the involvement in opportunity B at about 3/4 of the OWI, but has dropped the involvement in opportunity C to about 2/3 of its *OWI*—in line with E/C being smallest for opportunity C, so that the lower budget is forcing the working interest in opportunity C closer to zero first.

C. BUDGET OF B = $4 MM

In this case the threshold value of $E/C = 0.3$ has been crossed for opportunity C, which is therefore discounted completely. The budget is then split between opportunities A and B in the proportions:

$W(A) = 0.3765$; $RAV(A) = \$10.3$ MM; $C(A)W(A) = \$3.7647$ MM
$W(B) = 0.0024$; $RAV(B) = \$0.115$MM; $C(B)W(B) = \$0.2353$ MM

for a total *RAV* of $10.415 MM. In this case, because opportunity B has a lower E/C ($\equiv 0.5$) than opportunity A ($\equiv 5$), the lower budget forces less involvement in opportunity B. The total maximum *RAV* for all three opportunities is $12.9 MM, so that portfolio balancing yields 97.6% (at $B = \$11$ MM) and 80.9% (at $B = \$4$ MM) of the maximum by optimizing the budget fractions allocated to each opportunity.

IV. PROBABILISTIC PORTFOLIO BALANCING

For ranges of uncertainty on value V, cost C, success probability p_s, and risk tolerance RT for each opportunity, MacKay and Lerche (1996d) have shown how to construct a cumulative probability distribution for *RAV* at different values of working interest, W. Also shown in MacKay and Lerche (1996d) are (1) the construction of a cumulative probability for $RAV(i, W)$ when W is also allowed to vary within a prescribed range and (2) the cumulative probability for *OWI* as p_s, V, and C vary. A measure of the worth of the ith opportunity is taken to be the average value for *RAV*, written $\langle RAV(i, W) \rangle$ where angular brackets denote average values; a measure of uncertainty on this value is taken to be the volatility with

$$v_i(W) = \{RAV_{90}(i, W) - RAV_{10}(i, W)\}/\langle RAV(i, W) \rangle, \quad (4.27)$$

where RAV_{90} and RAV_{10} are the values of *RAV* at 90 and 10% cumulative probability, respectively. It is also taken that $\langle RAV(i, W) \rangle > 0$.

In this case a volatility weighted relative importance, $RI(i, W_i)$ is given through

$$RI(i, W_i) = \frac{[\langle RAV(i, W_i) \rangle / v_i(W_i)]}{\sum_{j=1}^{N} \{\langle RAV(j, W_j) \rangle / v_j(W_j)\}}, \quad (4.28)$$

while an unweighted relative importance can be given through

$$RI(i, W_i) = \frac{\langle RAV(i, W_i) \rangle}{\sum_{j=1}^{N} (\langle RAV(j, W_j) \rangle)}. \tag{4.29}$$

The contribution, P_i, to the total probable profit, P, at the average cumulative value is

$$P_i = \langle RAV(i, W_i) \rangle \quad \text{(unweighted)} \tag{4.30a}$$

or

$$P_i = \frac{(\langle RAV(i, W_i) \rangle / v_i(W_i))}{\sum_{j=1}^{N} 1/v_j(W_j)} \quad \text{(weighted)}, \tag{4.30b}$$

with

$$P = \sum_{i=1}^{N} P_i. \tag{4.30c}$$

The problem then is to determine the fractional working interests appropriate to maximize likely probability of profit given the uncertainties and given a fixed budget, B.

Consider the deterministic expressions given in Section II for fixed values of p_{si}, V_i, and C_i for each opportunity ($i = 1, \ldots, N$) together with the fixed values of RT and budget, B. Corporate budget and corporate risk tolerance are more nearly carved in stone than the variations known to occur in estimates of p_{si}, V_i, and C_i for each opportunity. Thus, if p_{si}, V_i, and C_i are allowed to vary within preset ranges for each opportunity and with known distributions, then Monte Carlo simulations can be run using the deterministic formulas of Section II and constructions made of the probable ranges of each W_i, together with the probable RAV distribution for each opportunity and for the total of all opportunities. One merely has to record frequency of occurrence of events and also the individual variations of the intrinsic parameters which contribute the largest degrees of uncertainty to the probabilistic outcomes, in a manner similar to that done elsewhere (MacKay and Lerche, 1996d) for a different problem.

In this way not only can portfolio balancing be done in a deterministic manner, but the balancing can also be done in a probabilistic framework. In addition, cost exposure is used to balance a fixed budget.

V. NUMERICAL ILLUSTRATIONS WITH VARIABLE PARAMETERS

A. VARIABLE VALUES

The value, V, ascribed to an opportunity is usually somewhat uncertain due to the two dominant factors of potential reserve assessments and potential future selling price. As a consequence there is some fluctuation to be expected in the optimum working interest that should be taken in each opportunity and in the total RAV.

To illustrate the effect of uncertainties in value, V, on the optimization, the values for the three opportunities presented in the previous section were used and the values allowed to vary uniformly by $\pm 10\%$ around central values. Thus for opportunity A, $100 MM $\leq V \leq$ $120 MM, with a central value of $110 MM; for opportunity B, $180 MM $\leq V \leq$ $220 MM, with a central value of $200 MM; for opportunity C, $270 MM $\leq V \leq$ $330 MM, with a central value of $300 MM. The Monte Carlo procedure then selected values at random from each of the three distributions in constructing cumulative probability distributions for total RAV and working interest for each opportunity using the analytic formulas for parabolic working interest given in Section II. Approximately 1500 trials were run on a small PC in approximately 6 minutes (the Simplex solution solver was not used in this exercise because it was far too slow except to spot-check individual runs). The computations were done for a budget of $11 MM, a case for which all three opportunities have some working interest at the center values, as exhibited in Section III.

Shown in Figure 4.1 is the cumulative probability plot for total RAV with the abscissa in units of $MM. In this case P_{68} (an approximate estimate

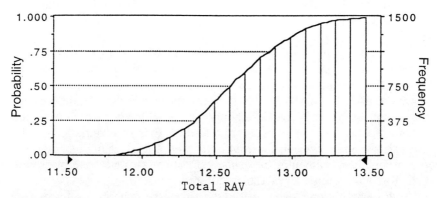

FIGURE 4.1 Cumulative probability distribution plot for total RAV when values for each of the three opportunities vary by $\pm 10\%$ around their central values, for a budget of $11 MM. The abscissa unit is $MM.

of the mean) occurs at $RAV_{68} = \$12.78$ MM; while P_{90} is at $RAV_{10} = \$13.06$ MM and P_{10} at $RAV_{10} = \$12.13$ MM, so the volatility of the 68% chance is only $v = 7.2\%$, indicating a high degree of confidence in an overall RAV of around \$12.78 MM. There is a 2/3 chance the total RAV will be less than \$12.78 MM but only a 10% chance of *less* than \$12.13 MM, and only a 10% chance of *greater* than \$13.06 MM. Hence, one can write $RAV = \$12.78^{+0.28}_{-0.65}$ MM to display directly the uncertainty at the 10 and 90% levels. Of interest also is the range of uncertainty on working interest for each opportunity, exhibited in Figures 4.2a–c for opportunities A, B, and C, respectively. Note for opportunity A that P_{68} occurs at $W_{68}(A) = 0.41$, while P_{90} and P_{10} occur, respectively, at $W_{90}(A) = 0.43$ and $W_{10}(A) = 0.38$, so that the volatility on working interest is $v(A) = 12.2\%$; equally, from Figure 4.2b, one has $W_{68}(B) = 0.045$, $W_{90}(B) = 0.0474$, and $W_{10}(B) = 0.0425$, for a volatility of $v(B) = 13.1\%$; while, from Figure 4.2c, one has $W_{68}(C) = 0.022$, $W_{90}(C) = 0.025$, and $W_{10}(C) = 0.015$ for a volatility of $v(C) = 45\%$. The increasing volatilities of the different opportunities reflect directly the fact that the closer the budget is to crossing over the threshold at which $H = E/C$ for each opportunity, the lower the working interest to be taken in that opportunity. Because opportunity C is the closest to this threshold, that fact is picked up in the greater degree of volatility on the likely working interest to be taken.

B. VARIABLE VALUES AND COSTS

In addition to the value of an opportunity being uncertain because of uncertainties in reserve estimates and future selling price, it is also usually the situation that opportunity cost estimates are also uncertain.

To illustrate how variations in cost influence total RAV and working interest in the portfolio, the costs of each opportunity were also allowed to vary randomly with a uniform distribution, centered on the deterministic values used in Section III and with 10% variation around the central values. Thus, for opportunity A, $\$9$ MM $\leq C \leq \$11$ MM; for opportunity B, $\$90$ MM $\leq C \leq \$110$ MM; and for opportunity C, $\$108$ MM $\leq C \leq \$132$ MM. The values for each opportunity were also taken to vary uniformly within 10% of their central values as for the previous example. Again a Monte Carlo set of 1500 random runs was made using a small PC with the following results. As shown in Figure 4.3, the total RAV is now changed due to the extra "spread" arising from the cost uncertainty. Thus, $RAV_{68} = \$12.8$ MM, $RAV_{90} = \$13.25$ MM, and $RAV_{10} = \$12.0$ MM, so that the volatility is now $v = 9.8\%$, an increase of 2.6% relative to the previous case of variable values only.

Equally the probable working interests for each opportunity also reflect the extra uncertainty caused by cost fluctuations. Shown in Figures 4.4a–c, for opportunities A, B, and C, respectively, are the cumulative probability

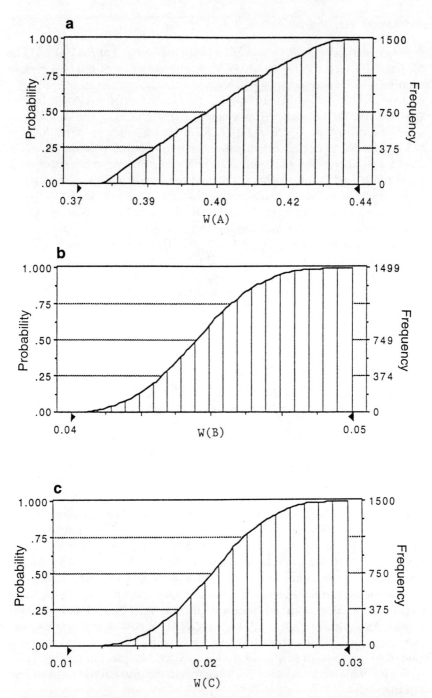

FIGURE 4.2 Cumulative probability distribution plot for working interest optimizing the total RAV when values for each of the three opportunities vary by ±10% around their central values, for a budget of $11 MM. The abscissa unit is fractional working interest (a) for opportunity A; (b) for opportunity B; (c) for opportunity C.

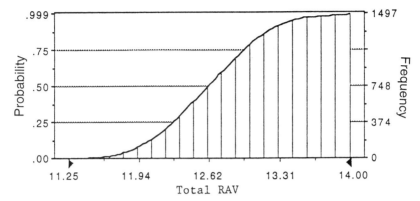

FIGURE 4.3 As for Figure 4.1 but now including the cost of each of the opportunities being uncertain by ±10% around their central values. The budget is at $11 MM.

distributions of working interest. Reading from Figure 4.4, we get:

$W_{68}(A) = 0.41$, $W_{90}(A) = 0.43$, $W_{10}(A) = 0.38$, $v(A) = 12.2\%$
$W_{68}(B) = 0.05$, $W_{90}(B) = 0.06$, $W_{10}(B) = 0.04$, $v(B) = 40\%$
$W_{68}(C) = 0.022$, $W_{90}(C) = 0.029$, $W_{10}(C) = 0.015$, $v(C) = 64\%$.

Thus, there is a significant increase produced in the range of uncertainty of the probable working interests for opportunities B and C by the 10% fluctuations in costs of each opportunity, representing the fact that low value and high cost, when taken together, both drive E/C to a smaller value and so influence the closeness of the working interest to the point where $H = E/C$ demands no involvement in the opportunity. Put another way, near the peaks of the individual RAV curves there is less influence of variations in input uncertainties because there is no slope at the peak of RAV with W. But the steepness of the slopes away from the peaks, particularly for opportunity C, causes a rapid slide away from significant investment in that opportunity relative to less steeply sloping opportunities.

C. VARIABLE VALUES, COSTS, AND SUCCESS PROBABILITIES

Apart from uncertainties in values and costs, it is also the case that the chance of success, p_s, is uncertain. To illustrate the additional effects on RAV and working interest probabilities caused by uncertainty in success probability, and to keep a balance with the prior two examples, a 10% uncertainty on p_s for each opportunity, drawn from a uniform population, is now included with the 10% uncertainties on V and C. Thus, for opportunities A and B, $0.45 \leq p_s \leq 0.55$; for opportunity C, $0.36 \leq p_s \leq 0.44$, with each centered at the deterministic values of Section III. A suite of 1500

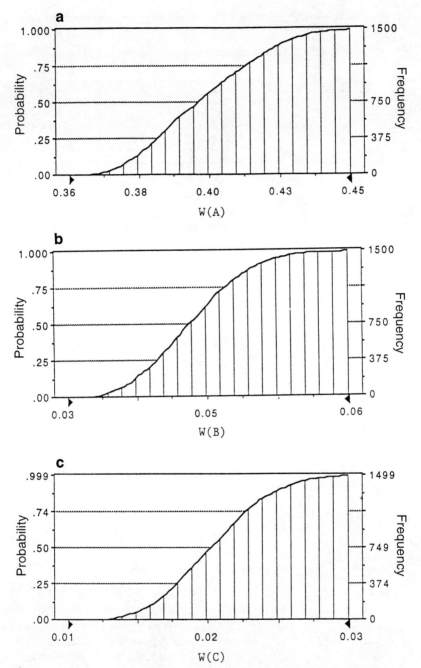

FIGURE 4.4 As for Figure 4.2, but now including the cost of each of the opportunities being uncertain by ±10% around their central values. The budget is at $11 MM. (a) For opportunity A; (b) for opportunity B; (c) for opportunity C.

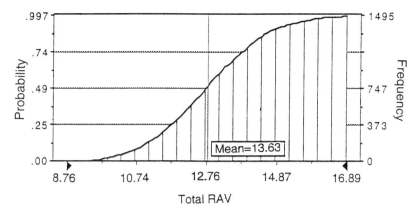

FIGURE 4.5 As for Figure 4.3 but now also including the probability of success for each of the opportunities being uncertain by ±10% around their central values. The budget is at $11 MM.

Monte Carlo runs was again done, with results for cumulative *RAV* probability shown in Figure 4.5 and for working interest probability in Figures 4.6a–c for opportunities A, B, and C, respectively. The budget was kept at $11 MM.

Note from Figure 4.5 that RAV_{68} = $13.63 MM, RAV_{90} = $14.87 MM, and RAV_{10} = $10.74 MM, for a volatility of v = 30%—a considerable increase compared to the prior two examples and reflecting the relative importance of uncertainty in success probability in determining *RAV*.

Reading from Figure 4.6, the working interests for each opportunity are:

$W_{68}(A)$ = 0.42, $W_{90}(A)$ = 0.45, $W_{10}(A)$ = 0.36, $v(A)$ = 9.4%
$W_{68}(B)$ = 0.05, $W_{90}(B)$ = 0.06, $W_{10}(B)$ = 0.03, $v(B)$ = 60%
$W_{68}(C)$ = 0.023, $W_{90}(C)$ = 0.032, $W_{10}(C)$ = 0.01, $v(C)$ = 96%.

Thus, the 10% fluctuations in success probability have a considerable effect on the uncertainties in working interest for the more risky opportunities B and C, with opportunity C clearly dominated by being closer to the critical threshold of no involvement.

D. VERY LOW BUDGET

The prior three examples were all carried out for a budget of $11 MM, at which partial involvement could be taken in all three opportunities.

Consider, then, the same uncertainty ranges on values, costs, and success probabilities for each opportunity, but when the budget is lowered to $4 MM. This situation was shown in Section III (the deterministic case) to lead to no involvement in opportunity C, and the budget was partitioned only between opportunities A and B. Again, a suite of 1500 Monte Carlo

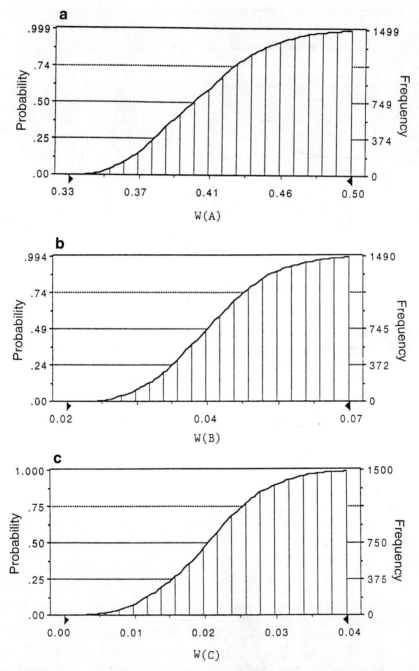

FIGURE 4.6 As for Figure 4.4, but now also including the probability of success for each of the opportunities being uncertain by ±10% around their central values. The budget is at $11 MM. (a) For opportunity A; (b) for opportunity B; (c) for opportunity C.

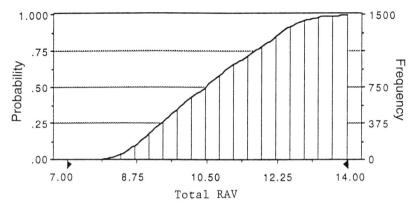

FIGURE 4.7 As for Figure 4.5 but with the budget reduced to $4 MM.

runs was done with the results exhibited in Figure 4.7 for total RAV probability and in Figures 4.8a–c for cumulative working interest probability for opportunities A, B, and C, respectively.

Reading from Figure 4.7 indicates that RAV_{68} = $11.3 MM, RAV_{90} = $12.45 MM, and RAV_{10} = $8.74 MM, with a volatility v = 33.5%; Figure 4.8 indicates that it is still the case that no involvement should be taken in opportunity C. Figures 4.8a and b indicate that:

$W_{68}(A) = 0.39; W_{90}(A) = 0.42; W_{10}(A) = 0.33, v(A) = 23\%$
$W_{68}(B) = 0.048; W_{90}(B) = 0.055; W_{10}(B) = 0, v(B) = 115\%$.

Thus, most of the budget is now committed to opportunity A, and the volatility of opportunity B makes that opportunity one of higher risk.

E. LOW TO HIGH BUDGET COMPARISON

As the budget varies, for given ranges of uncertainty on values, costs, and success probabilities, the relative importance is needed of each uncertainty in influencing the range of variation of the total RAV. The importance of this information lies in the fact that if one wishes to determine better the investment in each opportunity, then it is crucial to be aware of which parameter values, and their ranges of uncertainty, are causing the largest degree of variability in RAV. In this way the decision can be made as to which parameters to focus on in order to narrow their ranges of uncertainty. The relative importance (RI) of the contribution of each parameter to uncertainty in the total RAV cumulative probability distribution provides the requisite measure. Note that this procedure only prioritizes the relative dominance of individual parameters. Thus one has a measure that *if* one wishes to reduce uncertainty ranges than the RI provides the information

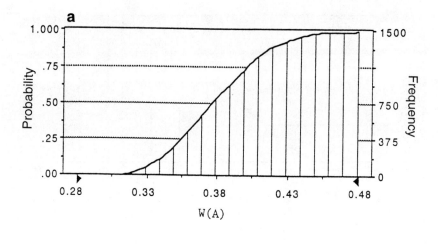
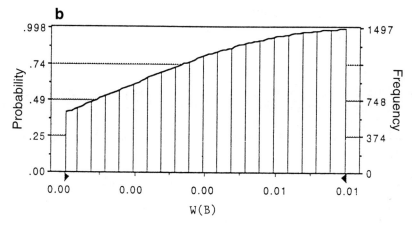
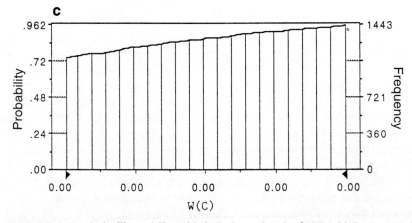

FIGURE 4.8 As for Figure 4.6 but with the budget reduced to $4 MM. (a) For opportunity A; (b) for opportunity B; (c) for opportunity C.

on where to expend effort to narrow uncertainty; the volatility provides a measure of *when* one needs to improve uncertainty ranges.

To illustrate the effect of each of the ranges of variability of values, costs, and success probabilities, Figures 4.9a and 4.9b plot the relative importances (in %) of each uncertainty factor used in the previous numerical examples for the cases of a budget of $11 MM (Figure 4.9a) and $4 MM (Figure 4.9b). Because investment can be taken in all three opportunities in the case of a budget of $11 MM (albeit with only small *RAV* values for opportunities B and C compared to opportunity A), Figure 4.9a shows that

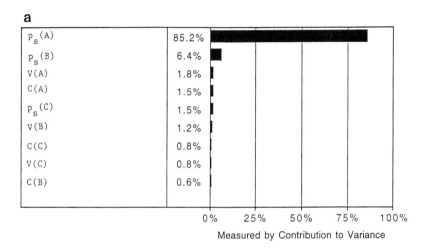

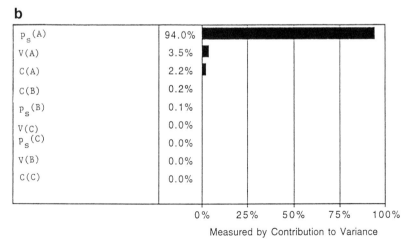

FIGURE 4.9 Relative importance diagram for individual parameters influencing the uncertainty on the total *RAV*. Note the dominance of the uncertainty in probability of success for opportunity A (a) for a budget of $11 MM; (b) for a budget of $4 MM.

there is some relative contribution to the total RAV variance of each uncertainty. However, what is clear is that the uncertainty in success probability for opportunity A is absolutely dominant in causing the majority (85%) of the variance in RAV, with about 7% caused by uncertainty in success probability for opportunity B. The remaining factors add only 8% cumulatively. Clearly, if the uncertainty on total RAV is to be more limited in the $11 MM budget case, then most effort (85%) has to be put on narrowing the range of uncertainty of success probability for opportunity A; all other factors are secondary to this goal.

Equally, lowering the budget to $4 MM removes opportunity C as a contributor to the total RAV, as reflected in Figure 4.9b where there is no contribution to the uncertainty in total RAV from parameter variations of opportunity C. Because a greater fraction of the budget now goes into opportunity A, the dominance of the uncertainty in success probability for opportunity A is increased in relative importance to 94%, with the uncertainties in the value of A and the cost of A adding another 5.7%. The uncertainty in the success probability and cost for opportunity B are now of extremely small relative importance. Thus, in this low budget situation, even more effort should be put into attempting to narrow down the range of uncertainty of the chance of success for opportunity A.

The advantage to having available the ability to determine which factors are causing the greatest uncertainty in estimates of total RAV in a portfolio is that one can quickly determine where to place effort to improve matters, without spending effort on factors that do not play a significant role. It is this fact that the relative importance figures provide in simple graphical form.

VI. COMPARISON OF PARABOLIC AND COZZOLINO RAV RESULTS

A. DETERMINISTIC RESULTS

The three opportunities in Sections III and V were arranged with opportunity A always dominant compared to opportunities B and C, so that it was easy to visualize the widely separated parameter ranges which showed the relative impact of changes influencing RAV results.

However, it is most often the case that the parameters describing different opportunities are much closer together than the simple examples used in the previous sections for illustrative purposes. Accordingly, in this section of the chapter a discussion is given of three opportunities with parameters as shown in Table 4.1, together with a fixed risk tolerance of $RT = \$300$ MM and a budget $B = \$100$ MM, so that OWI cannot be taken in each and every opportunity. Because the functional forms are different for the

TABLE 4.1 Parameters for Opportunities A, B, and C

Opportunity	V ($MM)	C ($MM)	p_s	p_f	E ($MM)	E/C
A	100	80	0.6	0.4	28	0.35
B	200	100	0.5	0.5	50	0.5
C	300	120	0.4	0.6	48	0.4

Cozzolino and parabolic RAV formulas, different *OWI* ensue, as shown in Table 4.2. At a budget of $100 MM and *RT* = $300 MM, the portfolio is optimized as shown in Table 4.3. There is, then, little to choose between the values of *RAV* or working interest from either the parabolic or Cozzolino formulas. Perhaps the point to be made is that the two formulas for RAV reflect a rugged insensitivity of the optimum *RAV* for the total portfolio to particular choices of functional *RAV* forms for each opportunity. Such stability of results is of great use when planning optimization strategies, because the results are not volatile, implying a statistical sharpness to financial decisions for involvement in the different opportunities.

The major part of the reason for this close parallel between the optimizations of the parabolic and Cozzolino risk-adjusted formulas is that the parabolic *RAV* formula is a Taylor series expansion (to quadratic order) in working interest of the Cozzolino exponential formula. Because all optimization values have positive slopes for $\partial RAV(W)/\partial W$, each *RAV* formula yields a working interest less than *OWI*, precisely the domain where the parabolic and Cozzolino *RAV* formulas are very close to each other.

B. STATISTICAL RESULTS

Because of the close parallel between the parabolic and Cozzolino *RAV* formulas in the region where $W < OWI$, it follows that one also anticipates a close parallel in probabilistic ranges of output when values, costs, and

TABLE 4.2 Comparison of Parabolic and Cozzolino Results

	Parabolic		Cozzolino	
Opportunity	OWI	C × OWI ($MM)	OWI	C × OWI ($MM)
A	1.0	80.000	1.00	80.000
B	0.67	66.667	0.693	69.315
C	0.34	40.816	0.365	43.785
Total		187.483		193.090

TABLE 4.3 Comparison of Parabolic and Cozzolino Results

	Parabolic			Cozzolino		
Opportunity	W	C × W ($MM)	RAV	W	C × W ($MM)	RAV
A	0.4793	38.345	10.443	0.4495	35.963	10.422
B	0.4071	40.706	14.139	0.419	41.937	14.470
C	0.1746	20.949	6.229	0.184	22.100	6.473
Total			30.811			31.365

success probabilities are allowed to vary for each opportunity. To demonstrate that such a close parallel does in fact occur, 1500 Monte Carlo simulations were run for both the parabolic and Cozzolino RAV formulas, respectively, when $\pm 10\%$ uncertainty is permitted on each of V, p_s, and C for each of the three opportunities, with uniform probability distributions for each parameter, and with centered values as for the deterministic case given above. The risk tolerance, RT, was held fixed at $300 MM, and the budget, B, was held constant at $100 MM for all runs.

Figure 4.10 shows the cumulative probability distributions for RAV from the parabolic (Figure 4.10a) and Cozzolino (Figure 4.10b) formulas. The difference between the two plots of less than 2% is minor, reinforcing the anticipation of only minor shifts.

Figure 4.11 shows the cumulative probability distributions for working interest in the three opportunities using both the parabolic (Figures 4.11a–c) and Cozzolino (Figures 4.11d–f) RAV formulas. Again there is little difference in the working interest probability distributions (less than about 2%), reflecting the close parallel of the two formulas for RAV in the domain where $W < OWI$.

The relative importance of the variations in the input parameters V, C, and p_s for each opportunity in contributing to uncertainty on the optimum RAV of the portfolio also reflects the close parallel of the parabolic (Figure 4.12a) and Cozzolino (Figure 4.12b) formulas. Both Figures 4.12a and 4.12b indicate that it is the uncertainty in success probabilities of opportunities B and A which dominate the relative importance, with about 34% from $p_s(B)$ and 25% from $p_s(A)$, for a total of around 59% of the total variance in RAV for the portfolio. The uncertainty in working interest for any opportunity is also closely parallel for both formulas, as exhibited in Figure 4.13 for opportunity A, with $p_s(A)$, $C(A)$, and $p_s(B)$ contributing about 75% of the variance in both Figures 4.13a (parabolic RAV) and 4.13b (Cozzolino RAV). Similar parallelism also occurs for opportunities B and C (not shown here).

The conclusion is clear: The expectation that the parabolic and Cozzolino

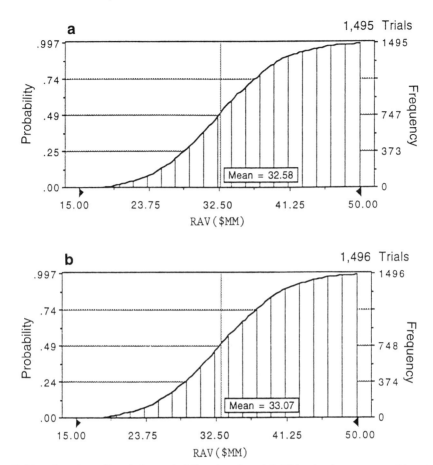

FIGURE 4.10 Cumulative probability plot of total *RAV* for the three opportunities of Section VI: (a) for the parabolic *RAV* formula; (b) for the Cozzolino formula. The abscissa is *RAV* in $MM.

RAV formulas provide closely parallel results for portfolio optimization is substantiated in both the deterministic and statistical treatments. Either formula will lead to a portfolio optimization that is ruggedly stable and insensitive to the precise details of the functional form, making for trustworthy participation decisions in multiple opportunities, and allowing identification of the sensitivity and relative importance of the various input variables in influencing both the uncertainty of the total *RAV*, and the uncertainty of the individual working interests for each opportunity. In this way priority can be assigned to improving individual factors influencing investment decisions without the concern that such priorities are *RAV* model dependent to a high degree—such is not the case.

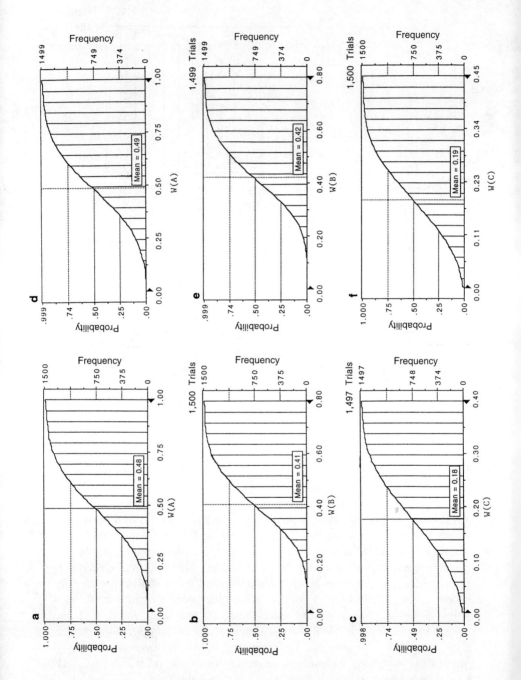

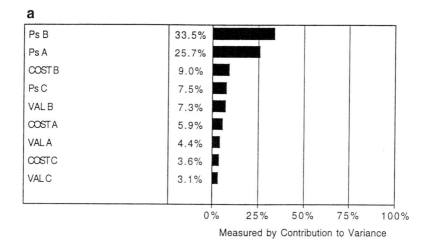

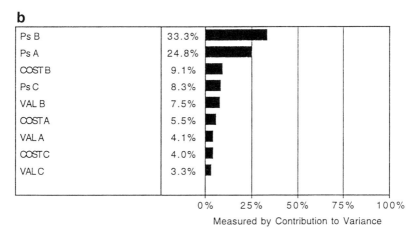

FIGURE 4.12 Relative importance diagram of ±10% variations in V, C, and p_s for each opportunity in contributing to the uncertainty on the total RAV of the portfolio: (a) for the parabolic RAV formula; (b) for the Cozzolino RAV formula.

VII. DISCUSSION AND CONCLUSIONS

This chapter has served two purposes: First was the need to obtain analytic expressions for optimizing total risk-adjusted value for a portfolio of opportunities in the face of a constrained budget; second was the need

←
FIGURE 4.11 Cumulative probability plot of fractional working interest for the three opportunities of Section V. The parabolic RAV formula was used to obtain the plots for (a) opportunity A; (b) opportunity B; (c) opportunity C. The Cozzolino RAV formula was used to obtain the plots for (d) opportunity A; (e) opportunity B; (f) opportunity C.

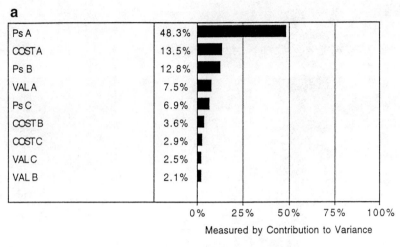

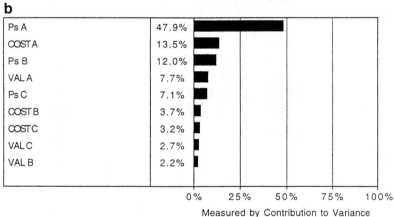

FIGURE 4.13 Relative importance diagram of ±10% variations in V, C, and p_s for each opportunity in contributing to the uncertainty on working interest for opportunity A: (a) for the parabolic RAV formula; (b) for the Cozzolino RAV formula.

to determine the effect of ranges of uncertainties of values, costs, and success chances for each opportunity on the total RAV, and to figure out a procedure to identify which parameters were causing the greatest uncertainty.

With respect to the first purpose, it was shown for a parabolic profile of RAV versus working interest for each opportunity that a closed-form expression could be written down exactly for the working interest that should be taken in each opportunity in order to maximize the total RAV under a constrained budget condition. It was also shown for the Cozzolino RAV formula (which uses exponential weighting) that the same procedure allows one to write down an analytic, but parametric, relation between working interest and total budget in terms of a single parameter. The procedure developed operates with any functional form chosen for *RAV* versus working interest. Numerical illustrations of the procedure indicated the pragmatic operation of the method (which is extremely fast numerically compared to the more conventional Simplex solution solver methods that have been used to date).

With respect to the second purpose, the ability to provide a rapid investigation of ranges of uncertainty of values, costs, and chances of success for each opportunity on the total *RAV* probability and probable working interest ranges for each opportunity means that one can quickly focus on which parameters need to be addressed if one is to narrow the uncertainty on total *RAV* for a given budget or, indeed, for a range of possible budgets.

The methods presented here can be utilized with any functional form of *RAV* dependence on working interest to provide simple expressions, either in closed form or involving only a single parametric representation, for the working interest to be taken to optimize a portfolio of opportunities. It is this fact that both the deterministic and probabilistic numerical illustrations have been designed to exhibit.

VIII. APPENDIX: WEIGHTED *RAV* OPTIMIZATION

In the body of the chapter the total RAV was maximized with each opportunity having its *RAV* added arithmetically to the total. However, corporations often weight the relative *RAV* contributions. The purpose of this appendix is to show that the optimization of the portfolio can equally be carried through with weighting factors.

Thus, if the weights assigned are $\sigma_i(>0)$ $(i = 1, \ldots, N)$ for $RAV_i(W_i)$, then the total weighted *RAV* is

$$RAV = \sum_{i=1}^{N} \sigma_i RAV_i(W_i) \bigg/ \sum_{j=1}^{N} \sigma_j. \quad (4.31)$$

The budget constraint could also be weighted if done more on a cash flow basis rather than a fixed budget basis. Let that weighting be $\rho_i > 0$ $(i = 1, \ldots, N)$ in the sense that the constraint to be applied is

$$\sum_{i=1}^{N} C_i \rho_i W_i = B. \quad (4.32)$$

Then, following the procedure of Section II, the optimum *RAV* occurs when

$$\frac{\sigma_i}{\rho_i C_i} \frac{\partial RAV_i(W_i)}{\partial W_i} = H = \text{constant, all } i. \quad (4.33)$$

For the parabolic RAV formula

$$\frac{\partial RAVi(W_i)}{\partial W_i} = E_i(1 - W_i/OWI_i), \qquad (4.34)$$

so that Eqs. (4.33) and (4.34) yield

$$W_i = OWI_i(1 - H\rho_i C_i/\sigma_i E_i)$$

with

$$\sum_{i=1}^{N} C_i \, OWI_i \, \rho_i - H \sum_{i=1}^{N} OWI_i \, \rho_i^2 C_i/\sigma_i E_i = B,$$

which determines H in terms of the budget B. A similar analysis can be carried through for any weighting applied to any functional form chosen for the dependence of RAV on working interest.

5

SIMILARITY, DEPENDENCE, AND CORRELATION CONSIDERATIONS FOR RISK-ADJUSTED VALUES AND WORKING INTEREST

I. INTRODUCTION

Consider two exploration opportunities, the first with a probability of success p_{s1} (and so a probability of failure $p_{f1} = 1 - p_{s1}$), potential gains, G_1, and costs C_1; the second with, correspondingly, p_{s2}, G_2, and C_2. A corporate risk tolerance of RT is in force and a fixed budget B available. The corporate risk tolerance is a critical ingredient in calculating the best working interest to take because otherwise one would either take all of an opportunity or none of it depending on whether the expected value, $p_s G - C$, is positive or negative, respectively. We want to figure out the best working interests, W_1 and W_2, respectively, to take in each of the opportunities in order to maximize the total risk-adjusted value (RAV) for the pair of opportunities. If the exploration opportunities are taken to be *independent* of each other, then it is possible to figure out precisely the optimum working interest to take in each, so that the total RAV is maximized. Now the budget requirement is made up of

$$B = W_1 C_1 + W_2 C_2. \tag{5.1}$$

Suppose the two opportunities are identical in every respect, with $p_{s1} = p_{s2}$, $G_1 = G_2$, and $C_1 = C_2$. Then, if any working interest is taken at all, it would seem from symmetry that equal working interest in each opportunity would be appropriate (i.e., $W_1 = W_2 = W$) so that the budget requirement alone[1] yields

$$W = B/2C \tag{5.2}$$

for a budget $B \leq 2C$. For a budget $B > 2C$ then 100% of working interest ($W = 1$) *could* be taken in each opportunity for a total cost of $2C$, thereby returning ($B - 2C$) of the budget to the corporate coffers. In this extreme situation note that no mention of risk-adjusted value has yet been made, nor is there any dependence mentioned of W on risk tolerance. One has to then check whether the maximum RAV occurs at $W = 1$; else one should take a working interest lower than 100% in order to maximize RAV, as is shown directly. Even with a risk tolerance, RT, low enough to impact the working interest decision, symmetry should prevail and so one expects $W_1 = W_2$. As will become clear in Section II, this diversification maximizes the total RAV of two opportunities even though total cost overheads are increased.

Clearly, then, there is a difference on investment decisions and working interest for each opportunity, depending on how dissimilar and/or dependent the opportunities are to each other. Three obvious questions arise:

1. How similar do two opportunities have to be in order that a decision to invest in one of the opportunities, at a working interest W_1, immediately provides an assessment of the working interest, W_2, to be taken in the second opportunity? And what is the characterization of "similar"?
2. As the opportunities become more and more dissimilar (or less dependent[2] on each other), how does one evaluate when the opportunities are best treated as unrelated to each other?
3. What effect does any correlation between opportunities have on working interest and risk-adjusted value?

Answers to these questions enable corporate decision makers to evaluate the likelihood of information being sufficient to make informed commitments for, say, multiple wells on a single structure, or for leasing positions for similar plays on a fairway, for wells on similar structures, etc. Here the above three questions are investigated using a comparison of just two exploration opportunities. The more general case of N opportunities can also be worked through following arguments similar to those presented here. While the general method of analysis does not change as the number of opportunities is increased, the technical details become more involved.

---

[1] There is a degeneracy ambiguity for *exactly* identical opportunities, for then *any* combination of working interest in both opportunities, such that the total budget is spent, is of equal value. This degeneracy is lifted for even slight differences in the opportunities, as is shown later.

[2] *Similarity* and *dependence* are different concepts. Two chances to gain $10 on separate coin tosses at a cost of $2 each are very similar but very independent.

The situation of two opportunities best illustrates, as simply as possible, the basic procedures in a manner consistent with clarity of presentation.

In addition, the calculations are carried through only for a parabolic *RAV* formula, which for any one opportunity takes the form

$$RAV = W\left(E - \frac{1}{2}p_s p_f W G^2 / RT\right), \quad (5.3a)$$
$$\text{in } 0 \leq W \leq \min\{1, 2ERT/(p_s p_f G^2)\},$$

where $2ERT/(p_s p_f G^2)$ is the value of W where $RAV = 0$, the expected value is $E = p_s V - p_f C$, $V = G - C$, G is the gain, C the cost, the probability of success is p_s, and $p_f (\equiv 1 - p_s)$ is the probability of failure, while *RT* is the corporate risk tolerance. The optimum working interest, *OWI*, is

$$OWI = \min\{ERT/(p_s p_f G^2), 1\}, \quad (5.3b)$$

and the maximum *RAV* at $W = OWI$ is then

$$RAV_{\max} = \frac{1}{2} E \times OWI, \quad \text{in } 0 \leq OWI \leq 1. \quad (5.3c)$$

Other functional forms for *RAV* dependence on working interest could equally be used [such as the Cozzolino (1977a, 1978) exponential dependence or the hyperbolic dependence], but such forms lead to nonlinear and transcendental equations which complicate the technical aspects of the problem without adding any particular insight. The parabolically dependent *RAV* formula leads to simple analytic expressions, which aid in the exposition.

II. GENERAL ARGUMENTS

Consider two opportunities with parameters p_{s1}, p_{f1}, V_1, G_1, and C_1, and p_{s2}, p_{f2}, V_2, G_2, and C_2, respectively. Let the working interests for the two opportunities be W_1 and W_2. For a budget, B, small enough that one cannot take the optimum working interest, *OWI*, in each opportunity [where $OWI = \min\{(ERT)/(G^2 p_f p_s), 1\}$], that is, for $B \leq C_1 OWI_1 + C_2 OWI_2$, the budget requirement is then

$$B = W_1 C_1 + W_2 C_2, \quad (5.4)$$

while the total *RAV* for both opportunities is

$$RAV = W_1 \left[E_1 - \frac{1}{2} p_{s1} p_{f1} W_1 G_1^2 / RT\right] \\ + W_2 \left[E_2 - \frac{1}{2} p_{s2} p_{f2} W_2 G_2^2 / RT\right]. \quad (5.5)$$

For a fixed budget, B, Eq. (5.4) can be used to replace W_2 in favor of W_1 in Eq. (5.5) with

$$W_2 = (B - W_1C_1)/C_2, \qquad (5.6)$$

provided that $0 \leq W_2 \leq \min\{1, 2E_2RTG_2^{-2}(p_{s2}p_{f2})^{-1}\}$ in order that the RAV for opportunity 2 is positive; and provided further that $B \geq W_1C_1$ so that $W_2 \geq 0$. Then the total RAV of Eq. (5.5) has a maximum when

$$W_1 = D^{-1}[E_1C_1^{-2} - E_2(C_1C_2)^{-1}\{1 - (B/C_2)p_{s2}p_{f2}E_2^{-1}G_2^2/RT\}] \qquad (5.7a)$$

and

$$W_2 = D^{-1}[E_2C_2^{-2} - E_1(C_1C_2)^{-1}\{1 - (B/C_1)p_{s1}p_{f1}E_1^{-1}G_1^2/RT\}], \qquad (5.7b)$$

where

$$D = (p_{s1}p_{f1}G_1^2C_1^{-2} + p_{s2}p_{f2}G_2^2C_2^{-2})/RT, \qquad (5.7c)$$

provided $0 \leq W_1 \leq \min\{1, 2E_1RTG_1^{-2}(p_{s1}p_{f1})^{-1}\} \equiv W_{1,\max}$; and $0 \leq W_2 \leq \min\{1, 2E_2RTG_2^{-2}(p_{s2}p_{f2})^{-1}\} \equiv W_{2,\max}$; and provided $RAV \geq 0$. If $W_1(W_2)$ of Eq. (5.7a) (5.7b) is negative then W_1 (W_2) is set to zero, while if W_1 (W_2) of Eq. (5.7a) (5.7b) exceeds $W_{1,\max}$ ($W_{2,\max}$) then W_1 (W_2) is set to $W_{1,\max}$ ($W_{2,\max}$). Note that if $E_1 = E_2$, $p_{s1} = p_{s2}$, $C_1 = C_2 \equiv C$ and $G_1 = G_2$, then Eqs. (5.7a) and (5.7b) yield

$$W_1 = W_2 = \frac{1}{2}B/C, \qquad (5.7d)$$

provided $B < 2C$, which is the value expected by symmetry considerations, as discussed previously.

A. SIMILAR OPPORTUNITIES

1. Equal Success Probabilities

Suppose, first, that the two projects are considered similar in the sense that $C_2 = \alpha C_1$, $G_2 = \alpha G_1$, and $p_{s2} = p_{s1}$, so that there is proportionate representation between the two opportunities with constant α. Note that because the failure probability for the second opportunity is $p_{f2} = 1 - p_{s2}$, there cannot be equal similarity of *every* parameter of opportunity 2 in relation to opportunity 1.

Then consider the working interest for opportunity 2 in relation to opportunity 1. One has

$$D = 2(p_{s1}p_{f1}G_1^2C_1^{-2}/RT), \qquad (5.8a)$$

so that

$$W_1 = D^{-1}E_1C_1^{-2}[1 - \{1 - (B/C_1)p_{s1}p_{f1}E_1^{-1}G_1^2/RT\}], \qquad (5.8b)$$

$$W_2 = D^{-1}E_1C_1^{-2}\alpha^{-1}[1 - (1 - (B/C_1)p_{s1}p_{f1}E_1^{-1}G_1^2/RT\}], \qquad (5.8c)$$

GENERAL ARGUMENTS

from which one obtains

$$W_2 = \alpha^{-1} W_1, \tag{5.9}$$

again provided $0 \leq W_1 \leq W_{1,\max}$ and $0 \leq W_2 \leq W_{2,\max}$. Thus, as both gain and cost rise in proportion, the working interest *decreases* in inverse proportion in order to maximize the total *RAV* with a fixed budget. From the budget requirement of Eq. (5.6), or directly from Eq. (5.8b), one then has

$$W_1 = (B/2C_1), \tag{5.10a}$$

so that

$$W_2 = \alpha^{-1}(B/2C_1). \tag{5.10b}$$

The maximum total *RAV* is then

$$RAV = 2W_1 \left(E_1 - \frac{1}{2} p_{s1} p_{f1} W_1 G_1^2 / RT \right), \tag{5.11}$$

which is precisely twice that of opportunity 1, so that the opportunities are similar because there is the *same RAV* for each opportunity, as long as costs and gains are proportional to each other for both opportunities, with the same success probability for each. The optimal working interests for the opportunities are then in *inverse* ratio to the proportionality of costs and gains.

2. Unequal Success Probabilities

Direct inspection of Eqs. (5.5) and (5.7) shows that, apart from the expected values E_1 and E_2, only the *products* $p_{s1}p_{f1}$ and $p_{s2}p_{f2}$ enter. Thus, one could have $p_{s1} = 1/4$, and $p_{s2} = 3/4$ to obtain $p_{s1}p_{f1} = p_{s2}p_{f2}$.

Two opportunities are then similar, in the sense that with $G_2 = \alpha G_1$, $C_1 = \alpha C_2$ it follows that $W_2 = \alpha^{-1} W_1$, as long as the success to failure probability *products* $p_{s1}p_{f1}$ and $p_{s2}p_{f2}$ are identical, provided however that

$$E_2 = \alpha E_1. \tag{5.12}$$

Equation (5.12) requires

$$p_{s2} = \alpha^{-1} p_{s1} - C_1 G_1^{-1}(1/\alpha - 1),$$

so that, with $p_{f1} = 1 - p_{s1}$, and with the requirement $p_{s1}p_{f1} = p_{s2}p_{f2}$, it follows that similarity, as defined, is achieved only when p_{s1} takes on the precise value

$$p_{s1} = \frac{1}{2}(\alpha + 1)^{-1}[\alpha + 2\beta - \{(2\beta - 1)^2 + (\alpha^2 - 1)\}^{1/2}] \tag{5.13}$$

where $\beta = C_1/G_1$, and provided $0 \leq p_{s1} \leq 1$; all other situations are dissimilar.

B. NEARLY SIMILAR AND DISSIMILAR OPPORTUNITIES

1. Nearly Similar Opportunities

For two opportunities with $p_{s1} = p_{s2}$, and in which $G_2 = \alpha G_1 + g$, $C_2 = \alpha C_1 + c$, where $g \neq 0 \neq c$, the opportunities can be said to be nearly similar when the influence of g and c on W_2 is small, in the sense that with

$$W_2 = \alpha^{-1} W_1 + \Delta W \tag{5.14}$$

then one requires $|\Delta W| \ll \alpha^{-1} W_1$ for near similarity. When $g = 0 = c$ it follows that $\Delta W = 0$, so that linear departures of g and c from zero are sufficient to evaluate a "nearly similar" statement. In addition, the departure of the total RAV from its maximum of twice the RAV for opportunity 1 also provides a measure of the near similarity of both opportunities.

Thus, use Eq. (5.14) to obtain

$$\begin{aligned} RAV = BC_1^{-1} &\left\{ p_{s1} G_1 - C_1 - \frac{1}{2} p_{s1} p_{f1} G_1^2 B/(RTC_1) \right\} \\ + \frac{1}{2} BC_1^{-1} &\left[g \left\{ p_{s1} G_1 - C_1 - \frac{1}{2} p_{s1} p_{f1} G_1^2 B/(RTC_1) \right\} \right. \\ &\left. - c \left\{ C_1 - \frac{1}{4} p_{s1} p_{f1} G_1^2 B/(RTC_1) \right\} \right] \end{aligned} \tag{5.15}$$

together with

$$W_2 = (B - W_1 C_1)(\alpha C_1)^{-1}(1 - c). \tag{5.16}$$

To first order in g and c, Eq. (5.14) yields

$$\Delta W/W_1 = 2\alpha^{-1}(g - c) E_1^{-1}[p_{s1} G_1 (1 + E_1 C_1 RT/(2 p_{s1} p_{f1} G_1^2 B)) - 3 C_1], \tag{5.17}$$

while the change, δRAV, in the total RAV is

$$\begin{aligned} \delta RAV = \frac{1}{2} BC_1^{-1} &\left[g \left\{ p_{s1} G_1 - \frac{1}{2} p_{s1} p_{f1} G_1^2 B/(RTC_1) \right\} \right. \\ &\left. - c \left\{ C_1 - \frac{1}{4} p_{s1} p_{f1} G_1^2 B/(RTC_1) \right\} \right], \end{aligned} \tag{5.18}$$

which should be small compared to the maximum total RAV occurring in the limit when $W_2 = \alpha^{-1} W_1$, in order that one would identify both the working interest W_2 and the total RAV as being nearly similar to the case of similar opportunities.

2. Dissimilar Opportunities

If the potential gains G_2 and/or the costs C_2 for opportunity 2 are sufficiently far removed from being proportional to the corresponding values for opportunity 1 that

$$|W_2 - W_1|/W_1 \geq 1 \tag{5.19a}$$

and

$$|RAV_2 - RAV_1|/|RAV_1| \geq 1, \tag{5.19b}$$

then the opportunities can be said to be dissimilar, since neither the working interests nor the respective RAVs for the different opportunities are at all close.

III. CORRELATED BEHAVIOR

A. CORRELATED VALUE AND COST FOR A SINGLE OPPORTUNITY

Quite often in estimating the worth of an opportunity to a corporation there is a correlated dependence (positive or negative) of value, V, and cost, C. Such a correlation can arise for a variety of reasons, an example of which is this: As one drills deeper, project costs arise, but it may be more likely that one will then encounter the dominant reservoir formations in a basin, so that value should then be positively correlated with cost. Of course, it is also possible that one is drilling away from the basin fairway so that there may be only a small chance of finding any oil at all; in that case, value is ngatively correlated with costs.

In addition, it is unlikely that precise numbers are available for costs, value, or even the success probability of finding hydrocarbons. Most often only ranges of C, V, and p_s are available. The problem then is to give not only the expected value $E(\equiv p_s V - p_f C)$ or RAV of the opportunity, but also the ranges of E and RAV as C, V, and p_s are allowed to vary in their assessed ranges.

To exemplify the pattern of behavior for each of these situations, numerical illustrations are considered here in which the success probability, p_s, can range uniformly between 10 and 30% with an average of 20%; costs, C, are estimated to range uniformly between $10 MM and $30 MM with an average of $20 MM; the net present value, V, is estimated to lie in the uniformly distributed range between $200 MM and $400 MM with an average of $300 MM; and the corporate risk tolerance, RT, is held fixed at $50 MM.

Three independent case histories are computed within the above parameter ranges. Case history 1 treats the parameter variations of p_s, V, and

C as independently randomly varying; a suite of random Monte Carlo simulations is then performed, extracting individual values for p_s, V, and C from the relevant uniform distributions. The cumulative probability distributions for optimum working interest, OWI, expected value, E, and risk-adjusted value, RAV, are then produced. Case history 2 repeats the logic path of case history 1 but takes the values of V and C to be *positively* correlated at the 90% level. Thus if C is selected above (below) its average value, then there is a 90% chance that V will be selected above (below) its average value. Case history 3 (the least likely) repeats case history 2 but with V and C taken to be 90% *negatively* correlated; a high (low) value of V is 90% certain to be associated with a correspondingly low (high) value of C.

These extreme values of correlated behavior have been chosen to maximize the contrast in results between the three situations of no correlation, 90% positive correlation, and 90% negative correlation. (Note negative correlations of cost and value are included to complete the illustration even though they are, in reality, rare.) For a comparison standard, note that if each of p_s, V, and C were to be held at their central values of $p_s = 0.2$, $V = \$300$ MM, and $C = \$20$ MM, then the optimum working interest would be at $OWI = 20.7\%$, for an expected value $\{(p_s V - p_f C) \times OWI\}$ of $9.09 MM at a working interest of OWI, and a risk-adjusted value of $3.8 MM at OWI. For 100% working interest, the expected value is $44 MM but then the corresponding risk-adjusted value is $-\$8.9$ MM at a risk tolerance of $50 MM.

Figures 5.1a–c show cumulative probability distributions for OWI, expected value at OWI, and RAV and OWI, for the suite of random Monte Carlo runs for case history 1 of no correlation. Figures 5.2a–c show the corresponding cumulative probability plots for case history 2 of 90% *positive* correlation between V and C; while Figures 5.3a–c show the probability plots for case history 3 of 90% *negative* correlation between V and C.

Two simple measures of the effect of the correlation are readily available from the results presented. First is the volatility, ν, of each result given by

$$\nu = \{P(90) - P(10)\}/P(50), \qquad (5.20)$$

which provides an indication of the stability of the 50% value, because the 90% cumulative probability value, $P(90) > P(50)$, and, in turn, $P(50) > P(10)$. If $\nu \ll 1$ then the $P(90)$ and $P(10)$ values are close to $P(50)$, indicating little uncertainty; while $\nu \gg 1$ implies implies significant uncertainty.

In addition, of interest is the question of which of $P(90)$ or $P(10)$ is departing most from $P(50)$, thereby providing a measure of whether the skew enhances the upside potential $\{P(90) - P(50)\}$ more than the downside $\{P(50) - P(10)\}$. A useful way to categorize this distortion is the skewness measure

$$\nu_+/\nu_- = \{P(90) - P(50)\}/\{P(50) - P(10)\}. \qquad (5.21)$$

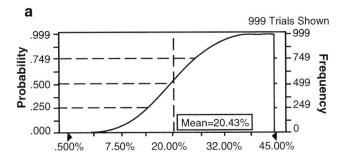

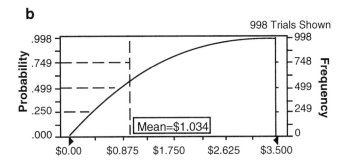

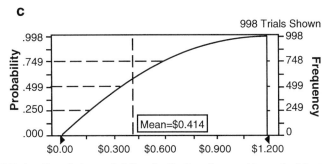

FIGURE 5.1 Cumulative probability distributions for case history 1 of the text, with no correlation between value, V, and costs, C: (a) for optimum working interest OWI; (b) for expected value, E, at OWI; (c) for risk-adjusted value, RAV, at OWI.

A value of $\nu_+/\nu_- = 1$ implies $P(90)$ is as far on the upside of $P(50)$ as $P(10)$ is on the downside {symmetric about $P(50)$}; while a value of $\nu_+/\nu_- \ll 1$ implies $P(90)$ is much closer to $P(50)$ than is $P(10)$; while a value of $\nu_+/\nu_- \gg 1$ implies just the opposite.

Table 5.1 records the values of $P(10)$, $P(50)$, and $P(90)$ for each of OWI,

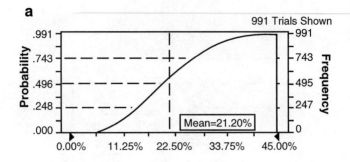

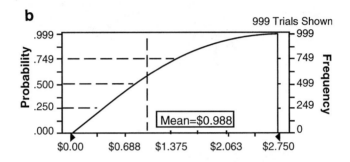

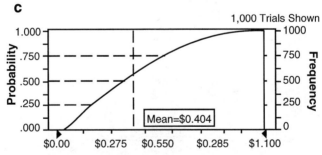

FIGURE 5.2 As for Figure 5.1 but for case history 2 of the text when value and cost are 90% positively correlated.

E at OWI, and RAV at OWI for each of case histories 1, 2, and 3. Using Eqs. (5.20) and (5.21) the volitility and skewness measures for each case history, and for each of OWI, expected value, and RAV can be computed as presented in Table 5.2. Note that for optimum working interest the volatility and skewness are both lowest (highest) for a negative (positive) correlation of costs and value, while for both expected value and risk-adjusted value the volatility and skewness are highest (lowest) for a positive

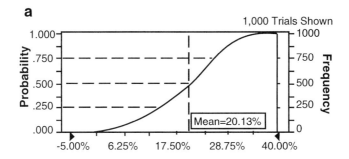

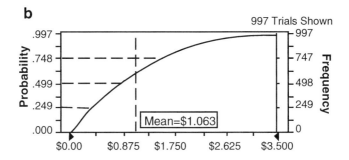

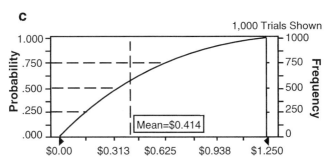

FIGURE 5.3 As for Figure 5.1 but for case history 3 of the text when value and cost are 90% negatively correlated.

(negative) correlation of costs and value. The implications are that a positive correlation permits a broader range of optimum working interests to produce positive expected values and risk-adjusted values, and the high skewness implies the range of *OWI* values is more biased to taking a larger share ($\nu_+ > \nu_-$) of the opportunity. For a negative correlation precisely the opposite is true: a narrow range of *OWI* values and a skewness biased to the downside of *OWI* for a smaller share.

TABLE 5.1 Cumulative Probability Values for Case Histories 1, 2, and 3

	No correlation	Positive correlation	Negative correlation
OWI(%)			
$P(10)$	9.79	10.36	9.20
$P(50)$	20.71	20.20	20.92
$P(90)$	31.04	33.59	29.56
Expected value, E, ($MM) at *OWI*			
$P(10)$	1.67	2.02	1.32
$P(50)$	8.67	9.05	8.72
$P(90)$	21.59	19.00	23.17
RAV ($MM) at *OWI*			
$P(10)$	0.76	0.91	0.61
$P(50)$	3.66	3.80	3.67
$P(90)$	8.21	7.63	8.60

For both expected value and *RAV*, a positive (negative) correlation produces the smallest (largest) volatility and the smallest (largest) skewness, implying that there is the least (greatest) uncertainty on the $P(50)$ values and the least (greatest) bias around the $P(50)$ values. Thus, a positive correlation gives the widest range of *OWI* values at which the opportunity will yield a profit, the lowest uncertainty on that profit, and the lowest bias; a negative correlation does precisely the opposite.

In addition, note from Table 5.1 that the $P(50)$ value for both *RAV* and expected value is highest for the positively correlated situation. The importance of correlated behavior is thus exhibited by these three case histories.

TABLE 5.2 Volatilities and Skewnesses for Case Histories 1, 2, and 3

	No correlation	Positive correlation	Negative correlation
OWI			
Volatility, ν	1.03 (M)	1.15 (H)	0.97 (L)
Skewness, ν_+/ν_-	0.95 (M)	1.16 (H)	0.74 (L)
Expected value			
Volatility, ν	2.30 (M)	1.88 (L)	2.51 (H)
Skewness, ν_+/ν_-	1.85 (M)	1.42 (L)	2.09 (H)
Risk-adjusted value			
Volatility, ν	2.04 (M)	1.77 (L)	2.18 (H)
Skewness, ν_+/ν_-	1.57 (M)	1.42 (L)	1.61 (H)

Key: L, lowest; M, middle; H, highest.

One further aspect of the results for the case histories is of interest. Because each of V, C, and p_s is uncertain, within their prescribed ranges, and because optimum working interest, OWI, expected value at OWI, and risk-adjusted value at OWI are all nonlinearly dependent on V, C, and p_s, it is *not* the case that equal fractional variations of V, C, and p_s around their respective mean values will cause the same degree of variation of OWI, E, or RAV; such would be the case only if each of OWI, E, or RAV were linearly dependent on V, C, and p_s taken separately (i.e., of the form $aV + bp_s + fC$ where a, b, and f are constants).

A question of interest is to evaluate the relative contributions of the ranges of uncertainties in V, C, and p_s to the uncertainties in OWI, E, and/or RAV, for then one could determine where to place the main emphasis of effort in any attempt to narrow the range of uncertainty of OWI, E, and/or RAV. Illustrations of this relative importance procedure were given earlier in this volume, as were the technical details, which do not need to be repeated here. Perhaps of novelty, however, is to consider the relative importance of variations in V, C, and p_s to, say, OWI from a rank correlation viewpoint rather than being viewed as causing fractional contributions to the variance around the mean OWI, as has already been done.

Such a depiction is given in Figures 5.4a–c for the noncorrelated, positively correlated, and negatively correlated case histories. The relative contributions of uncertainties in p_s (labeled chance), costs (labeled cost), and value (labeled *NPV* for net present value) to the ranges of optimum working interest uncertainty are given by the numbers opposite each entry (thus, in Figure 5.4a, p_s, C, and V each contributes to the uncertainty in OWI in the ratios 74:63:4), while the sign and rank correlation indicate the direction of influence. A negative rank correlation implies a lowering of OWI for an increase in the variable, and vice versa. Notice the shift in contributions of V occurring as the case history is changed from uncorrelated (Figure 5.4a), through positively correlated (Figure 5.4b), to negatively correlated (Figure 5.4c). Notice also how the relative strengths of the different contributions change in the three cases, reflecting directly the correlated dependence (positive or negative) of V and C.

These examples serve to exhibit the different patterns of behavior for a single exploration opportunity depending on how the particular components are correlated.

B. GEOLOGICALLY CORRELATED OPPORTUNITIES

Geological correlation is often used to indicate that different opportunities have values to a corporation which are correlated. For instance, two opportunities on the same structure would be expected to be correlated, because the geological conditions responsible for one opportunity likely

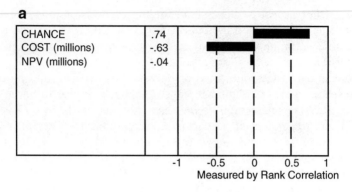
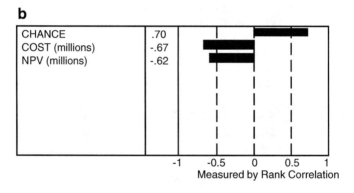
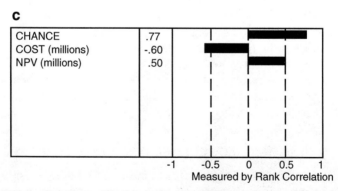

FIGURE 5.4 Relative contributions to the range of optimum working interest from a rank correlation viewpoint of uncertainties in p_s, costs, and value: (a) case history 1, uncorrelated; (b) case history 2, V and C 90% positively correlated; (c) case history 3, V and C 90% negatively correlated.

being hydrocarbon bearing probably define conditions for the other opportunity as well.

As the opportunities are displaced further away from each other, the geological conditions responsible for one opportunity likely being hydrocarbon bearing have less and less to do with another opportunity being hydrocarbon bearing. The correlation of dependent behavior then diminishes.

The analysis of correlated behavior of opportunities in a basin then depends on whether one is evaluating potential opportunities on (1) a prospect scale, (2) a regional fairway scale (and one then also has the job of attempting to identify the fairway direction and/or dip angle), or (3) a basinal intercomparison scale, in which case one has to identify similar geologic developments of source rock; migration pathways; structural, tectonic, and stratigraphic conditions; and of sealing capabilities through to the present day for each basin.

Further, the correlation effects mentioned above are related to intercomparison of different opportunities. There are also correlation effects that occur for a *single* opportunity. There is the fact that the cost of drilling increases (approximately linearly) with increasing depth to a target. There is the fact that the historical distribution of oil reserves in a basin with increasing subsurface depth first increases and then, at large depth, decreases in every basin so far investigated. Thus, there must be some correlation between reserve size R, anticipated at subsurface depth z, and costs to drill to that depth (Porter, 1992). Equally, the probability distribution of success in finding a reserve of size R at depth z must first increase at small z, and eventually decrease at large z, so that field size, costs, and success probability are also correlated. The point is that there are, quite often, situations in which significant correlations of value, costs, and success probability are to be expected to be the rule rather than the exception.

1. Correlation Effects for a Single Opportunity

To illustrate how the effects of drilling depth can impact the correlated interplay of potential gains, costs, and success probability for a *single* opportunity, suppose that costs increase linearly with depth z as

$$C(z) = C_0 + C_1(z/z_*), \tag{5.22}$$

where C_0 is the fixed cost of rig hiring, towing, etc., and C_1 is the cost to drill to a target depth z_*.

Suppose that the hydrocarbon reserve size (in barrels of oil equivalent) in a basin varies with depth as

$$R(z) = R_0[(z/z_*) + (z_*/z)]^{-1} \tag{5.23}$$

so that reserves that could be found at shallow depths ($z \ll z_*$) are

$$R(z) \cong R_0(z/z_*) \tag{5.24a}$$

and at great depths ($z \gg z_*$) are

$$R(z) \cong R_0 z_*/z \tag{5.24b}$$

with a peak value of $R = R_0/2$ on $z = z_*$.

The gain, G, if one were to find a reserve of size R at depth z is, at a fixed selling price \$$u$ per barrel, directly proportional to the reserve size given through Eq. (5.23); that is,

$$G(z) = G_0(z/z_* + z_*/z)^{-1}, \tag{5.25}$$

where $G_0 = uR_0$. The maximum of $G(z)$ is $1/2 G_0$ occurring on $z = z_*$. The probability of actually encountering a hydrocarbon reserve is small near the surface and initially increases with depth to a maximum value of, say, ε. In frontier basins where little information is available, ε is often as low as 1/20, while in well-developed basins ε can occasionally rise to as high as 1/3. The success probability will eventually decrease at great depths in a basin because the probability of finding hydrocarbon pools is related to the ability to image the subsurface. This resolution of the subsurface image is reflected in the empirically determined finding success. The empirical form

$$p_s = \varepsilon(z/z_*) \exp(1 - z/z_*) \tag{5.26}$$

is a fairly good representation of the behavior of finding success probability with increasing drill depth, z, and has a maximum of $p_s = \varepsilon$ on $z = z_*$. The minimum budget, $B_{\min}$, required to drill to a depth z is then

$$B_{\min}(z) = OWI(z)C(z), \tag{5.27}$$

where $OWI(z)$ is the optimum working interest given by Eq. (5.3b).

These correlated patterns of behavior can be used to assess the working interest that a corporation should take in a given opportunity for a fixed risk tolerance, RT, as shown in the next section of this chapter.

2. Correlation Effects between Opportunities

For different opportunities, each with its own values of the parameters C_0, C_1, G_0, ε, and z_*, the correlated behavior of the opportunities is now manifested through the drill depth z, regarded as a variable. Thus, if one were to drill to depth z_1 for opportunity 1, and to depth z_2 for opportunity 2, then the correlation of the two opportunities is most easily given in terms of a fixed total budget, B, using Eqs. (5.5), (5.7a), and (5.7b). In principle, one would like to drill the lowest cost opportunity first and then decide on drilling the second opportunity. Such an ideal situation seldom occurs due to the problems of mandated drilling as work obligations, expiration times of leases, government controls, etc. For instance, one division of a corporation could be drilling a well in the state-controlled waters of offshore Alabama, while a different division could be drilling in federal-controlled deeper waters of offshore Alabama at the same time. There is a geological correlation between the wells if both have the Norphlet, aeolian sand trend as a target, but both wells can be drilled independently.

Equally, one division of a corporation may want to drill an opportunity in the Gulf of Thailand and a different division may have an opportunity in West Texas. If the corporation has a fixed budget, it must choose the amount to be expended on the drilling opportunity for each division. In the first case the opportunities are directly geologically correlated; in the second case they are only financially correlated because of the fixed budget.

In either case, if one were to drill both opportunities only to very shallow depths, then both the success probabilities and gains are small, so that the total RAV is low or negative. If one were to drill both sets of opportunities to large depths, then the costs increase so that the total RAV would again be negative. One could drill opportunity 1 to some intermediate depth and opportunity 2 to shallow or great depth, or vice versa of course, but then the RAV of one of the opportunities would be less. Thus, there must be a combination of drill depths, z_1 and z_2 for the two opportunities, that maximizes the total RAV for a fixed budget, B. It is also possible to use the total risk-adjusted value per total unit of cost as a basis for maximization, in which case one is interested in the maximum of

$$M = \frac{\sum_{i=1}^{N} RAV_i}{\left(\sum_{i=1}^{N} C_i W_i\right)} \tag{5.28}$$

for a fixed budget B. But, because

$$B = \sum_{i=1}^{N} C_i W_i, \tag{5.29}$$

the maximum of M occurs when the total RAV is maximized. An alternative maximization control function is

$$F = \sum_{i=1}^{N} \{RAV_i/(C_i W_i)\}, \tag{5.30}$$

which maximizes RAV for each opportunity relative to amount spent ($C_i W_i$) on each, with the budget constraint of Eq. (5.29) in force.

The correlation of geological information is then related to how the parameters z_*, ε, C_0, C_1, and G_0 vary as the opportunities increase in distance from each other. This correlated behavior between two opportunities is illustrated in the next section of this chapter.

IV. NUMERICAL ILLUSTRATIONS

A. CORRELATION EFFECTS FOR A SINGLE OPPORTUNITY

To illustrate patterns of behavior and results as drill depth increases, consider an exploratory basin and set the overall chance of success at $\varepsilon = 0.1$. Take the majority of reserves as occurring at a typical depth of about 10,000 ft ($\equiv z_*$). In addition, choose a maximum scaling value of producible reserves as 10^8 bbls ($\equiv R_0$) and a selling price per barrel of \$20/bbl ($\equiv u$) so that if no costs at all were to be included, a naive estimate of maximum total present-day value is $G_0 = \$2 \times 10^9$ ($= uR_0$).

If drilling costs are taken to increase at \$1000/ft, then to drill to 10,000 ft costs approximately $\$10^7$ ($\equiv C_1$). The fixed costs, C_0, are taken to be about \$5 MM, equivalent to about \$0.1 MM per day for rig rental, and a drill time to 10,000 ft of about 50 days.

The corporate risk tolerance, RT, is dependent on a variety of internal corporate factors. For the purposes of illustration the calculations have been run with high, intermediate, and low corporate risk tolerance, in order to visualize better the influence of shifts in corporate attitude to risk as drill depths to targets increase.

With the parameters set, the optimal working interest, maximum RAV, and minimum budget for the opportunity were then evaluated with increasing depth, z. For example, shown in Figures 5.5a–c are the optimum working interest versus depth, risk-adjusted value versus depth, and minimum budget required (B_{min}) versus depth for $\varepsilon = 0.1$, $C_0 = \$5$ MM, $C_1 = \$10$ MM, $G_0 = \$2000$ MM, $z_* = 10^4$ ft, for the three cases of $RT = \$10$ MM, $RT = \$100$ MM, and $RT = \$1000$ MM, respectively; Figures 5.5d–f show the corresponding values when $\varepsilon = 0.2$. Note from Figure 5.5 that the OWI shows an abrupt increase from zero at the depth at which $p_s G$ first exceeds C and returns to zero at the greater depth where $p_s G$ again drops below C. The RAV variation with depth shows a similar behavior, but note that the depth of the maximum RAV is different than that for the maximum OWI, reflecting the spatial variability differences of costs, success probability, and gains. The minimum budget to achieve an OWI at a given depth has a dependence that is a product of the increasing costs with depth and the variations of OWI with depth. For instance, if OWI decreases over a range of depths faster than costs rise, then a lower budget is required to drill to that target depth than to a shallower target. This is true because lower working interest is taken in the deeper target.

As the corporate risk tolerance increases, the OWI that can be taken also rises until, at $RT = \$10^9$, as shown in Figure 5.5, there are depth ranges where 100% working interest is appropriate.

Note also from Figure 5.5 that, because the reserve distribution and the success probability are both distributed around a most likely depth of $z =$

z_*, it is possible to strike oil both shallower and deeper than z_*, accounting for the broad range of depths that can lead to positive RAV values for the opportunity.

The point of this example is to show how geologic correlation influences OWI, RAV, and the budget required to explore an opportunity. Of course, one could argue that the functional forms chosen for $p_s(z)$, $C(z)$, and $G(z)$ as illustrative devices are not those pertaining to a particular basin or basinal fairway; and one could equally argue that the choices of parameters entering the illustrations are either not relevant for a particular exploration area or are not precisely known, but only bounded in some broad ranges. But the logic of the procedure given operates with any functional choices for $p_s(z)$, $C(z)$, and $G(z)$, so that one can insert what are considered relevant behaviors for p_s, C, and G for a particular area and perform calculations similar to those provided here to obtain the corresponding shapes of OWI, RAV, and B_{min} with depth; the general patterns of behavior will remain the same. Equally, if it is considered that parameter values (such as ε, C_0, C_1, z_*, G_0, RT, etc.) are not too well known, then an obvious strategy is to use a suite of Monte Carlo simulations (1) to obtain the cumulative probabilities for OWI, RAV, and B_{min} at each depth; and (2) to provide relative importance measures so that one can determine which of the parameters most needs its range of uncertainty to be addressed in order to narrow the range of uncertainty of OWI, RAV, and B_{min}.

B. CORRELATION EFFECTS BETWEEN OPPORTUNITIES

Consider, once again, two opportunities. If the values of C_0, C_1, z_*, ε, and G_0 are the same for both opportunities, then both are similar, in the sense that one would then set $W_1 = W_2$ and drill to the same depths, $z_1 = z_2$. However, if one includes geological information it may be that both opportunities are only correlated but very dissimilar. For instance, suppose that the scale depth, z_*, to the largest oil pools is related to a structural trend, so that z_* is around 10^4 ft for opportunity 1, but z_* is around 1.5×10^4 ft for opportunity 2, perhaps reflecting a flexural trend, as in the Gulf of Mexico. If all other parameters, C_0, C_1, ε, and G_0, are identical then it is the systematic increase of z_* for opportunity 2 that ends up making that opportunity less worthwhile, because costs escalate as one has to drill to greater depth to obtain the same reserves of oil as one anticipated obtaining for opportunity 1. Clearly, the two opportunities will be similar only if the RAVs and appropriate working interests follow the similarity prescriptions given in Section II. The opportunities can be extremely dissimilar but still highly correlated as we now show.

Consider the same parameter values as in the previous example for *both* opportunities with the single exception of scaling depth z_*, which is retained

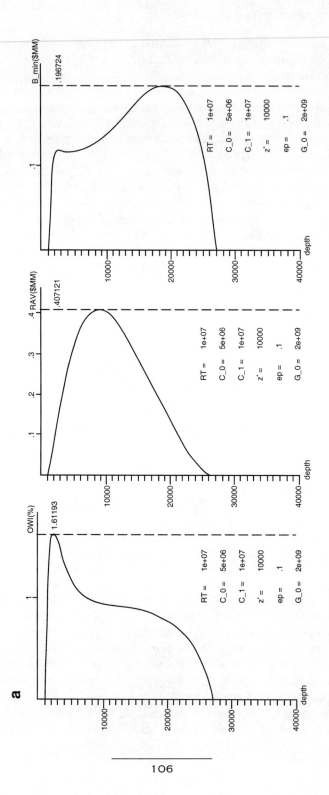

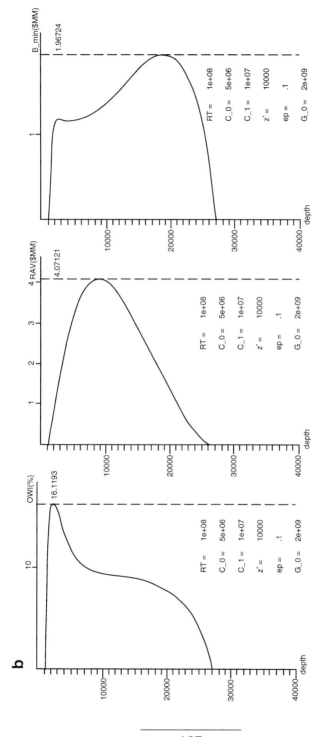

FIGURE 5.5 Triplet panels of *OWI*, *RAV*, and required budget versus depth, respectively, for a single opportunity for corporate risk tolerance values of (a) $10 MM, at maximum success probability of $\varepsilon = 0.1$; (b) $100 MM at $\varepsilon = 0.1$; (c) $1000 MM at $\varepsilon = 0.1$. parts (d), (e), and (f) repeat the triplet panels of parts (a), (b), and (c), respectively, but with the maximum success probability increased to $\varepsilon = 0.2$. Other parameter values are given in text and on each panel.

107

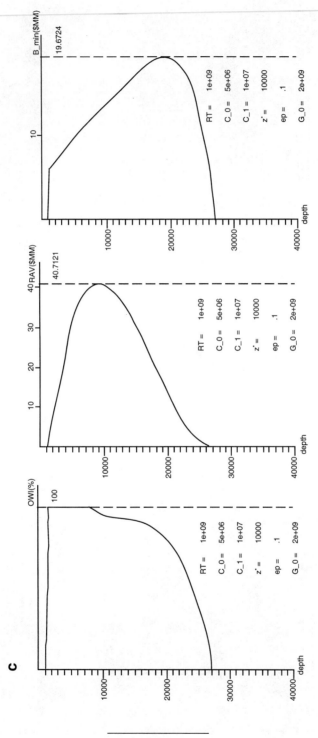

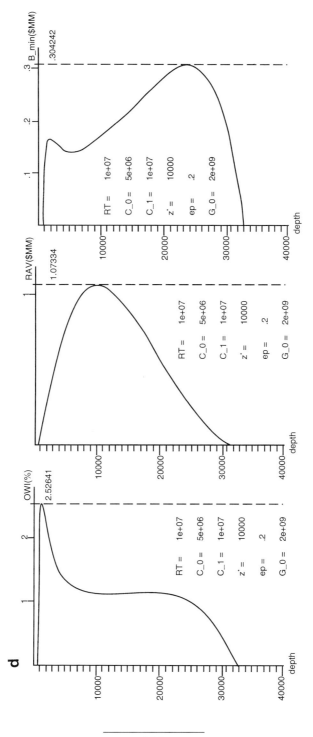

FIGURE 5.5 (*Continued*)

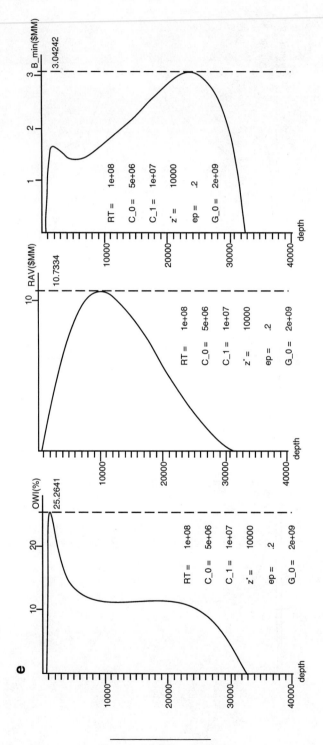

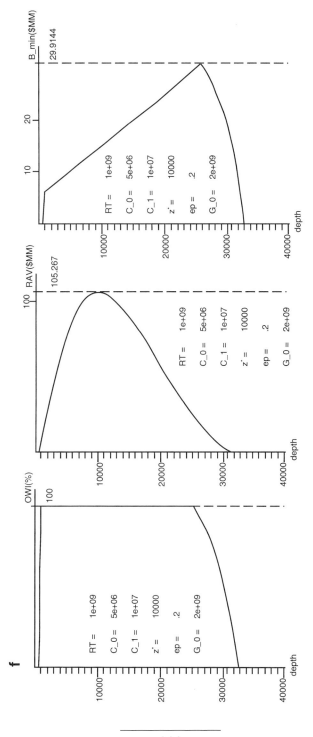

FIGURE 5.5 (*Continued*)

at 10^4 ft for opportunity 1; but for opportunity 2 take z_* to systematically increase from 10^4 ft through to 2×10^4 ft. Then the total RAV and optimal working interests to take in each opportunity are characterized as follows.

In the range where working interest is $0 \leq W \leq W_{max}$ for each opportunity, the RAV of opportunity 1 is positive (under the parabolic RAV formula) when

$$0 \leq W_1 \leq \min\{1, 2E_1RT/(p_{s1}p_{f1}G_{1}^{2})\} \equiv W_{1,max}, \quad (5.31)$$

provided $E_1 \geq 0$, with a similar expression for W_2.

For a fixed budget B, evaluate the range of depths within which $0 \leq W \leq W_{max}$ and RAV is positive for each opportunity. Then the correlated behavior at depth z_1 (>0) for opportunity 1 and depth z_2 for opportunity 2 (in those depth ranges yielding $RAV \geq 0$) can be described by contours of constant total RAV on a (z_1, z_2) plot. Plots can also be given of RAV_1, RAV_2, W_1, and W_2 for each opportunity on similar axes to illustrate the correlated behavior. As z_1 and z_2 vary independently, the correlated behavior of the two opportunities will have maxima in $\infty \geq z_1 \geq 0$, $\infty \geq z_2 \geq 0$, which provide the maximum joint values for correlated drilling depths, working interest, and maximum total RAV. Pragmatically, the numerical procedure for addressing this correlated problem proceeds as follows. Suppose, first, one were to regard the two opportunities as independent. Then the minimum budget needed to take an optimum working interest in each opportunity, with the first to be drilled to depth z_1 and the second to depth z_2, is

$$B_{min}(z_1, z_2) = OWI_1(z_1)C(z_1) + OWI_2(z_2)C(z_2). \quad (5.32)$$

Now as z_1 and z_2 vary independently in $0 \leq z_1 \leq \infty$ and $0 \leq z_2 \leq \infty$, the minimum budget has a maximum value when $OWI_1(z_1)C(z_1)$ and $OWI_2(z_2)C(z_2)$ are, independently, as large as possible. For fixed values of all other parameters let this maximum be B_{crit}. If the actual budget, B, is greater than B_{crit} then one can always take the optimum working interest in *both* opportunities at all depths z_1 and z_2, and one can return the amount $B - B_{min}(z_1, z_2)$ to the corporate coffers. However, when the actual budget B is less than B_{crit} then one cannot take the OWI in both opportunities at *all* depths z_1, z_2; one must settle for less than the optimum working interest in each opportunity in order to satisfy the budget constraint. As $OWI_1(z_1)$ heads to zero at the two depths where $p_s(z_1)G(z_1) \leq C(z_1)$, and similarly for $OWI_2(z_2)$, then $B_{min}(z_1, z_2)$ will certainly drop below the actual budget, B. Thus, there is a range of shallow depths and also a range of deeper depths where the OWI in opportunities 1 and 2 can again be taken. But there is also a central region of depths for z_1 and z_2 where the actual budget, B, is less than $B_{min}(z_1, z_2)$, and in that region the OWI cannot be taken in both opportunities.

The smaller the actual budget in comparison to B_{crit}, the wider the range of central values of z_1 and z_2 where less than optimal working interest can be taken in each opportunity.

Numerically, then, for fixed values of C_0, C_1, G_0, z_*, ε, and RT for each opportunity, one first computes B_{crit} in comparison to B. If $B \leq B_{crit}$, then the shallow and deep depth ranges for both z_1 and z_2 are computed for which $B \geq B_{min}(z_1, z_2)$ and OWI for each opportunity is taken in those ranges. In the intervening (z_1, z_2) ranges, where $B \leq B_{min}(z_1, z_2)$, the appropriate working interest which maximizes the *total RAV* for both opportunities taken together, as well as the RAV and budget fraction for each opportunity, are computed from Eqs. (5.4), (5.5), and (5.7) for all other parameters held fixed.

To illustrate the patterns of behavior two main developments are given here. First, the scale depth, z_*, for opportunity 2 is systematically increased with respect to that for opportunity 1. In this way one can, for instance, adjust the lateral spacing, x, between the two opportunities in a manner dependent on the difference between z_{*2} and z_{*1}, so that different trends of reservoirs can be addressed. For example, if x is taken to increase linearly with $(z_{*2} - z_{*1})$ then a sloping reservoir fairway is constructed, while addition of a parabolic dependence would produce a curved trend, representative perhaps of the flex trend of South Louisiana. In this way, a variety of fairway paths can be constructed.

In addition, the potential gains, G_0, for each opportunity can be chosen to be very similar or very different. For instance, if opportunity 1 is thought to be a gas structure then the potential gains would likely be less than for a deeper (larger z_*) oil structure for opportunity 2. While the drilling costs to reach the most likely target depth of z_* for opportunity 1 are less than for opportunity 2, the gains from opportunity 2 are higher. For a fixed budget and fixed corporate risk tolerance, it is then not obvious which opportunity is going to have the higher RAV. So several cases are evaluated here to illustrate the difference in behaviors: Opportunity 1 is taken to be a gas or oil reservoir and, independently, opportunity 2 an oil or gas reservoir, as the scale depth z_{*2} increases with respect to z_{*1}.

The functional dependence with depth of $p_s(z)$ and $G(z)$ for both opportunities is retained as in the previous illustration, although one can, of course, let each vary in a different manner with depth depending on empirical information gleaned from historical precedent in a given basis. Consider each case history in turn.

1. Case History 4: Oil at the Sites of Opportunities 1 and 2

For total RAV, RAV_1, and RAV_2 for opportunities 1 and 2, respectively, working interest for each opportunity (labeled W_1 and W_2), and total budget required, B_{min}, together with the budget required for each opportunity (B_1 and B_2), relevant variations are presented in Figures 5.6 through 5.9 as

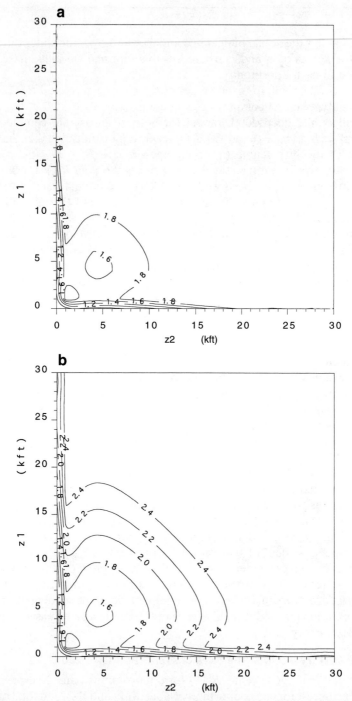

FIGURE 5.6 Correlated pattern of minimum budget needed to drill two oil field opportunities to depths z_1 and z_2, respectively, as limited by a fixed corporate budget of (a) $2 MM; (b) $2.5 MM; (c) $3 MM; (d) $4 MM. All other parameters are from case history 4 of the text.

NUMERICAL ILLUSTRATIONS   115

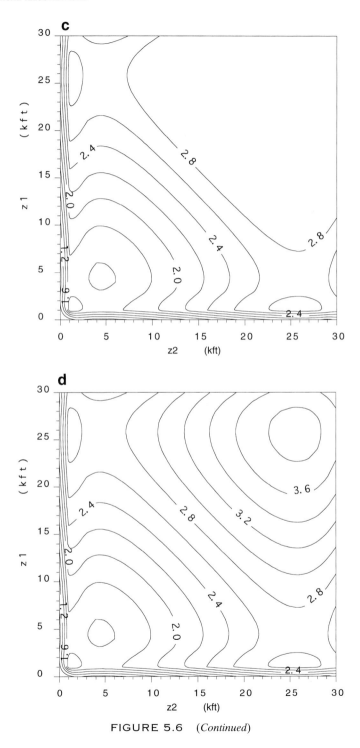

FIGURE 5.6   (*Continued*)

target depths z_1 and z_2 are increased independently for each opportunity, with parameter values set at $\varepsilon_1 = \varepsilon_2 = 0.3$, $G_0(1) = G_0(2) = \$2 \times 10^9$, $z_{*1} = 10^4$ ft, $C_0 = \$0.5 \times 10^6$, $C_1 = \$10^3$/ft, and a corporate risk tolerance of $RT = \$5 \times 10^7$. The corporate budget, B, is variable. The three cases considered are for $z_{*2} = z_{*1}$, $z_{*2} = 1.5z_{*1}$, and $z_{*2} = 2z_{*1}$ so that one is considering similar gains from both opportunities, with the scaling depth, z_{*2}, for opportunity 2 systematically increased with respect to opportunity 1.

Figures 5.6a–d provides contours of the budget distribution to drill to different depths (z_1 and z_2) for opportunities 1 and 2, respectively, when the total budget available is \$2 MM (Figure 5.6a), \$2.5 MM (Figure 5.6b), \$3 MM (Figure 5.6c), and \$4 MM (Figure 5.6d). Symmetry prevails between the two opportunities, as can be seen by the symmetric reflection around the 45° line of all contours in Figure 5.6, and as expected because both opportunities are identical.

Consider now a low corporate budget B of \$1.6 MM, so that one cannot take the optimum working interest in both opportunities at all depths that return a positive total RAV. Then for the case where the scaling depths are equal ($z_{*1} = z_{*2}$) both opportunities have identical characteristic parameters so that one again expects symmetry to prevail and, as shown in Figures 5.7a–c for W, RAV, and budget fraction for each opportunity as z_1 and z_2 are independently varied, there is indeed complete symmetry.

As the scale depth, z_{*2}, for the second opportunity increases to $1.5z_{*1}$, the symmetry of Figure 5.7 is broken. Because the costs to drill deeper increase, and because the same oil reserves are peaked at a greater depth, the RAV and W for opportunity 2 are less than for opportunity 1 at the same drill depth. As shown in Figures 5.8a–c for W, RAV, and budget fraction for each opportunity, the shallow opportunity 1 generally has a greater RAV and W, and takes less of the budget to return a positive gain.

The contrast between W, RAV, and budget fraction for the shallow opportunity 1 and the deeper opportunity 2 increases as the scale depth z_{*2} increases even further to $2z_{*1}$, as shown in Figures 5.9a–c for W, RAV, and budget fraction for each opportunity.

The point to make from these illustrations is that the two opportunities are, or can be, both geologically correlated (if the increase in z_* represents a deepening trend on a single fairway) and are certainly correlated because of the financial restriction of a fixed budget. The two opportunities may also be completely uncorrelated geologically (one could be in the Gulf of Thailand, say, and the other in West Texas) but are still correlated because of the financial budget limitations.

2. Case History 5: Gas at Opportunity 1, Oil at Opportunity 2

Here the same parameter values as for case history 4 are taken, with the single exception of the gains for opportunity 1, which are taken to be $G_0(1) = 0.1G_0(2)$, so that a mimic of a shallow gas field for opportunity 1

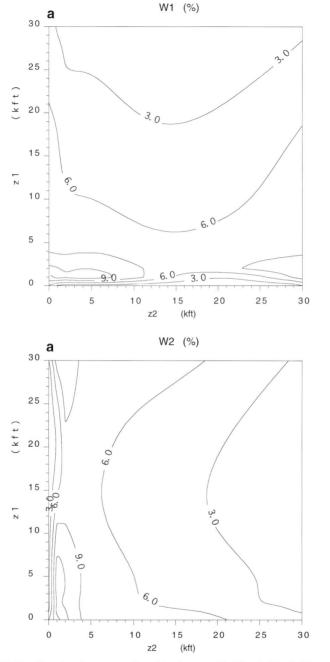

FIGURE 5.7 Correlated patterns of working interest W_1, W_2, RAV_1, RAV_2, total RAV, B_1, and B_2 for the two oil field opportunities when both have identical characteristics, as given in case history 4 of the text, and a corporate budget of $1.6 MM so that optimum working interest in both opportunities cannot be taken at all depths: (a) panels of W_1 and W_2 (%) versus z_1 and z_2; (b) panels of RAV_1, RAV_2, and total RAV (in $MM) versus z_1 and z_2; (c) panels of B_1 and B_2 (in $MM) versus z_1 and z_2.

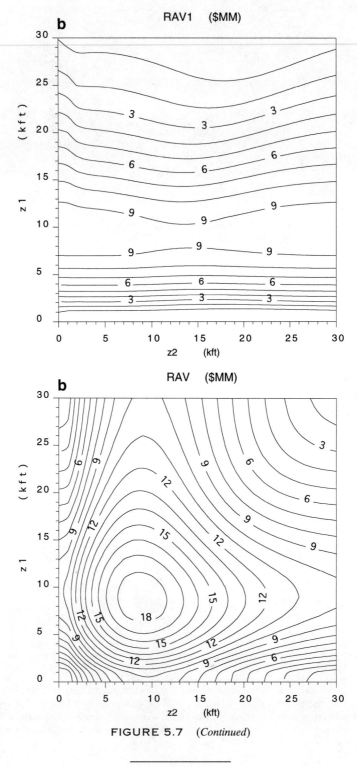

FIGURE 5.7 (*Continued*)

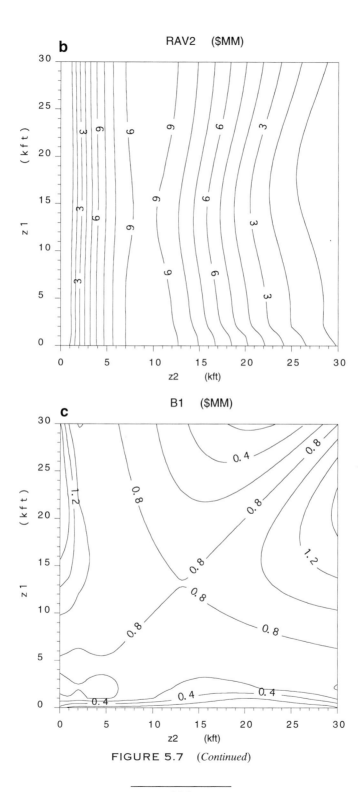

FIGURE 5.7 (*Continued*)

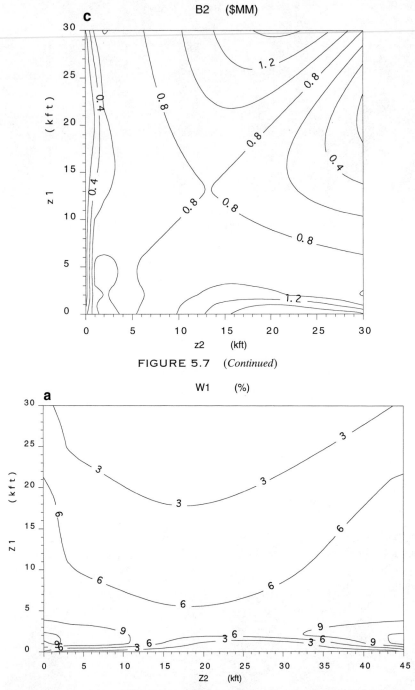

FIGURE 5.8 As for Figure 5.7 but with the scale depth of opportunity 2 increased to $z_{*2} = 1.5 z_{*1}$, rather than having $z_{*1} = z_{*2}$ as for Figure 5.7: (a) panels of W_1 and W_2 (in %) versus z_1 and z_2; (b) panels of RAV_1, RAV_2, and total RAV (in $MM) versus z_1 and z_2; (c) panels of B_1 and B_2 (in $MM) versus z_1 and z_2.

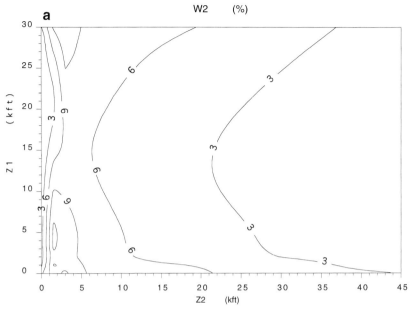

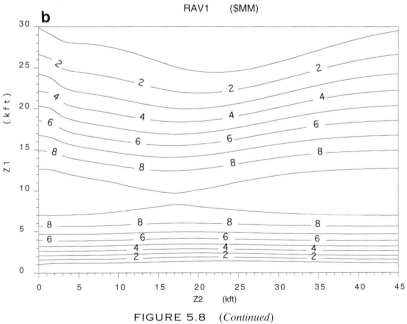

FIGURE 5.8 (*Continued*)

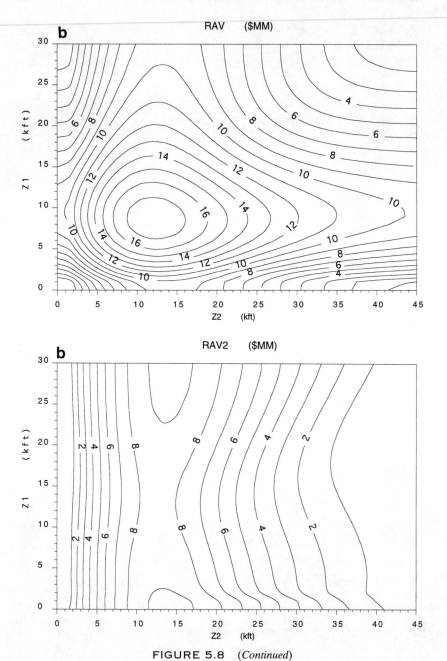

FIGURE 5.8 (*Continued*)

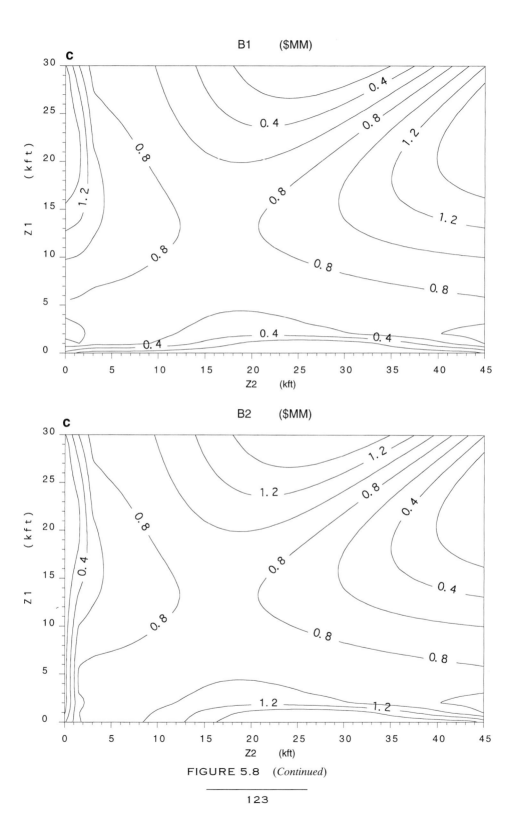

FIGURE 5.8 (*Continued*)

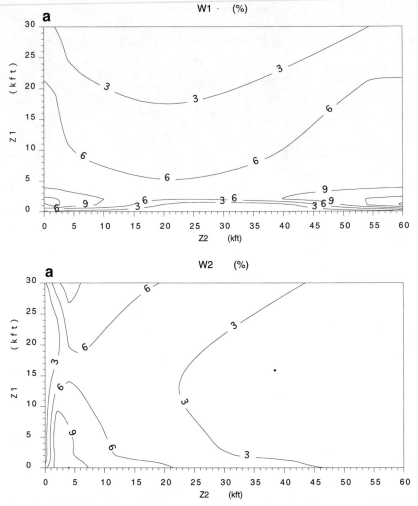

FIGURE 5.9 As for Figure 5.7 but with the scale depth of opportunity 2 increased to $z_{*2} = 2z_{*1}$: (a) panels of W_1 and W_2 in (%) versus z_1 and Z_2; (b) panels of RAV_1, RAV_2, and total RAV (in \$MM) versus z_1 and z_2; (c) panels of B_1 and B_2 (in \$MM) versus z_1 and z_2.

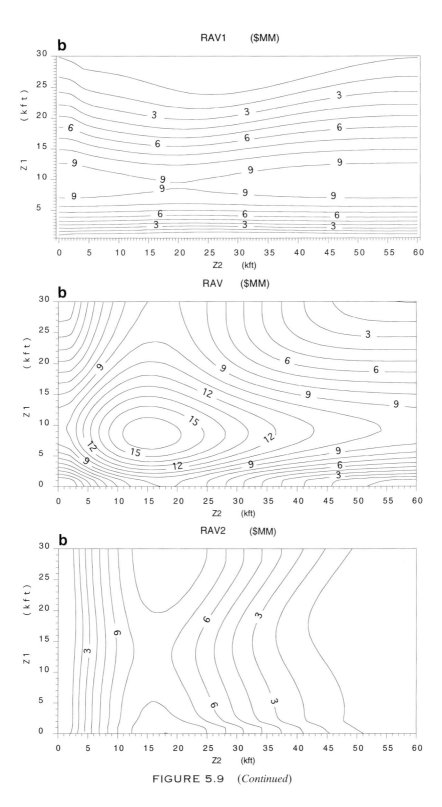

FIGURE 5.9 (*Continued*)

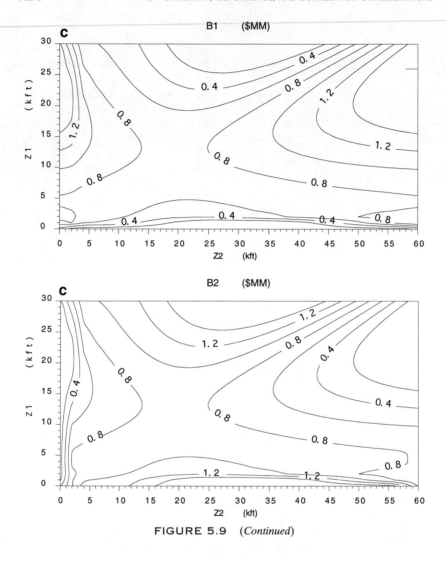

FIGURE 5.9 (*Continued*)

is compared against an ever deepening oil field (opportunity 2) as the scale depth z_{*2} increases relative to that for z_{*1}. Two cases are examined. First, z_{*1} is set equal to z_{*2} so that the gas field and oil field characteristics are identical, except for the 10:1 ratio in potential gains of the oil field relative to the gas field. (Practically, one has to include the operation and development costs as well, which will lower this 10:1 advantage. For ease of illustration such costs are not considered here.) If both opportunities have the same scale depth ($z_{*1} = z_{*2}$) then, because the likelihood of finding oil is the same as finding gas, and the costs of drilling to the same depth for either opportunity are identical, almost all of the limited budget goes to

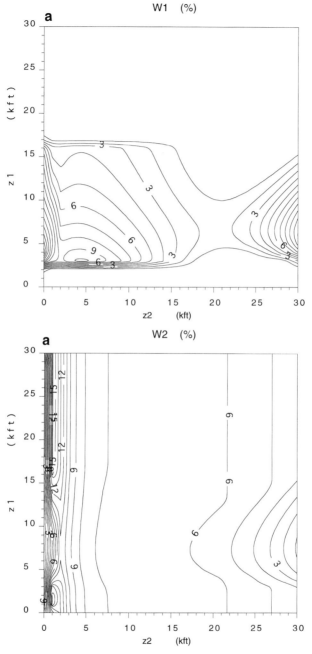

FIGURE 5.10 Correlation patterns of working interest W_1 and W_2, RAV_1, RAV_2, total RAV, B_1, and B_2 for the case of opportunity 1 being a gas field and opportunity 2 an oil field, with gains from the gas field estimated at 0.1 of the gains from the oil field. All other parameters are taken from case history 5 of the text, and the scale depths of the oil and gas opportunities are identical ($z_{*2} = z_{*1} = 10^4$ ft): (a) panels of W_1 and W_2 (in %) versus z_1 and z_2; (b) panels of RAV_1, RAV_2, and total RAV (in $MM) versus z_1 and z_2; (c) panels of B_1 and B_2 (in $MM) versus z_1 and z_2.

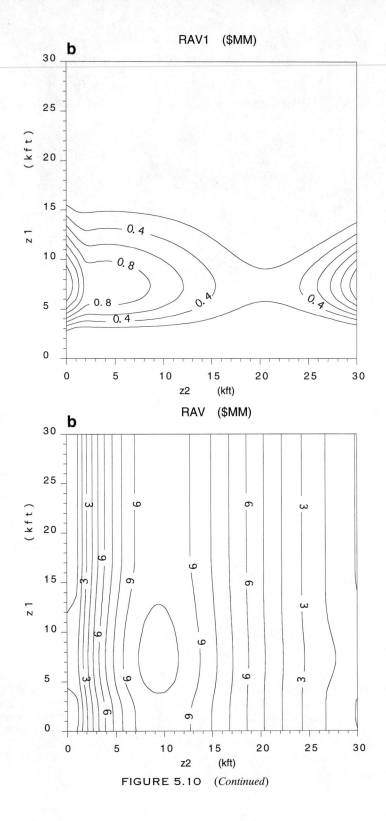

FIGURE 5.10 (*Continued*)

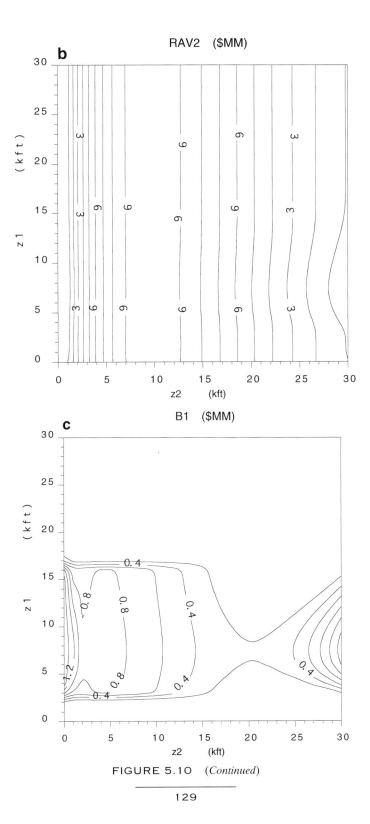

FIGURE 5.10 (*Continued*)

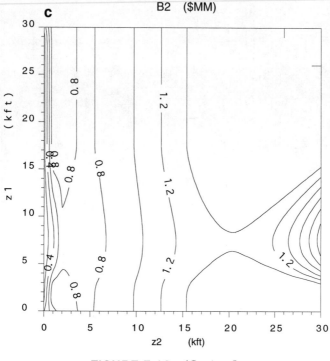

FIGURE 5.10 (*Continued*)

the potentially more lucrative oil opportunity. This situation is clearly seen in the results for W, RAV, and budget fraction depicted in Figures 5.10a–c.

Second, as the scale depth, z_{*2}, to the likely oil accumulation increases, the cost of drilling increases so that the profitability of a deeper oil find decreases. Eventually, at a great enough depth, opportunity 2 becomes less worthwhile than the shallow gas opportunity 1; more of the limited budget is then spent on opportunity 1 and the working interest taken in opportunity 2 is decreased. Figures 5.11a–c show these patterns of variation as the drill depths z_1 and z_2 for opportunities 1 and 2 are allowed to increase independently, for scale depth $z_{*2} = 2z_{*1}$.

The point about this case history is to show that there is not just a geologic correlation as a consequence of relative depth of scale for two opportunities of interest, but also a combined geologic/financial correlation as a consequence of anticipated type of product (gas/oil) to be found, in concert with both a fixed budget and also a shift in the reserve scale-depth, z_*.

For each opportunity one could, of course, vary the functional forms of $p_s(z)$, $R(z)$, $G(z)$, and $C(z)$ independently, as well as the parameters appropriate to each opportunity; and such variations will color the pictures

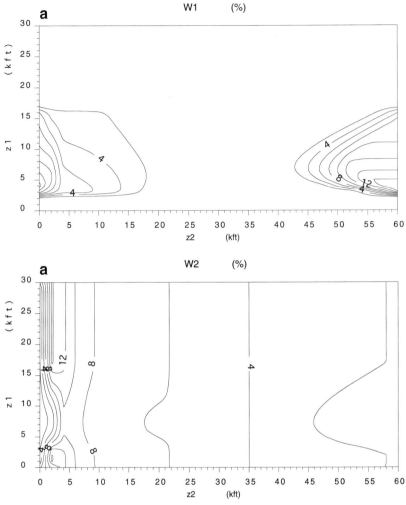

FIGURE 5.11 As for Figure 5.10 but with the oil field opportunity scale depth, z_{*2}, now at $2z_{*1}$. Note that the shallow gas field opportunity now becomes more enticing than it was when both oil and gas field opportunities had identical geological characteristics: (a) panels of W_1 and W_2 (in %) versus z_1 and z_2; (b) panels of RAV_1, RAV_2, and total RAV (in $MM) versus z_1 and z_2; (c) panels of B_1 and B_2 (in $MM) versus z_1 and z_2.

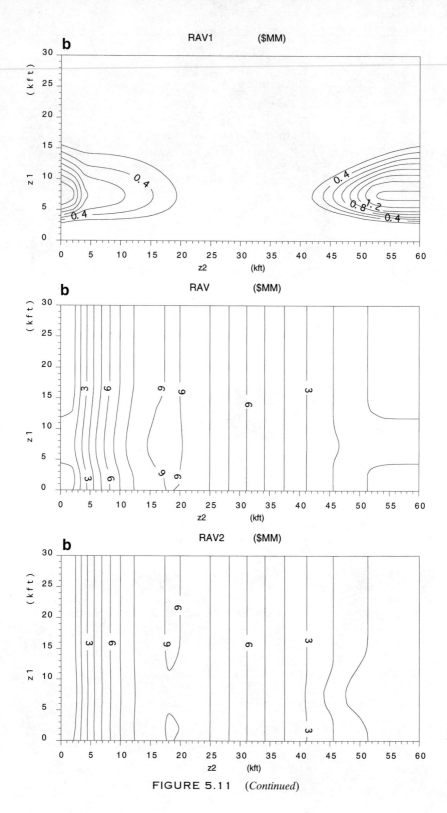

FIGURE 5.11 (*Continued*)

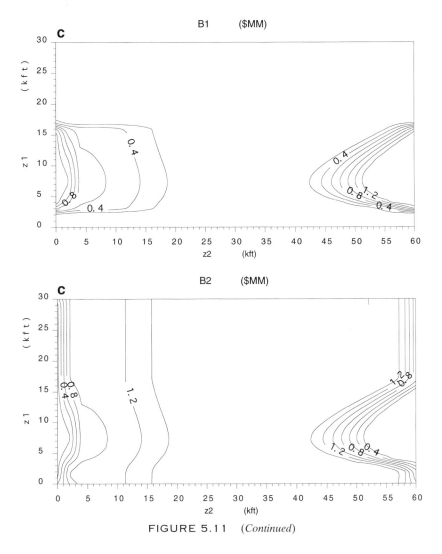

FIGURE 5.11 (*Continued*)

painted here in different ways. But the patterns for the shallow and deep regions will retain their overall topology; only the relative distortions of the already sketched contours will change.

V. DISCUSSION AND CONCLUSIONS

The purpose of this chapter has been to illustrate how similar, dissimilar, and correlated exploration opportunities can be addressed within the framework of both a fixed corporate risk tolerance and also a fixed corporate budget, in terms of providing effective working interest to be taken in a given opportunity and the risk-adjusted value to be expected given the constraints of corporate risk tolerance and a fixed budget. The definitions of similar and dissimilar opportunities can be made quantitatively precise, as has been shown.

In addition, the correlated nature of a single opportunity with respect to a pure fiscal correlation (with values correlated or anticorrelated with costs) can be provided, as can the dominant variations of uncertainty of inputs in controlling the fluctuations in both RAV and OWI. How this correlated behavior influences both the volatility of estimated worth of an opportunity, and whether the correlation skews the uncertainty more to upside potential or downside loss, as well as which factors are most important in causing the variations, is important to illustrate so that one can evaluate which factors need better control.

Apart from the fiscally correlated behavior for a single exploration opportunity, there is also a geologically correlated component, illustrated by considering that the reserve distribution, finding success probability, and drilling costs are all to be correlated with depth drilled.

Numerical illustrations show how the risk-adjusted value, optimum working interest, and required budget are all interconnected through the depth dependencies chosen for reserves, success probability, and costs.

The combined interaction of geological correlation and independent fiscal correlation can then be intertwined for any given exploration opportunity. One can then easily also allow all relevant parameters to vary within preselected ranges of variation to provide a probabilistic range of uncertainty.

For more than one exploration opportunity, the correlated behavior depends on a couple of main factors: First is the total budget available to be split in some proportion between the opportunities—a fiscal constraint that forces a correlated behavior on two or more opportunities. Second is the geological correlation of successfully finding hydrocarbons (oil or gas) for one opportunity at a greater depth than for another opportunity. Depending on the intrinsic product worth of each reserve (oil or gas), one may be led to minimize involvement in a shallow gas opportunity (with small RAV) in favor of a greater involvement in a deep oil opportunity. The contrast pattern is between likely depth to each opportunity, versus the expected gains of each, versus the success probability and fixed budget constraints. Simple numerical examples illuminate the interesting patterns of behavior for RAV, W, and budget fraction that can be coaxed out of seemingly mundane situations.

The general conclusion is that attention must be paid to *both* fiscal and geologic correlations in exploration opportunities of both similar and dissimilar sorts. It is this dominant point that the examples presented here have been designed to illustrate as simply as possible.

VI. APPENDIX: CUMULATIVE DEPTH CONSIDERATIONS FOR A SINGLE OPPORTUNITY

Quite often it is not so much the risk-adjusted value and optimum working interest associated with a likely finding success for an opportunity at depth which are relevant. If the total depth to be drilled is given as $z = TD$, most often the question is reduced to determining the best *constant* working interest that one should take in the opportunity. The point here is that one is not interested solely in potential reserves that are expected to be found at TD, but rather one is interested in the cumulative reserves that are found at any drill depth less than or equal to TD. In addition it is not often the case that fractional participation in an opportunity is allowed to be anything except a constant fraction. (A few exceptions do occur, such as contracts written to give, say, 10% participation to 10,000 ft; 8% participation from 10,000 to 20,000 ft; and 0% at greater depths. Such contracts are presumably written to attempt to follow roughly the OWI versus depth curve for a corporate assessed behavior.)

For reserves estimated to be found at depth z, the parabolic $RAV(z)$ and working interest, W, formula is

$$RAV(z) = W\left\{E(z) - \frac{1}{2}WG(z)^2 p_s(z)p_f(z)/RT\right\}, \quad (5.33)$$

where

$$E(z) = p_s(z)G(z) - C(z). \quad (5.34)$$

The average RAV ($CRAV$) to depth z is

$$CRAV = \frac{\left\{\int_0^z RAV(z)\,dz\right\}}{z}, \quad (5.35)$$

which, for a constant working interest, W, is given by

$$z\,CRAV = W\left(\int_0^z E(z')\,dz'\right) - \frac{1}{2}\frac{W^2}{RT}\int_0^z p_s(z')p_f(z')G(z')^2\,dz'. \quad (5.36)$$

A maximum of $CRAV$ with respect to W occurs at $W = OWI$ (in $0 \leq OWI \leq 1$) given by

$$OWI = \min\left\{RT\left[\int_0^z E(z')\,dz'\right]\left[\int_0^z dz'\,p_s(z')p_f(z')G(z')^2\,dz'\right]^{-1}, 1\right\} \quad (5.37)$$

with the maximum value of $CRAV_{max}$ provided through

$$z\,CRAV_{max} = \frac{\frac{1}{2}RT\left[\int_0^z E(z')\,dz'\right]^2}{\left[\int_0^z p_s(z')p_f(z')G(z')^2\,dz'\right]}, \quad \text{in } 0 \le OWI \le 1; \quad (5.38a)$$

$$z\,CRAV_{max} = \int_0^z E(z')\,dz' - \frac{1}{2}\int_0^z p_s(z')p_f(z')G(z')^2\,dz'/RT, \quad \text{on } OWI \equiv 1; \quad (5.38b)$$

and

$$z\,CRAV_{max} = 0, \quad \text{in } OWI < 0. \quad (5.38c)$$

Thus, by plotting OWI from Eq. (5.37) versus z ($\equiv TD$) and also $CRAV_{max}$ versus z ($\equiv TD$) for increasing TD, one can determine both the likely average RAV to be achieved for any hydrocarbon accumulation found at depth TD or shallower, as well as the constant optimum working interest one should take in the opportunity, which allows for the cumulative probability of finding hydrocarbons at any depth less than and including TD.

A similar argument can, of course, be carried through for multiple, correlated opportunities.

6

Modifications to Risk Aversion in High Gain Situations

I. INTRODUCTION

As noted in Chapter 2, one of the standard devices for estimating a risk-adjusted value (RAV) for a petroleum exploration opportunity is the utility function, originally developed within the petroleum exploration business by Cozzolino (1977b, 1978) using utility theory with exponential weighting. In terms of a corporate risk tolerance, RT, success probability, p_s (and so failure probability $p_f \equiv 1 - p_s$), costs, C, and potential gains, G, if the opportunity is successful, the Cozzolino formula for RAV is given by

$$RAV = -RT \ln\{p_s \exp(-WV/RT) + p_f \exp(WC/RT)\}, \quad (6.1)$$

where W is the fractional working interest in the opportunity ($0 \le W \le 1$) and the value $V = G - C$.

As given in Chapter 2, RAV has a maximum with respect to working interest at an optimum working interest, OWI, given by

$$OWI = RT \ln(p_s V/p_f C)/(V + C), \quad (6.2)$$

provided $0 \le OWI \le 1$ and, if $OWI > 1$, then OWI is set to unity, while if $OWI < 0$ (i.e., when $V \le p_f C/p_s$) then OWI is set to zero. At the value $W = OWI$, the maximum RAV is given by

$$RAV_{max} = -RT \ln[p_s (1 + V/C)] + V \times OWI \quad (6.3a)$$
$$= RT\{V(V + C)^{-1} \ln(p_s V/p_f C) - \ln[p_s(1 + V/C)]\}. \quad (6.3b)$$

Now OWI of Eq. (6.2) itself has a maximum with respect to value, V, at $V = V_{max}$ given by[1]

$$\ln(p_s V_{max}/p_f C) = 1 + C/V_{max} \qquad (6.4)$$

and then

$$OWI_{max} = RT/V_{max} \qquad (6.5)$$

so that, for $V_{max} > RT$, $OWI_{max} < 1$.

But then there is the interesting problem that for extremely high gains G of an opportunity ($G \to \infty$ and so $V \to \infty$), OWI from Eq. (6.2) is *less* than the maximum from Eq. (6.5) and, as $V \to \infty$,

$$OWI \to \frac{RT}{V} \ln(p_s V/p_f C) \to 0. \qquad (6.6)$$

It is not difficult to show that as $V \to \infty$ then $RAV \to -RT \ln p_f$, a finite limit. Hence, the paradox: As potential value in an opportunity tends to very large values (a good risk situation), one should apparently take a smaller and smaller working interest in the opportunity, irrespective of its success or failure probability. Even low-risk (p_s close to unity) opportunities with high potential gains should, apparently, not be invested in to any significant degree ($OWI \to 0$), which seems contrary to anticipated behavior. One would anticipate that taking a high working interest in a low-risk, high-gain opportunity would be the appropriate strategy to follow. This paradox is not unique to the RAV derived from the Cozzolino formula; it also occurs for the hyperbolic RAV formula and for the parabolic RAV formula exhibited earlier.

The question addressed here is how one resolves the anticipated behavior at high values, V, relative to the behavior predicted by the Cozzolino RAV formula.

Major pressure to consider this problem came (Rose, 1996) from a corporation. In the process of evaluating the economic aspects of the corporation's opportunity it was requested that a high-gain possibility be considered. The paradoxical result came through that the optimum working interest should be *reduced* relative to a lower gain scenario—much to the corporation's disbelief and to individual bemusement.

Accordingly, an immediate concern became one of determining good risk versus bad risk (increased uncertainty due to increased cost or decreased chance) situations within a strategy utilizing risk-adjusted value/optimum working interest.

---

[1] Appendix A at the end of this chapter provides the method for constructing accurate numerical solutions to Eq. (6.4) for V_{max}.

One might think that the specific case which triggered this investigation is itself suspect, for it can be argued that it is rare in the real world to have an enormous discrepancy between reward and cost (the ratio was 100:1 in the specific case) and so such situations may, perhaps, give one cause to be suspicious of the accuracy of the number anyway and therefore to reduce working interest. Perhaps so, but the paradox stands because *any* value, V, even slightly greater than V_{max}, argues that the working interest should be lower than the maximum optimum working interest occurring on $V = V_{max}$; and V_{max} does not have to be enormously discrepant compared to costs to force such a paradox. For example, with $p_s \cong 47\%$, the ratio of V_{max}, to costs, C, is the modest value 4 from Eq. (6.4); so for this case the paradox occurs for any V greater than $4 \times$ costs, a common occurrence in the real world. Also in the real world corporations often do not have the luxury of choosing the optimum working interest; nevertheless, the modifications to the classical Cozzolino formula are of interest because they indicate what would be optimal, thereby helping corporations in their exploration decisions.

The procedure given in this chapter not only provides a pragmatic method for identifying the change from a bad risk to a good risk scenario, but also specifies the transition in terms of the values, V_{max}, determined from the Cozzolino formula by Eq. (6.4), at which the optimum working interest has its maximum value.

The detailed technique development is carried out in Appendix A for the Cozzolino (exponential weighting) risk-adjusted value formula, and in Appendix B for the corresponding formula for *RAV* for parabolic weighting.

This relegation of the technical details to appendixes at the end of the chapter is done for a variety of reasons: First, the details of the technical manipulations are collected in one place, and we also show that the paradox is not a consequence solely of the functional form used for the Cozzolino *RAV* but is generic in character. Second, the body of the text and the flow of the chapter are not overly cluttered with detail. An exploration risk analyst is unlikely to wade through the technical details to arrive at final, useful, modified formulas for applications to business, while a geomathematician is unlikely to appreciate the practical business applications as a consequence of a detailed, expansive, mathematical treatment. Third, those whose job it is to convert mathematical risk analysis methods to computer-driven applications can read the appendixes directly without being forced to bounce backwards and forwards through the chapter looking for a particular point or condition to ensure numerical accuracy of a computer code.

In the body of the text we give only the pertinent results for modifying the Cozzolino *RAV* formula; the corresponding modified results for the parabolic *RAV* formula are detailed in Appendix B.

II. MODIFICATIONS TO THE COZZOLINO *RAV* FORMULA

A. OVERVIEW

The mean value, E, of the choices between the success and failure channels of Figure 6.1 is given by $E = p_s V - p_f C$, while the variance, σ^2, is given by $\sigma^2 = p_s p_f (V + C)^2$. A measure of accuracy of the mean is customarily given by the volatility, $v = \sigma/|E|$, which measures the signal-to-noise fluctuations around the mean. A small value of volatility ($v \ll 1$) means low uncertainty in the expected mean value E, whereas a large value ($v \gg 1$) implies considerable uncertainty in E. Note that

$$v^2 = p_s p_f (V + C)^2 / (p_s V - p_f C)^2 \qquad (6.7)$$

has the property that $v^2 \to p_f/p_s$, as $V \to \infty$, which is a fixed limit for v^2, independent of the value V. Thus the uncertainty of the mean, σ, does not decrease to zero as $V \to \infty$. Because the uncertainty of the mean is *not* decreasing at high values, the usual central limit theorem on increasing accuracy of the mean as value, V, increases fails to be appropriate. This uncertainty in the mean should therefore be incorporated somehow in the relevant *RAV* formula.

As shown in Appendix A at the end of the chapter, the modified Cozzolino-type formula for *RAV*, which includes the corresponding variance in the mean, is given by

$$RAV = -RT \ln\{p_s \exp(-VW/RT_a) + p_f \exp(WC/RT_a)\} \qquad (6.8)$$

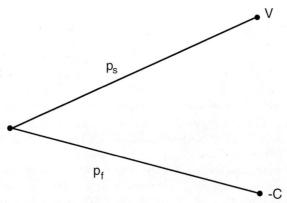

FIGURE 6.1 Chance node diagram illustrating value, V, costs, C, probability of success p_s, and probability of failure p_f ($\equiv 1 - p_s$). The expected value at the chance node is $E = (p_s V - p_f C)W$ at a working interest $W < 100\%$.

with

$$RT_a = [(RT^{2a} + \{(p_s p_f)^{1/2} (V + C)\}^{2a}]^{1/2a}, \qquad (6.9)$$

where a is a positive number greater than about 2 (see Appendix A). Note that with this inclusion of variance, RT_a reduces to the usual corporate risk tolerance, RT, as long as the standard error (square root of variance), σ, is much less than the corporate risk tolerance, RT. However, when the standard error is large compared to the corporate risk tolerance, then RT_a tends to $(p_s p_f)^{1/2}(V + C)$.

While there is no influence on the maximum RAV as given through Eq. (6.3b), by this choice (or any other) for RT_a, there is a large effect on the optimum working interest, OWI_a, because now

$$OWI_a = RT_a \ln\{p_s V/p_f C\}/(V + C), \qquad \text{for } 0 \leq OWI_a \leq 1 \quad (6.10)$$

and, as $V \to \infty$, OWI_a from Eq. (6.10) tends to the limiting form $(p_s p_f)^{1/2} \times \ln(p_s V/p_f C)$, very different from the conventional Cozzolino formula. So this sort of variance inclusion procedure is clearly pointing in the correct direction of resolving the high-V paradox.

B. PRACTICAL CONSIDERATIONS

From a practical viewpoint there are two problems with the above development, which includes the use of the variance to identify a shift of OWI more in accord with anticipated behavior.

First, there is no particularly good way of identifying the value of the parameter a (except to note that $a = 1$ if one can really argue that the factors entering RT_a can be added in a Gaussian least squares manner). One could use past historical precedent for case histories to determine empirically an approximate range of values of a. This process tends to give a value of $a \cong 1$–3, somewhat more than the value of $a = 1$ appropriate to the Gaussian assumption. In addition, there is no guarantee that RAV analysis on a particular opportunity is going to fit against such a statistical empirical range of a values.

Second, at low working interest values it is a tried-and-true fact that the conventional Cozzolino formula does an acceptable job of providing a dependence of RAV on value V; one would like to preserve that part of the curve of OWI versus V at values of V less than $V_{\max}$.

A practical procedure for addressing both of these concerns at once is as follows. Take the optimum working interest, OWI, at a value V to be given by

$$\text{OWI} = (RT/V_{\max}) [\tanh\{[OWI_a/(RT/V_{\max})]^b\}]^{1/b}, \qquad (6.11)$$

where RT/V_{max} is just OWI_{max} of the Cozzolino formula [Eq. (6.5)] and b is a numerical coefficient greater than 8 (see Appendix A). If $b \geq 8$ then on $OWI_a = RT/V_{max}$, Eq. (6.11) yields an OWI within 3% of RT/V_{max}. For values of $OWI_a \ll RT/V_{max}$ Eq. (6.11) returns $OWI = OWI_a$, while for values of $OWI_a \gg RT/V_{max}$ Eq. (6.11) returns the value $OWI = RT/V_{max}$; and Eq. (6.11) is restricted to $0 \leq OWI \leq 1$. Note that if one has determined RT/V_{max} for the Cozzolino formula (see Appendix A) then all that is needed now are Eqs. (6.10) and (6.11).

The corresponding value of RAV at $W = OWI$ at a value V is then given directly by insertion into the modified Cozzolino-type formula:

$$RAV = -RT \ln\{p_s \exp(-OWI \times V/RT_a) + p_f \exp(OWI \times C/RT_a)\}. \quad (6.12)$$

Note that RAV from Eq. (6.12) tends to the limiting value $RAV \rightarrow -RT \ln[p_f + p_s \exp\{-(RT/V_{max})(p_s p_f)^{-1/2}\}]$ as $V \rightarrow \infty$—a slightly smaller limiting value than the high-V limit of $-RT \ln p_f$ for the classical Cozzolino formula.

This procedure honors the best aspects of both the Cozzolino formula and also of the variance inclusion without having to be unduly concerned with a precise determination of the parameter a from some statistical distribution model, although a rough range for parameter a is needed, and a must exceed 2 (see Appendix A).

The reason for wanting to preserve, to the extent possible, the low-V part of the Cozzolino RAV formula is to keep a separation between "bad risk" and "good risk" opportunities. Here bad risk represents situations where $p_s V$ is only slightly larger than $p_f C$ so that the mean expected value $E \equiv p_s V - p_f C$ is only slightly positive, in which case the uncertainty $\pm \sigma$ on E, given by $\sigma = \pm (p_s p_f)^{1/2}(V + C)$, can easily lead to a negative result within one standard error. For values of $V < V_{max}$ the Cozzolino formula correctly predicts that one should take a smaller and smaller working interest in an opportunity as V tends from above toward $p_f C/p_s$. On the other hand, a good risk situation is one in which the uncertainty on expected mean value is sufficiently small, and E is itself sufficiently high that even one standard error, $-\sigma$, on the low side of the mean will likely yield a positive return. This situation is encompassed by the region $V \gtrsim V_{max}$ in the classical Cozzolino formula and is clearly a good risk despite the Cozzolino formula indicating that a lower and lower optimum working interest should be taken as value increases.

In summary, the practical procedure recommended above cleanly separates good risk from bad risk situations and provides optimum working interest for both situations, which honors the basic structure anticipated for the opportunity. (If one wishes to be more rigorous, one can always use the modified RT_a formula for corporate-defined values of the parameters a

and b.) In addition, the practical solution always gives an *OWI* that is extremely close to $RT/V_{\max}$ at all $V \geq V_{\max}$.

C. "CROSSOVER" FROM THE CLASSICAL COZZOLINO FORMULA

The "crossover" from the point at which the classical Cozzolino formula for *OWI* provides a good representation of real situations to when it does not requires that several criteria be checked in order, as depicted in the flow diagram of Figure 6.2, from which one sees that twin requirements must be in force:

$$RT \gtrsim (p_s p_f)^{1/2} (V + C) \equiv \sigma, \qquad (6.13a)$$

and, when inequality (6.13a) is obeyed, one must also have in force:

$$\ln(p_s V / p_f C) \lesssim (V + C)/V_{\max}, \qquad (6.13b)$$

if the classical Cozzolino formula is to depict accurately *RAV* and *OWI*. Otherwise, the modified procedure provided here is a more appropriate representation of *RAV*; and then only if $RT \leq V_{\max}$ is $OWI \leq 1$.

Simply put, (1) if the standard error $\sigma\{\equiv(V + C)(p_s p_f)^{1/2}\}$ of the mean value $E(\equiv p_s V - p_f C)$ is either comparable to E or is large compared to the corporate risk tolerance, *RT*, or both, then modifications to the classical

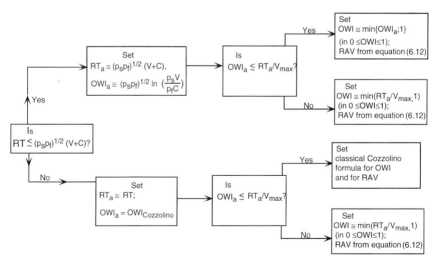

FIGURE 6.2 Flow diagram indicating the regimes of values for *RT* and *V* when the classical Cozzolino formula needs modifications.

Cozzolino formula certainly need to be made; and (2) even if σ is not large compared to RT, modifications to the classical Cozzolino formula still need to be made (a) whenever the project value, V, is larger than V_{max}, where V_{max} is the value at which the optimum working interest occurs, given by $\ln(p_s V_{max}/p_f C) = 1 + C/V_{max}$ for the Cozzolino RAV formula; and (b) whenever $RT \geq V_{max}$ and $V \geq V_{max}$, independently of whether RT is large or small compared to σ.

D. NUMERICAL ILLUSTRATIONS

To show when the modified procedure provides that a necessary change be made in the optimum working interest and associated RAV that would otherwise be calculated from the classical Cozzolino RAV formula, two groups of numerical calculations have been made. The first group exhibits the resulting modifications to the OWI formula [Eq. (6.11)] for different values of the parameters a and b, and is followed by two case histories done with $a = 2$ and $b = 8$. The second group of numerical illustrations considers the behavior of working interest in six potential opportunities and illustrates the range of results for RAV and OWI which are influenced by the non-Cozzolino nature of the opportunities.

1. Group 1 Illustrations

To illuminate the behaviors of the various different patterns of dependence for different values of the parameters a and b, we consider several cases where $p_s = 0.2$ ($p_f = 0.8$), with costs $C = \$2$ or $\$8$ MM; and corporate risk tolerance $RT = \$6$ or $\$20$ MM.

Plots of OWI [from Eq. (6.11)] versus value, V, are presented in Figures 6.3 through 6.6, for different values of C and RT, and with both parameters a and b varying. The corresponding plots of OWI versus V for the Cozzolino formula are shown by dashed lines on Figures 6.3 through 6.6. Several numerical cases have shown that the modified formula [Eq. (6.11)] with $a = 1$ is too aggressive in its rise relative to the bad risk portion of the conventional Cozzolino formula at values of $V < V_{max}$. Increasing the value of the parameter a to 2 from unity allows a less aggressive dependence of OWI on V in $V < V_{max}$ as shown in Figures 6.3 through 6.6 A value of $b = 8$ produces curves on Figures 6.3 through 6.6 not too dissimilar from the bad risk portion of the Cozzolino curve, while increasing the value to $b = 16$ shows an improvement. The overall results from the investigation

FIGURE 6.3 Ordinate shows fractional optimum working interest versus increasing value, V (in $MM), on the abscissa for both the Cozzolino formula (dashed curve) and the modified OWI formula (solid curve); for $RT = \$20$ MM, $p_s = 0.2$, $p_f = 0.8$, and costs $C = \$2$ MM. Shown in clockwise order, starting at the upper left panel, are parameter value pairs: $a = 1$, $b = 8$; $a = 1$, $b = 16$; $a = 2$, $b = 16$; $a = 2$, $b = 8$.

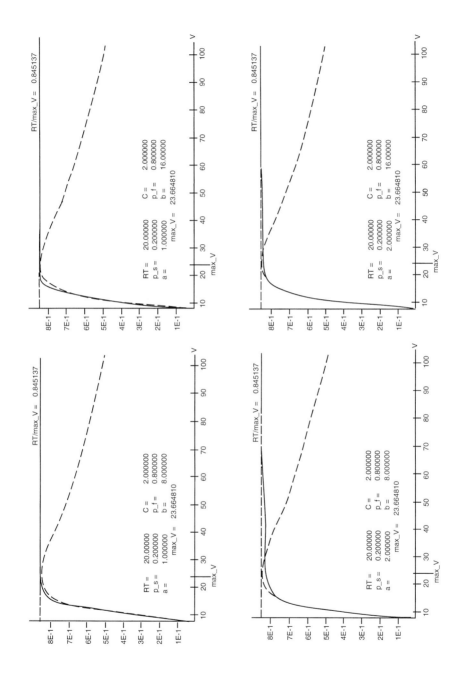

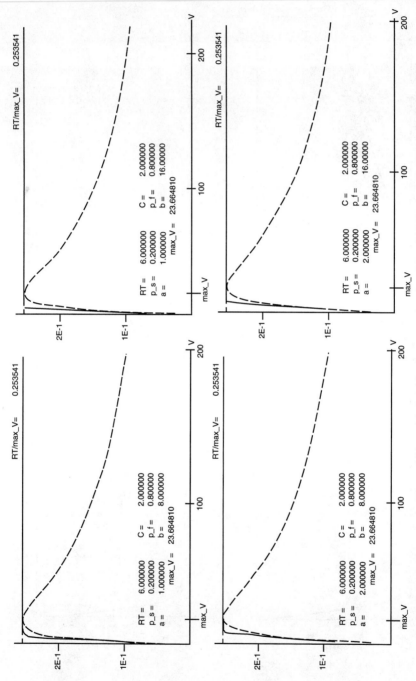

FIGURE 6.4 As for Figure 6.3 but with the risk tolerance reduced to $RT = \$6$ MM, and all other values held fixed.

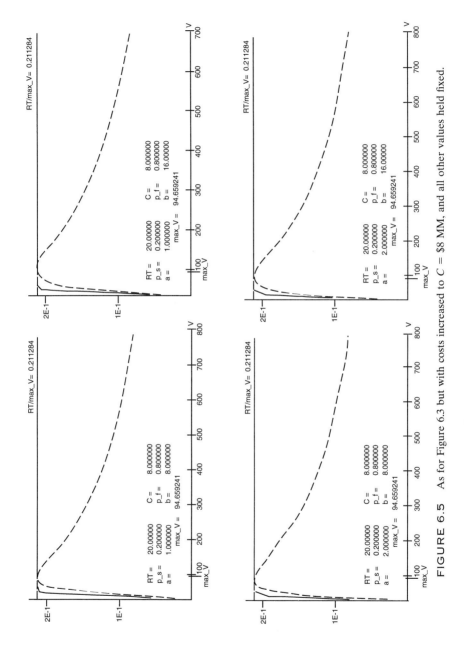

FIGURE 6.5 As for Figure 6.3 but with costs increased to $C = \$8$ MM, and all other values held fixed.

147

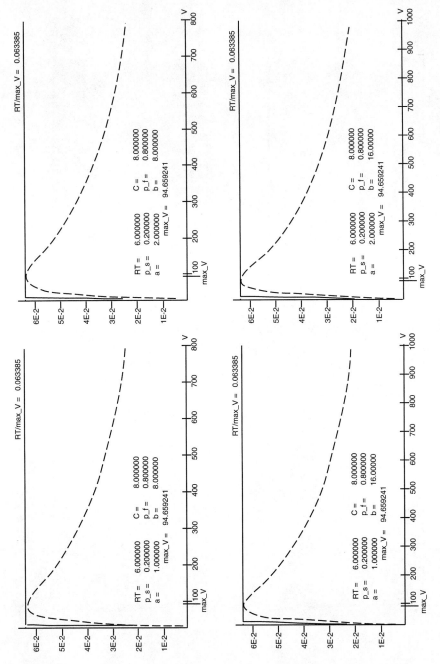

FIGURE 6.6 As for Figure 6.3, but with costs increased to $C = \$8\text{MM}$, risk tolerance reduced to $RT = \$6$ MM, and all other values held fixed.

suggest that parameter values of $a \geq 2$ and $b \geq 8$ do an adequate job of paralleling OWI versus V from the Cozzolino formula at $V \leq V_{max}$, and do a much better job of adequately describing the good risk region $V \geq V_{max}$. Thus the good risk aspects of the high-gain situation are more sharply emphasized by the modified procedure, while the bad risk situations remain clearly identified by the low-V end of the conventional Cozzolino formula.

a. Case History for a "Low-Gain" Situation

Consider an opportunity with the following parameters: $p_s = 0.2$, $V = \$5$ MM, $C = \$1$ MM, $RT = \$10$ MM. In this case the expected value is $E = \$0.2$ MM, while the standard error is $\sigma = \$0.24$ MM $\ll RT$, so that the volatility, v, (σ/E) is 1.2, implying a significant uncertainty on the expected value. The optimum working interest for the Cozzolino formula is 37.19%, while the modified optimum working interest, per Eq. (6.11), is 37.22%. The corresponding curves of risk-adjusted value versus working interest are given in Figure 6.7, indicating that the Cozzolino formula does an acceptable job of defining OWI. The change in optimum working interest as the estimated value, V, increases is given in Figure 6.8 for both the Cozzolino formula and the modified formula of Eq. 6.11, showing that both are remarkably close until the estimated value is in excess of about \$25 MM.

Thus for any value of V less than \$25 MM there is no need in this case to modify significantly the standard Cozzolino results.

b. Case History for a "High-Gain" Situation

Consider an opportunity with the following parameters: $p_s = 0.2$, $V = \$50$ MM, $C = \$1$ MM, $RT = \$10$ MM. In this case the expected value is $E = \$9.2$ MM, while the standard deviation is $\sigma = \$20.4$ MM, so that $\sigma/E \cong 2.2$ implying nearly a doubling of the volatility compared to the previous case history. Further, $\sigma \cong 2RT$ in this case, so that modifications to the conventional Cozzolino formula are surely going to be required.

The optimum working interest from the Cozzolino formula is 49.52% in this case, while the modified formula [Eq. (6.11)] indicates that an OWI of 84.51% should be taken.

Shown in Figure 6.9 are the OWI values for both the Cozzolino formula and the modified formula as value increases, indicating that at any value $V \gtrsim \$15$ MM modifications are needed to the Cozzolino formula and the estimated value of $V \cong \$50$ MM is in this region. This point is further reinforced if a plot is made of risk-adjusted value versus working interest for both the Cozzolino formula and the modified formula for the input parameters, as given in Figure 6.10, where the shift in RAV is clearly exhibited.

The point of these examples has been to illuminate the behavior of RAV and OWI for fixed values of costs, success probability, and risk tolerance, as the estimated project value increases, and to use the approximate rules

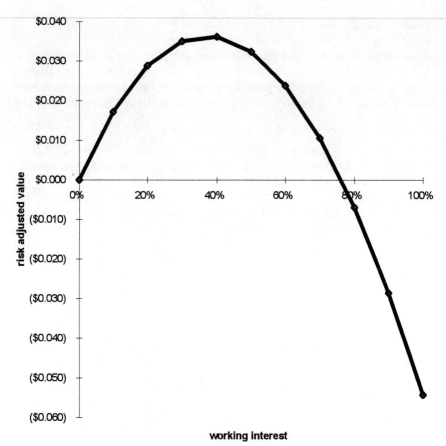

FIGURE 6.7 Risk-adjusted value versus working interest for both the basic Cozzolino formula and the modified formula, for $p_s = 0.2$, $V = \$5$ MM, $C = \$1$ MM, $RT = \$10$ MM; both curves almost overlie each other. Ordinate in $MM.

given in Figure 6.2 to examine when modifications to the Cozzolino formula play significant roles.

2. Group 2 Illustrations

As part of the information that can be gleaned from the modified RAV formula, it is of interest to consider the six opportunities presented in Table 6.1, where the estimated values, costs, and success probabilities are held fixed at the entries listed in Table 6.1 but the risk tolerance, RT, is variable.

From the data of Table 6.1 it is a simple matter to construct the standard error σ, volatility v ($\equiv \sigma/E$), the optimum working interest per $MM of RT (i.e., OWI/RT), the value V_{max} corresponding to the maximum OWI, and the risk-adjusted value, RAV, at the optimum working interest per $MM

MODIFICATIONS TO THE COZZOLINO RAV FORMULA

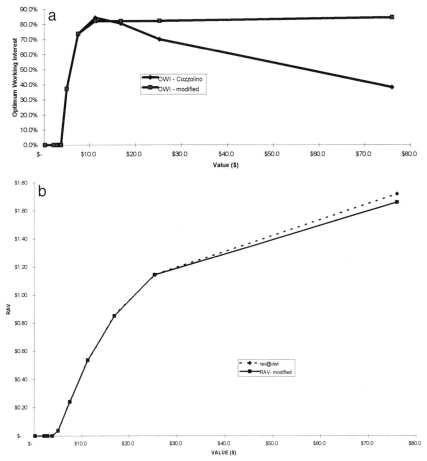

FIGURE 6.8 (a) Change in *OWI* as the value, *V*, increases for both the Cozzolino formula and the modified formula using the same parameters as for Figure 6.7, ordinate in %, abscissa in $MM. (b) *RAV* versus value, *V*, for both the Cozzolino formula and the modified formula using the same parameters as in Figure 6.7, ordinate and abscissa in $MM.

TABLE 6.1

Opportunity	Value, V ($MM)	Costs, C ($MM)	Success probability, p_s	Failure probability, p_f	Expected value, E ($MM)
A	1,800	30	0.15	0.85	244.5
B	500	10	0.20	0.80	92.0
C	140	3	0.25	0.75	32.8
D	40	1	0.30	0.70	11.3
E	10	0.3	0.35	0.65	3.3
F	2.5	0.1	0.40	0.60	0.9

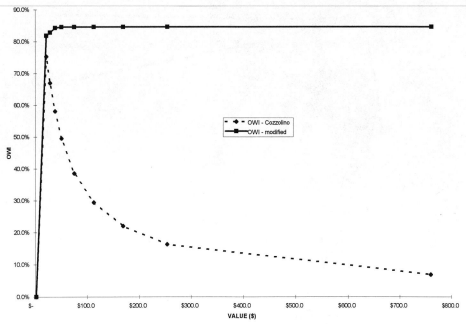

FIGURE 6.9 OWI versus value, V, for both the Cozzolino formula and the modified formula, for parameters $p_s = 0.2$, $v = \$50$ MM, $C = \$1$ MM, $RT = \$10$ MM. Ordinate in %, abscissa in \$MM.

of RT (i.e., RAV/RT), for each of the opportunities using the Cozzolino formula. This information is given in Table 6.2. Plotted on Figure 6.11 are the curves of RAV versus RT for each of the six opportunities A–F and, superimposed on each curve, are the points where $RT \cong \sigma$ through which are drawn the hatched lines defining domains (shaded) in RAV versus RT space at which the Cozzolino formula needs to be modified for each of the opportunities; these areas are labeled "non-Cozzolino."

TABLE 6.2

Opportunity	Standard error, σ (\$MM)	Volatility, $v \equiv \sigma/E$	$OWI(\%)/RT$ (\$MM)$^{-1}$	$RAV(\$MM)/RT(\$MM)$	V_{max} (\$MM)
A	253.1	1.04	1.26×10^{-1}	0.06	480
B	204.0	2.22	4.9×10^{-1}	0.20	135
C	61.9	1.89	1.9	0.24	29
D	18.8	1.66	6.8	0.31	8
E	4.9	1.48	2.74×10^{1}	0.39	2
F	1.3	1.44	1.06×10^{2}	0.41	0.48

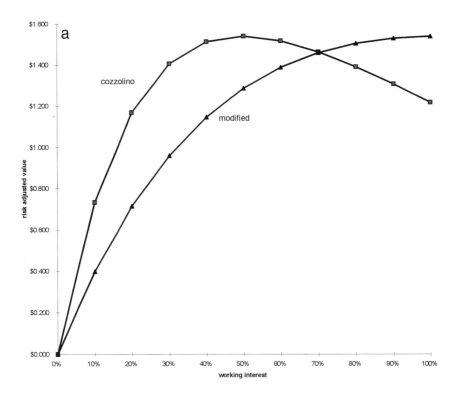
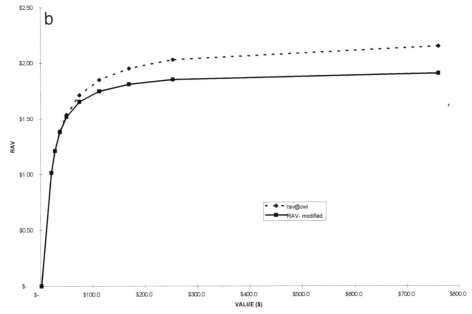

FIGURE 6.10 (a) Risk-adjusted value versus working interest for both the Cozzolino formula and the modified formula using the same parameters as in Figure 6.9. Ordinate in $MM. (b) RAV versus value, V, for both the Cozzolino formula and the modified formula using the same parameters as is Figure 6.9. Ordinate and abscissa in $MM.

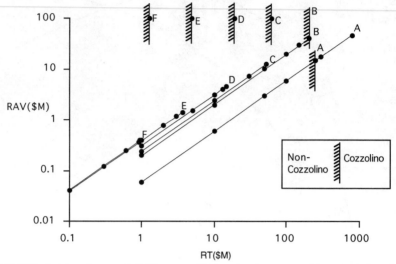

FIGURE 6.11 Curves of *RAV* versus *RT* for the six opportunities, A–F, described in the text, with the regions where the Cozzolino formula needs modification labeled as "non-Cozzolino."

Plotted on Figure 6.12 are the corresponding curves of *OWI* versus *RT* for each of opportunities A–F and, superimposed on each curve, are the points where $RT \cong \sigma$; hatched lines again show the regions where the Cozzolino formula needs to be modified. Also given in Figure 6.12 are the values for *RT* for each opportunity corresponding to an optimum working interest of 100%, as determined from the Cozzolino formula.

As can be seen from Figures 6.11 and 6.12 the regions where the Cozzolino formula needs modification are not obtained solely under situations where there is a large discrepancy between reward and cost, but occur for all six of the cases in regimes of parameters commonly occurring in real-world situations—this is the major point to be made for the six illustrative situations given.

E. WEIGHTING THE UNCERTAINTIES

Perhaps one of the major unaddressed concerns is illustrated by considering the worth of the estimated expected values for each of the opportunities in Table 6.1. In addition to the actual expected value for each opportunity, there are also standard errors, σ, of the expected value, as given in Table 6.2.

The volatility factor is useful with more than one category of potential opportunity resource for two reasons: If the volatility is small, then the estimate is fairly accurate because the $P(84)$ and $P(16)$ values, which bracket

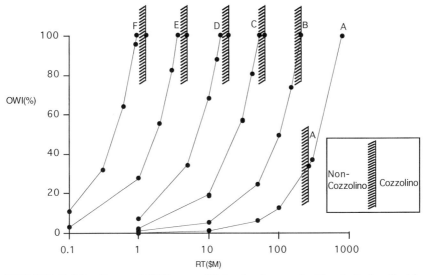

FIGURE 6.12 Curves of OWI versus RT for the six opportunities, A–F, described in the text, with the regions where $RT \lesssim \sigma$ for each opportunity labeled "non-Cozzolino," implying the need for modification of the classical Cozzolino formula.

the $P(68)$ value, are then close, so the fractional error in $P(68)$ is small. If the volatility is large, then a large fractional error exists in the $P(68)$ estimate, and that estimate is then not as reliable as one with a low volatility. Here $P(16)$, $P(68)$, and $P(84)$ represent the cumulative probability values and are approximately given by $E - \sigma$, E, and $E + \sigma$, respectively, typical of a moderately skewed log-normal distribution.

Two calculations can then be performed simply on a set of opportunities. Suppose N categories are being investigated, each with its expected value $E(x_i)$, corresponding to about $P(68)$,[2] and each with its volatility, $v \cong \{P(84) - P(16)\}/P(68)$.

Then a volatility-weighted measure, R_i, of potential expected value can be made for the ith opportunity as

$$R_i = \{E(x_i)/v_i\}, \tag{6.14a}$$

which weights the contribution of each opportunity to the total according to its relative volatility, with a total worth from all opportunities of

$$R = \sum_{i=1}^{N} R_i \tag{6.14b}$$

---

[2] A value more precise than 68% is $P(50) + 1/2\mu\{P(84) - P(50)\} \equiv (50 + 17\mu)\%$, where $\mu = 0.5 \ln\{P(84)/P(16)\}$.

The second calculation that is available is the volatility-weighted relative importance, $RI(i)$ (in %) of the ith opportunity in relation to the other opportunities, with

$$RI(i) = \frac{100\{E(x_i)/v_i\}}{\sum_{i=1}^{N} E(x_i)/v_i}. \qquad (6.15)$$

This relative importance factor measures the likelihood of a particular opportunity contributing potential worth compared to all opportunities. As such, RI is a useful measure of where most exploration effort should be put to enhance the likelihood of resource appreciation being realized, with allowance made for the speculative nature of individual opportunities. Such a device is very useful in the decision-making process of where to commit capital investment and other resources.

F. VOLATILITY-WEIGHTED ESTIMATES OF POTENTIAL OPPORTUNITIES

Each individual opportunity given in Table 6.1 enables one to construct the volatility, v, of each opportunity given in Table 6.3.

Inspection of Table 6.2 shows that opportunity A has the lowest volatility. However, the volatility is nearly as high as 104%, and the remaining opportunities have between about 144 and 222% volatilities—suggesting that they are extremely uncertain relative to opportunity A—a factor of between about 1.4 to 2.2 more uncertainty based on relative volatilities. It would seem that expected value assessments of the opportunities with high volatility are less than accurate.

Table 6.3 gives the volatility-weighted assessments of the opportunities and the relative importance of each opportunity, indicating that the volatility-weighted values are considerably less optimistic than an unweighted assessment of expected values.

TABLE 6.3 Volatility-Weighted Assessments and Relative Importances

Opportunity	Weighted value ($MM)	Relative importance (%)
A	235.1	77
B	41.9	14
C	21.2	6
D	6.8	2
E	2.2	0.7
F	0.6	0.2
Total	307.8	99.9

Thus, instead of adding all the opportunities as though each expected value estimate had the same level of certainty as all others, if one allows for the relative worth of each estimated opportunity, then a more representative measure can be given of what amount each opportunity is *likely* to contribute to the total estimated revenue. It would seem that opportunity A is relatively more important (77%) than any of the remaining opportunities, and that considerably less attention and resources should be spent on the remaining opportunities. The suggestion, then, is that not all opportunities are relatively attractive based on the magnitude of the expected value.

G. RELATIVE IMPORTANCE OF EACH CATEGORY

As well as estimating the total potential worth likely to be found for each opportunity, it is of equal importance to know which opportunities carry the most certainty of maximizing gains. Such assessments then provide information on where to concentrate activities and capital to maximize the probability of actually finding the resources.

Table 6.3 provides the volatility-weighted relative importance (RI) for each opportunity respectively, using the data of Table 6.1 and the volatility measures given in Table 6.2. Each RI is evaluated using values at $P(68)$, $P(16)$, and $P(84)$, so that one has an appreciation of the stability and rank of each category's relative importance.

Inspection of Table 6.3 strongly suggests that, at any level, the relative importance is dominated by opportunity A, with opportunity B a distant second and the remaining opportunities of much lower importance.

H. OTHER WEIGHTING MEASURES

Alternative weighting measures are available. For instance, it is often the case that a corporation chooses to analyze projects with the $P(50)$ value, $E_{1/2}$, rather than the mean, E. For a log-normal distribution the exact relation is

$$E_{1/2} = E/(1 + \nu^2)^{1/2}. \tag{6.16}$$

Thus the weighting of each opportunity is then done in an absolute sense rather than in a relative sense. For the six opportunities, the corresponding values are given in Table 6.4. Note also, from Table 6.4, that a relative importance, defined for the ith opportunity by

$$RI_i(\%) = \frac{100 \, E_{1/2}(i)}{\sum_{i=1}^{N} E_{1/2}(i)}, \tag{6.17}$$

yields similar values to those for weighting inversely with volatility.

TABLE 6.4 $P(50)$-Weighted Assessments and Relative Importances

Opportunity	Weighted value ($MM)	Relative importance (%)
A	169.1	73.5
B	37.9	16.4
C	15.3	6.6
D	5.7	2.5
E	1.8	0.8
F	0.5	0.2
Total	230.3	100.0

The major difference between the two weighting procedures is in the assessment of absolute worth versus relative worth as shown in Tables 6.3 and 6.4.

III. DISCUSSION AND CONCLUSIONS

The original motivation for addressing the problem of high-gain (high-V) situations was provided by the dichotomy of a real case history in which the conventional Cozzolino formula recommended a lower OWI as the gain increased, bringing the difficulty sharply to the fore. Rapid inspection of other functional formulas for risk-adjusted value and working interest made clear that the problem was of a generic nature and not peculiarly beholden to the Cozzolino formula alone.

Resolution to the problem is achieved here by incorporating the uncertainty in the expected value directly into RAV, so that corporate risk tolerance and the intrinsic uncertainty both contribute. In this way, for low-gain ("bad risk") situations the modifications reduce to the conventional Cozzolino formula, while for high-gain ("good risk") situations the modifications yield a higher working interest than does the Cozzolino formula.

Numerical illustrations of how the modifications adjust both RAV and OWI for a single exploration opportunity provide an idea of the numerical scale of values, while an intercomparison of six exploration opportunities provides not only a pattern of behaviors indicating when the modifications should be used, but also permits the relative contribution and relative importance of each opportunity to be assessed in terms of the intrinsic uncertainty of the expected value, independently of the corporate risk tolerance.

The major point to be garnered from the work reported here is that care must be exercised in applying any risk-adjusted value formula to an exploration opportunity without being aware of its fundamental strengths and weaknesses.

IV. APPENDIX A: MODIFYING RISK FORMULAS

A. MODIFICATIONS TO THE COZZOLINO (EXPONENTIAL WEIGHTING) FORMULA

1. General Considerations

Consider again the Cozzolino formula of the text, Eq. (6.1), but rewritten in the form

$$RAV = -RT \ln\{p_s \exp(-WV/RT_0) + p_f \exp(WC/RT_0)\}, \quad (6.18)$$

where the factor RT outside the logarithmic term is written separately from the value RT_o inside the logarithm in order to trace the effects of both factors.

Then one has the optimum working interest given through

$$OWI_o = RT_o \ln(p_s V/p_f C)/(V + C) \quad \text{for } 0 \leq OWI_o \leq 1 \quad (6.19)$$

and, at this optimum working interest, the maximum RAV is given through

$$\begin{aligned} RAV_{\max} &= V(RT/RT_o) \, OWI_o - RT \ln\{p_s(1 + V/C)\} \\ &= V \ln(p_s V/p_f C)/(V + C) - RT \ln\{p_s(1 + V/C)\} \end{aligned} \quad (6.20)$$

The important point to note is that the maximum RAV is *independent* of the value chosen for RT_o, whereas OWI_o is directly *dependent* on the choice of RT_o. Thus the question of the behavior of OWI_o as $V \to \infty$ is determined by the dependence chosen for RT_o on V, whereas the high-V behavior of $RAV_{\max}$ does not depend on this choice and always tends to $-RT \ln p_f$ as $V \to \infty$.

2. Variance Considerations

One way to argue for the inclusion of variance is from analogy. Consider Figure 6.1 but now regard the problem as two end members, one with "energy" V, the other with "energy" $-C$. If we were to measure the average energy of the resultant system, we would obtain the value $p_s V - p_f C$. But the corresponding spread in energy around the mean, σ, corresponds to an intrinsic temperature, $T(\text{intrinsic})$, of the resultant system with

$$RT(\text{intrinsic}) \equiv (p_s p_f)^{1/2}(V + C) \equiv \sigma, \quad (6.21)$$

where R is the gas constant. Thus, even when the original system is cold (i.e., has no temperature), the resultant "mix" of two end-member products *does* have a temperature.

If the original system also has a temperature, $T(\text{system})$, which is taken to be independent of the intrinsic temperature of the final mix, then the final temperature of the products is given by

$$(RT_o)^2 = \{RT(\text{system})\}^2 + \{RT(\text{intrinsic})\}^2, \quad (6.22d)$$

provided the variances due to temperature of the original system and of the products are taken in Gaussian least squares addition. For a more general procedure of addition, we would write

$$RT_o = \{[RT(\text{system})]^{2a} + [RT(\text{intrinsic})]^{2a}\}^{1/2a}, \quad (6.22b)$$

where a is a positive constant related to the statistical properties of the addition.

If, then, we wanted to write an energy distribution for the resultant mix, written with respect to the temperature, $T(\text{system}) \equiv T$, we would write

$$\exp(-RAV/RT) = p_s \exp(-WV/RT_o) + p_f \exp(WC/RT_o), \quad (6.23)$$

where RAV is the energy of the mix, and W is the fractional contribution of each energy state. Then,

$$OWI_a = (V + C)^{-1} \ln(p_s V/p_f C) \{(RT)^{2a} + [(p_s p_f)^{1/2}(V + C)]^{2a}\}^{1/2a} \\ \equiv (V + C)^{-1} \ln(p_s V/p_f C) RT_o, \quad (6.24)$$

which, as $V \to \infty$, now tends toward

$$OWI_a = (p_s p_f)^{1/2} \ln(p_s V/p_f C) \to \infty. \quad (6.25)$$

Thus the pattern of behavior of optimum working interest as V increases is changed from the basic Cozzolino formula, and allows for the fact that the uncertainty of the mean is not decreasing as the mean increases. At high V, the implication is that one should take a greater working interest in the opportunity, rather than the ever smaller working interest that would be suggested by the conventional Cozzolino formula; but both formulas provide the same maximum RAV of $-RT \ln p_f$ as $V \to \infty$.

B. AN ALGORITHM FOR DETERMINING V_{max} OF EQUATION (6.4)

Write $V_{\text{max}}/C = u$ when Eq. (6.4) of the text can be put in the form

$$\ln u = \{1 - \ln(p_s/p_f)\} + 1/u. \quad (6.26)$$

(a) If $p_s < ep_f$ then $S \equiv 1 - \ln(p_s/p_f) > 0$ and the iterative procedure for solution to Eq. (6.26) is

$$u_{n+1} = \exp\left(S + u_n^{-1}\right) > 1 \quad (6.27)$$

with initial estimate $u_o = \exp(S) > 1$.
(b) If $p_s = ep_f$, then $S = 0$ and Eq. (6.26) reduces to

$$u \ln u = 1 \quad (6.28)$$

with approximate solution $u = 1.75$.

(c) If $p_s > ep_f$, then $S < 0$ and an iterative procedure for the solution to Eq. (6.26) splits into two parts:

(i) If $e^2 p_f \geq p_s > ep_f$ then $u \geq 1$, with $u = 1$ only when $e^2 p_f = p_s$. Then write

$$u_{n+1} = \exp\left(u_n^{-1} + S\right), \qquad (6.29)$$

but with initial estimate $u_o = 1$. This iterative procedure guarantees to keep each iteration for $u_n > 1$.

(ii) If $p_s > e^2 p_f$ then $p_f e / p_s < u < 1$ and $S < -1$. In this case a suitable iteration is obtained by first writing $u = 1/y$, with $y > 1$. Then Eq. (6.26) is equivalent to

$$y = \ln(p_f/p_s) - 1 - \ln y. \qquad (6.30)$$

A suitable iteration procedure is as follows. Let $y_0 = \ln(p_s/p_f) - 1 \geq 1$, then write (for $n = 0, 1, 2, \ldots$)

$$y_{n+1} = y_o - \ln y_n \qquad (6.31)$$

provided $y_{n+1} > 1$. If $0 < y_{n+1} < 1$, then the next iteration is bound to increase y_{n+2} above unity because $\ln(y_{n+1}) < 0$, which then increases the right-hand side of Eq. (6.31). If $y_{n+1} < 0$ at any iteration, and because $y_n \geq 1$, then immediately replace y_{n+1} by $y_n/2 > 0$. If y_{n+2} is negative using Eq. (6.31), then again immediately replace y_{n+2} by $y_{n+1}/2 \equiv y_n/4$, and so on, eventually obtaining an iterate y_{n+m} which is less than unity but greater than zero. Then the next iteration is bound to produce y_{n+m+1}, which is positive and greater than unity. In practice, convergence occurs extremely rapidly. In addition, specific values can be used to construct a look-up table. For instance, on $p_s = 2e^3 p_f$, $u = 1/2$; on $p_s = 4e^5 p_f$, $u = 1/4$; and so on, with $u = 2^{-m}$ satisfying exactly Eq. (6.26) when $p_s = p_f 2^m \exp(1 + 2^m)$ for arbitrary values of m.

V. APPENDIX B: MODIFICATIONS TO THE PARABOLIC RISK AVERSION FORMULA

Instead of exponential weighting as per the Cozzolino *RAV* formula, a corresponding parabolic *RAV* formula with respect to working interest, W, is

$$RAV = EW\{1 - (E\nu^2 W/RT)/2\} \qquad (6.32)$$

where $E = p_s V - p_f C$ is the expected value, and ν^2 is the square of the volatility given by

$$\nu^2 = \{(p_s V^2 + p_f C^2)/E^2\} - 1. \qquad (6.33)$$

RAV of Eq. (6.32) has a maximum on $W = OWI$ where

$$OWI = RT/(\nu^2 E) \qquad \text{For } OWI \leq 1 \qquad (6.34)$$

and at $W = OWI$, the RAV is

$$RAV(W = OWI) = 1/2 \, E \times OWI. \qquad (6.35)$$

Note that as $V \to \infty$, Eq. (6.34) yields $OWI \to 0$ but Eq. (6.35) gives $RAV \to 1/2RT(p_s/p_f)$. Thus the same character of problem occurs for the parabolic RAV formula as occurred for the Cozzolino RAV formula.

As V increases from a minimum of $V = p_f C/p_s$ (where $E = 0$), so OWI increases from zero, reaching a largest value of

$$OWI_{\max} = 2p_s RT(1 + 2p_f)^{-2} C^{-1} \qquad \text{for } OWI_{\max} \leq 1, \qquad (6.36a)$$

for

$$V_{\max} = 3p_f C/p_s \qquad (6.36b)$$

For $V = V_{\max}$ and $OWI = OWI_{\max}$ one has

$$RAV_{\max} = 2RT p_s p_f (1 + 2p_f)^{-2} \qquad (6.37a)$$
$$\equiv p_f C \, OWI_{\max} \leq 1/2RT p_s/p_f. \qquad (6.37b)$$

In the same manner as the Cozzolino formula was modified to allow for the variance in the mean value E, introduce

$$OWI_a = [(RT)^{2a} + \{(p_f p_s)^{1/2}(V + C)\}^{2a}]^{1/2a} \qquad (6.38)$$

and then set OWI to be OWI_1 with

$$OWI_1 = OWI_{\max}[\tanh\{(OWI_a/OWI_{\max})^b\}]^{1/b} \qquad (6.39)$$

with the corresponding RAV of

$$RAV = 1/2E \times OWI_1. \qquad (6.40)$$

As for the Cozzolino formula, Eq. (6.39) reduces to $OWI_1 \cong OWI_a$ for $OWI_a \ll OWI_{\max}$, and tends to the constant value $OWI_1 \cong OWI_{\max}$ for $OWI_a \gg OWI_{\max}$. Again, for $RT \gg \nu E$ the modified result of Eq. (6.39) reduces to the result of Eq. (6.34) for the conventional parabolic RAV formula, provided $OWI_a \ll OWI_{\max}$, so that the flow diagram of Figure 6.2 still provides the conditions under which the parabolic RAV formula needs modification.

7

Corporate Funding Requests, Fixed Budgets, and Cost Balancing

I. INTRODUCTION

In any major oil company there are usually thousands of potential projects that desire funds, ranging from new software requests, to applications of basin models, to improved field size delineation by use of 4-D seismic, to leasehold extensions in time and/or increased acreage requests.

Some of the requests are field specific (e.g., 4-D seismic to trace remaining oil in a particular reservoir), while others are championed as of more general utility (e.g., applications to many exploration areas of a new seismic stratigraphy model or code).

Usually, each request needs to be accompanied by a quantitative cost analysis to the corporation if the project is to go ahead, but it is also usually the case that the gains of the project are described in rather vague terms (e.g., an increase in productivity, improved decision accuracy, a small percentage increase of oil in place, etc.) but without any quantitative estimates.

Because the funding requests do not all arrive for evaluation at the same time, it is difficult to intercompare the requests. Indeed, it seems more as though requests arrive at a roughly steady rate with time.

What is needed is a consistent merit-based process that allows funds to be allocated objectively to each worthwhile project. The problem is to provide an objective definition of "worthwhile," and then to figure out an objective procedure for allocating funds. In addition, one has to ensure that "threshold" conditions are honored; for instance, if a 4-D seismic survey has a minimum cost of $20 MM then there is little point in allocating $2 MM to that survey because nothing can then be done.

One must also ensure that the sum of all requests for funds in a given funding period does not exceed the total budget for that period. The difficulty here is that because funding requests arrive roughly steadily with time, there is no way of knowing if a highly worthwhile request is going to arrive in the late part of the budget period. In some situations a fraction of the budget is retained in the off chance that such a highly worthwhile project does arrive. The problem is that if such a project is not then forthcoming, one has a fraction of the budget left that one could have spent on other projects, perhaps of lesser worth but still of value to the corporation. What is also needed is a quantitative way to estimate the likelihood of a highly worthwhile request arriving in a given time period so that one can provide a likely "waiting time" estimate between highly worthwhile requests. In this way the fraction of the budget to retain, which itself depends on the residual time left in the budget period, can be assigned objectively.

The purposes of this chapter are to address these concerns.

II. SPECIFIC PROJECT REQUESTS

Start by considering a specific project request. As an example, consider a producing field, for which a specific request arrives to shoot 4-D seismic to improve estimates of field drainage area with time. Some cost δC is estimated for the acquisition, but only a vague claim is advanced that it will help to determine better the total remaining reserves in the field.

Prior to the request arriving, one has some expected value, E_0, for the remaining field potential based on criteria dealing with specific productivity, field costs, selling price with time, estimated remaining reserves in place, etc. Equally, because each of the component criteria had some range of uncertainty, one also has available an estimate of the volatility, v, of the expected value, in the sense that, with 90% confidence, one can write the uncertainty, σ, in the expected value as $\sigma = \pm vE_0$. This particular problem will be addressed in more detail in Chapter 8.

Suppose, initially, that a *perfect* funding request arrives for which are provided the extra costs, δC, the additional increase in gains, δG, and the increase in the chance, δp_s, that the request will lead to the additional gains.

Because both the expected value, E_0, and the standard error uncertainty, σ, of the project are known prior to the funding request arriving, one can write equivalent gain, G, cost, C, and success probability, p_s, values using the standard relations

$$E_0 = p_s G - C \tag{7.1}$$

and

$$\sigma^2 = p_s p_f G^2, \tag{7.2}$$

where $p_f = 1 - p_s$, to obtain

$$p_s = (E_0 + C)^2/[\sigma^2 + (E_0 + C)^2] \qquad (7.3a)$$

and

$$G = [\sigma^2 + (E_0 + C)^2]/(E_0 + C), \qquad (7.3b)$$

which expresses the equivalent values of G and p_s in terms of the known values σ, E_0, and the estimated project costs, C.

If the funding request is granted, the project will then have a new expected value of

$$E_1 = (p_s + \delta p_s)(G + \delta G) - (C + \delta C) \qquad (7.4a)$$

and a new standard error, σ_1, given through

$$\sigma_1^2 = (p_s + \delta p_s)(1 - p_s - \delta p_s)(G + \delta G)^2. \qquad (7.4b)$$

On average, the funding request is worth granting if $E_1 > E_0$, which can be written as an inequality:

$$p_s \delta G + \delta p_s G + \delta p_s \delta G > \delta C. \qquad (7.5)$$

Inequality (7.5) directly connects both the increase in gains, success probability, and costs of the request in terms of the original project costs.

But the improvement in expected value may be less than worthwhile if the increase in costs from C to $C + \delta C$ makes the venture less profitable in terms of estimated returns per unit of cost invested. Thus one should more appropriately require that

$$E_1/(C + \delta C) > E_0/C \qquad (7.6)$$

in order to improve the value per dollar spent. Inequality (7.6) can be written in the form

$$\frac{\delta P_s}{p_s} + \frac{\delta G}{G}\left(1 + \frac{\delta p_s}{p_s}\right) \geq \left(\frac{\delta C}{C}\right). \qquad (7.7)$$

Suppose now, as is most often the case, that only a vague proposal for funding arrives, in which the extra costs, δC, are well specified. (Usually this specification is a corporate mandated requirement if the request is to have any chance at all of being funded.) The increase in gains, δG, and the likely increase in success probability, δp_s, are usually where the funding requests become particularly vague in their description of benefits to the corporation of funding the request. The question then is how to assess the likely improvement in worth to a corporation of a project if the vague funding request is approved.

Two extreme possibilities suggest avenues for evaluation. First, suppose that no increase in gains ($\delta G = 0$) were to occur as a consequence of funding the proposal. Then the only motive for funding would be if the proposal were to increase p_s, the success probability. Such an increase would imply that either the expected value, E_0, is increased or the variance, σ^2, is decreased, or both. And the changes in E_0 and/or σ^2 would have to be sufficient to offset the increased costs. For instance, if one takes the requirement of inequality (7.7) as appropriate then, with $\delta G = 0$, one requires

$$\delta p_s / p_s \geq \delta C / C. \tag{7.8}$$

From Eq. (7.3a), one has

$$\frac{\delta p_s}{p_s} = \frac{\sigma^2 p_s}{(E_0 + C)^2} \left\{ \frac{2 \delta E_0}{(E_0 + C)} - \frac{\delta \sigma^2}{\sigma^2} \right\} \tag{7.9}$$

so that the requirement for funding the request is

$$\frac{2 \delta E_0}{(E_0 + C)} - \frac{\delta \sigma^2}{\sigma^2} \geq \frac{\delta C}{C} \frac{(E_0 + C)^2}{p_s \sigma^2}. \tag{7.10a}$$

Now, because $\delta G = 0$, it follows from Eq. (7.3b) that

$$\frac{\delta \sigma^2}{\sigma^2} = \frac{\delta E_0}{(E_0 + C)} \{1 - (E_0 + C)^2 / \sigma^2\} \tag{7.10b}$$

so that inequality (7.10a) can be written more simply as

$$\frac{\delta E_0}{(E_0 + C)} \geq \frac{\delta C}{C}. \tag{7.10c}$$

Suppose, for example, that $E_0 \gg C$, so that costs are negligible in relation to $p_s G$, the expected gains. Then, if one is to fund the project the required behavior is

$$\delta p_s \gtrsim \left(\frac{\delta C}{C} \right) \frac{(E_0 + C)}{G}. \tag{7.11}$$

For a highly successful project, $p_s \to 1$ (and so $p_f \to 0$) the increment in expected return $\delta E_0 / E_0$ would have to be extremely high and/or the decrease in expected variance, $-\delta \sigma^2 / \sigma^2$, extremely large if inequality (7.11) is to be obeyed.

Second, suppose that there is no change in the success probability ($\delta p_s = 0$) as a consequence of the funding request, but that the request argues for an increase in gains. Then inequality (7.7) would require

$$\delta G / G \gtrsim \delta C / C \tag{7.12}$$

if the project is to be funded.

These two extreme cases make it clear that it is not just the information in the funding request that has to be taken into account, but also the prior available information on the project in terms of G, C, σ^2, E_0, and p_s. Only by comparing the parameters of the project with the changes to those parameters that would be brought about by funding the specific request can one properly evaluate the worth of funding the project.

III. GENERAL PROJECT REQUESTS

A specific project request is relatively clear in what is proposed, namely, to improve the worth to the corporation of a particular prospect or producing field. By way of contrast, a general project request is less clear. For example, it might be that a funding request arrives to purchase a computer code from an outside vendor. While not germane to a particular business venture of the corporation, the requester proposes that the code will speed up the ability to process any and all seismic data so that the delivery efficiency of migrated seismic sections to geologists is improved, with the geologists then able to spend more time—and hence be more thorough—in their evaluation of the likely hydrocarbon proneness of different regions. In this case there is no one unique, specific, identifiable project which benefits solely from the funding request, but rather all projects are said to be positively influenced. Of course, as is usual, no details of the dollar value of the positive influence are provided in the request, merely a broad, vaguely worded, statement of the "exciting benefits" to the corporation.

A. PREFERENCE INDICES

One way to address the general request is to take the total costs, δC, of the proposal and divide it in some manner between all the projects, say, N, which it is said to benefit. If one could do so, assigning a fractional cost δC_n to the nth project (with $\delta C = \sum_{i=1}^{N} \delta C_n$), then one could evaluate the worth of the request to the total portfolio of opportunities that the proposal claims to benefit. For instance, one could determine if the increase in expected value, δE_n, is such that $\delta E \equiv \sum_{i=1}^{N} \delta E_n > 0$ for the sum of all projects, and use this criterion as a measure of worth of the general project request. But assigning fractional costs, δC_n, to the nth project is not completely straightforward. One could just simply take the total cost δC and arbitrarily assign the amount $\delta C_n = \delta C/N$ to each of the N projects. However, such a choice is likely to drive marginal projects (with very low expected values) negative, turning them from small, but positive, cash flows to the corporation to drains on the corporate coffers.

An alternative is to take the expected value, E_n, for each project *prior* to the funding request and rank all the N projects by a cumulative preference index, CPI_n, with

$$CPI_n = E_n \Big/ \sum_{i=1}^{N} E_i. \tag{7.13}$$

Costs of the request, δC, are then assigned in proportion to CPI_n, so that the costs assigned to the nth project are

$$\delta C_n = \delta C \left(CPI_n \bigg/ \sum_{i=1}^{N} CPI_i \right). \tag{7.14}$$

In the worst case, if no extra gains or increase in success probability were to accrue to any project as a consequence of granting the funding request, then one has diminished the expected value of each and every project to E'_n with

$$E'_n = E_n - \delta C_n. \tag{7.15}$$

Clearly, if $\delta C_n^2 < \sigma_n^2$, where the variance σ_n^2 is on the estimate of E_n prior to the request being approved, then, while the mean value of each project is decreased, no discernible difference arises because the decrease in E_n is within the "noise" of the uncertainty.

Other preference indices are available to use as constraints. One could use risk-adjusted value, RAV_n, for each project rather than E_n and construct a CPI based on relative importance of RAV; one could construct a CPI for each project weighted in some manner according to the volatility v_n ($\equiv \sigma_n/E_n$), such as $E_n/(1 + v_n^2)^{1/2}$, and so on.

However, if $\delta C_n > \sigma_n$, then the systematic decrease in E_n is larger than the statistical fluctuation of $\pm \sigma_n$. In such a situation, unless one is extremely careful, or unless $\delta C_n \ll E_n$, it is possible that the apportioned costs could negatively impact marginal projects to the point of making them unprofitable.

One way to operate is as follows: For those projects for which $E'_n \leq 0$, reassign the costs, δC_n, to the remaining projects by recalculating the CPI values based on E'_n for those projects with $E'_n > 0$ and based on E_n for those projects with $E'_n < 0$. Then reapportion costs, δC, according to the new CPI and again go through the iteration with CPI until fractional costs have been assigned to all projects appropriately, so that each is still a worthwhile venture. But this procedure again introduces arbitrary conditions. One would prefer to have an objective, reproducible criterion; value received for dollars spent provides such a measure.

B. VALUE PER DOLLAR SPENT ("BANG FOR THE BUCK")

If expected value divided by cost is used as a basis for maximizing the return per unit of cost, then the determination of cost splitting of a request is facilitated. For N projects said to benefit from the request, of cost δC, let a fraction f_i of δC be applied to the ith project. The cumulative preference index, *CPI*, for the *N* projects is then

$$CPI = \sum_{i=1}^{N} \frac{p_{si}G_i}{(C_i + \delta Cf_i)} \tag{7.16}$$

subject to the constraints $f_i \geq 0$; and $p_{si}G_i \geq (C_i + \delta Cf_i)$ in order to retain $E_i \geq 0$; and also

$$\sum_{i=1}^{N} f_i = 1. \tag{7.17}$$

Use constraint (7.17) to write $f_j = 1 - \sum_{i=1}^{N}{}' f_i$ where the prime means omit the term $i = j$. Then the *CPI* of Eq. (7.16) has an extremum with respect to f_i when

$$\frac{p_{si}G_i}{(C_i + \delta Cf_i)^2} = \frac{p_{sj}G_j}{(C_j + \delta Cf_j)^2} \quad \text{for all } i, j. \tag{7.18}$$

But *i* and *j* were arbitrary choices; the only way equality (7.18) can be valid for *all i* and *j* is if each side is an absolute positive constant, H^2, independent of *i* and *j*. Then

$$(p_{si}G_i)^{1/2} = H(C_i + \delta Cf_i). \tag{7.19}$$

One then requires $H \geq (p_{si}G_i)^{-1/2}$ to satisfy the constraint $p_{si}G_i \geq (C_i + \delta Cf_i)$ for the *i*th project. The constant *H* can be determined by summing both sides of Eq. (7.19) and using the constraint (7.17). Then

$$H = \frac{\sum_{i=1}^{N} (p_{si}G_i)^{1/2}}{\left(\delta C + \sum_{i=1}^{N} C_i\right)} > 0. \tag{7.20a}$$

Thus the first requirement for the *i*th project is that

$$\sum_{j=1}^{N} (p_{sj}G_j)^{1/2} \geq \frac{\left\{\delta C + \sum_{j=1}^{N} C_j\right\}}{(p_{si}G_i)^{1/2}}. \tag{7.20b}$$

It follows that the fraction of δC to be applied against the *i*th project is

$$f_i = [(p_{si}G_i)^{1/2} - HC_i]/H\delta C, \tag{7.21a}$$

provided that $0 \leq f_i \leq 1$ and provided that inequality (7.20b) is satisfied. If either $f_i \leq 0$ or if inequality (7.20b) is not satisfied, the correct inference is that the *i*th term should not be included in apportioning the costs, δC. If $f_i \geq 1$ the correct inference is that all the costs, δC, should be borne by the *i*th project. If f_i is greater than unity for more than one project, then one computes a suite of *CPI* values, with each calculated using $f_i = 1$ for those projects that gave $f_i \geq 1$. The highest CPI then provides the information on which project should bear the cost. When $f_i \leq 1$ for all projects, then the N projects against which the costs, δC, are to be fractionally apportioned are those for which $0 \leq f_i \leq 1$; all other projects with $f_i \leq 0$ are not made to bear any fraction of the request of δC.

Note that $f_i > 0$ for the ith project when

$$(p_{si}G_i)^{1/2}/C_i > \frac{\sum_{j=1}^{N}(p_{sj}G_j)^{1/2}}{\left(\sum_{j=1}^{N}C_j + \delta C\right)}, \tag{7.21b}$$

while $f_i \leq 1$ for the ith project when

$$\delta C\left[\sum_{j=1}^{N}(p_{sj}G_j)^{1/2} - (p_{si}G_i)^{1/2}\right] \geq \tag{7.21c}$$

$$\left(C_i \sum_{j=1}^{N} C_j\right)\left[\{(p_{si}G_i)^{1/2}/C_i\} - \sum_{j=1}^{N}(p_{sj}C_j)^{1/2}\Big/\sum_{j=1}^{N}C_j\right].$$

When f_i is given by Eq. (7.21a), the CPI is reduced from its value of $CPI_0 \equiv \sum_{i=1}^{N}(p_{si}G_i/C_i)$, obtaining in the absence of the request ($\delta C = 0$), to the value

$$CPI_1(\delta C) = \left\{\sum_{i=1}^{N}(p_{si}G_i)^{1/2}\right\}^2\left\{\sum_{i=1}^{N}C_i + \delta C\right\}^{-1}. \tag{7.22}$$

C. ANALYTIC EXAMPLE FOR TWO PROJECTS

To illustrate the variation, consider a request of δC to be apportioned between just two projects with parameters p_{s1}, G_1, C_1, and p_{s2}, G_2, C_2, respectively, and with $p_{s1}G_1/C_1^2 > p_{s2}G_2/C_2^2$. Let the fraction $f_1 = \sin^2\theta$, $f_2 = \cos^2\theta$ so that $f_1 + f_2 = 1$ automatically. Then

$$CPI = \frac{p_{s1}G_1}{(C_1 + \delta C \sin^2\theta)} + \frac{p_{s2}G_2}{(C_2 + \delta C \cos^2\theta)}. \tag{7.23}$$

Differentiating CPI with respect to θ and setting the derivative to zero yields

$$0 = \frac{\partial CPI}{\partial \theta} = -2\sin\theta\cos\theta\left[\frac{p_{s1}G_1}{(C_1 + \delta C \sin^2\theta)^2} - \frac{p_{s2}G_2}{(C_2 + \delta C \cos^2\theta)^2}\right]. \tag{7.24}$$

Hence, from Eq. (7.24) one has three possible values of θ: (1) $\theta = 0$, (2) $\theta = \pi/2$, or (3) $\theta = \theta_c$ with

$$\sin^2\theta_c = \{[(p_{s1}G_1)/(p_{s2}G_2)]^{1/2}(C_2 + \delta C) - C_1\}\{1 + [(p_{s1}G_1)/(p_{s2}G_2)]^{1/2}\}^{-1}\delta C^{-1}. \tag{7.25}$$

Note that $\sin^2\theta_c$ is intrinsically positive when $p_{s1}G_1/C_1^2 \geq p_{s2}G_2/C_2^2$; and note that $\sin^2\theta_c < 1$ when

$$\delta C \geq \{(p_{s1}G_1)/(p_{s2}G_2)\}^{1/2} C_2 - C_1 \equiv \delta C_* > 0. \tag{7.26}$$

If $\delta C < \delta C_*$ then $\sin^2\theta_c > 1$ in which case only the solutions $\theta = 0$ and $\theta = \pi/2$ are viable. For each of the three values $\theta = 0$, $\theta = \pi/2$, and $\theta = \theta_c$ (when $\sin^2\theta_c < 1$) the corresponding values of CPI are

$$CPI(\theta = 0) = p_{s1}G_1/C_1 + p_{s2}G_2/(C_2 + \delta C), \tag{7.27a}$$

$$CPI(\theta = \pi/2) = p_{s1}G_1/(C_1 + \delta C) + p_{s2}G_2/C_2, \tag{7.27b}$$

and

$$CPI(\theta = \theta_c) = (p_{s1}G_1)/(C_1 + \delta C \sin^2\theta_c) + (p_{s2}G_2)/(C_2 + \delta C \cos^2\theta_c) \tag{7.27c}$$

provided further that

$$p_{s1}G_1 \geq C_1 + \delta C \sin^2\theta \tag{7.27d}$$

and

$$p_{s2}G_2 \geq C_2 + \delta C \cos^2\theta. \tag{7.27e}$$

for each θ value $(0, \pi/2, \theta_c)$, in order to keep the expected values for each project positive. Different values of p_{s1}, p_{s2}, G_1, G_2, C_1, and C_2 can be inserted into the three formulas (7.27a, b, c) for CPI to determine which has the highest CPI subject to the constraints just outlined. Thus, the apportionment of the request costs, δC, is determined. Then return to the procedure of Section II for a single opportunity and determine if the request is worthwhile funding for each and every opportunity that would be impacted by the cost sharing.

D. NUMERICAL EXAMPLES FOR TWO PROJECTS

1. Increasing Extra Costs

Consider, for numerical illustration purposes, that opportunity 1 has the parameters $p_{s1} = 0.4$, $G_1 = \$50$ MM, $C_1 = \$10$ MM, so that the expected values $E_1 = p_{s1}G_1 - C_1 = \$10$ MM, and that opportunity 2 has the parameters $p_{s2} = 0.25$, $G_2 = \$80$ MM and also costs $C_2 = \$10$ MM so that $E_2 = p_{s2}G_2 - C_2 = \$10$ MM. Suppose the cost, δC, of acquiring further data is $\$5$ MM. Then $\sin^2\theta_c = 0.5$ so that

$$CPI(\theta = 0) = 10/3; \quad CPI(\theta = \pi/2) = 10/3; \quad CPI(\theta = \theta_c) = 3.2;$$

indicating that it is a better strategy to assign all of the $\$5$ MM cost of data acquisition either to the first opportunity ($\theta = 0$) or to the second opportunity ($\theta = \pi/2$) rather than splitting the costs.

As the extra costs rise, eventually at $\delta C > \$10$ MM, one can no longer satisfy inequalities (7.27d) or (7.27e) if one assigns *all* the costs to one project. But, by assigning half the extra cost to each project ($\sin^2\theta_c = 0.5$), one has $p_{s1}G_1/(C_1 + 1/2\delta C) \geq 1$ as long as $\delta C < \$20$ MM. Thus, at $\delta C = \$10$ MM one has the *CPI* ratio at 1.33 for each project. One can still keep each project profitable up to a maximum extra cost of $20 MM. Beyond $20 MM both projects fail to be profitable.

2. High Success and Marginal Success Opportunities

For the situation of one highly successful, low-cost opportunity, and a second, marginally successful, high-cost opportunity, with parameters $p_{s1} = 0.9$, $G_1 = \$100$ MM, $C_1 = \$10$ MM for the highly successful opportunity, and parameters $p_{s2} = 0.2$, $G_2 = \$50$ MM, $C_2 = \$9$ MM for the marginally successful opportunity, and with the additional data cost at $\delta C = \$46$ MM one has $\sin^2\theta_c = 0.842$ and

$CPI(\theta = 0) = 9.18$, which keeps expected values for both projects positive;
$CPI(\theta = \pi/2) = 2.72$, but then $p_{s2}G_2 < C_2 + \delta C$ so that this solution is forbidden;
$CPI(\theta = \theta_c) = 2.46$, but $p_{s2}G_2 < C_2 + \delta C \cos^2\theta_c$ so that this solution is also forbidden.

In this case the high additional data cost should be borne solely by the high success, low-cost opportunity, if the decision to acquire the data is made at all.

IV. MULTIPLE REQUESTS, FIXED BUDGETS, AND FUNDING STRATEGIES

A. GENERAL ANALYSIS

While the paradigms of the preceding two sections are often of benefit for determining *if* a particular proposal is worthwhile funding, one of the major difficulties faced by a corporation is that thousands of internal proposals are received over the course of a funding cycle (usually a year in duration) and one has the problems not only of deciding if a particular proposal is worth funding, but also of figuring out what fraction of the fixed budget, B, should be reserved at any given point in time for worthwhile, but possibly high-cost, projects that *might* arrive for consideration in the remainder of the budget cycle.[1]

Of course, historical precedent from prior years can be used to evaluate the yearly patterns of behavior of funding requests, the fractions approved,

---

[1] The other sort of budgeting is a continually rolling funding of budget, so that at any given time one always has the budget "topped up."

and distributions with time in the year. But it would be beneficial if a more general framework were available so that one could plan budget expenditures systematically and strategically in a given year, without having to be beholden to the vagaries and fluctuations of the historical record. The purpose of this section of the chapter is to show how to achieve such a goal.

Because thousands of funding proposals arrive throughout the funding period, it is appropriate to break the distribution of proposals into dollar intervals of width $\Delta\omega$, with $\Delta N/\Delta\omega$ the average number of proposals requesting funds in the dollar range ω to $\omega + \Delta\omega$ that arrive by time t in the funding cycle. The cumulative cash amount, $\Delta S/\Delta\omega$, requested in that range is then

$$\Delta S/\Delta\omega = \omega \Delta N/\Delta\omega. \tag{7.28}$$

If proposals arrive at a roughly steady rate, R, throughout the year, and if the dollar-requested interval distribution of proposals is also roughly independent of time, then over the funding cycle time of T, one has

$$R = T^{-1} \int_0^\infty \frac{\Delta N}{\Delta\omega} d\omega \tag{7.29}$$

and the average total amount requested per unit time, dS/dt, is

$$\frac{dS}{dt} = T^{-1} \int_0^\infty \omega \frac{\Delta N}{\Delta\omega} d\omega. \tag{7.30}$$

One is interested in the probability distribution, $p(B, t)$, that one will have left a budget between B and $B + dB$ at time t, having started with a budget B_0 at time $t = 0$. The probability, P, of $N(\omega)$ independent requests occurring in the dollar interval $\Delta\omega$ centered around a value ω_k is taken to be Poisson-distributed, with P then given by

$$P(N, \omega_k) = (N_k)^N \exp(-N_k)/N!, \tag{7.31}$$

where $N_k = \Delta\omega \, (\Delta N/\Delta\omega)_{\omega=\omega_k}$.

Because the budget for each request is ω_k, the probability, f_k, of residual funds being at B is

$$f_k(B) = \sum_{N=0}^\infty P(N, \omega_k) \, \delta(B_0 - B - N\omega_k), \tag{7.32}$$

where $\delta(x)$ is the Dirac δ function, because $B_0 - B$ is spent on the N_k requests.

The probability distribution, $p(B, t)$, of having funds left in the range B to $B + dB$, with requests arriving in intervals of width $\Delta\omega$ centered around $\omega_1, \omega_2, \omega_3, \ldots, \omega_k, \ldots$, is the sum of probabilities for $N_1, N_2, N_3, \ldots, N_k, \ldots$, requests in the corresponding request intervals, with the budget B then being given by

$$B_0 - B = \sum_{k=0}^{N} N_k \omega_k. \tag{7.33}$$

Thus one has

$$p(B;t) = \sum_{N_1,N_2\ldots=0}^{\infty} P_1(N_1,\omega_1)P_2(N_2,\omega_2)\ldots P_k(N_k,\omega_k)\ldots \delta\left(B_0 - B - \sum_{i=1}^{\infty} N_i\omega_i\right). \tag{7.34}$$

Rewrite the δ function in Eq. (7.34) using the standard integral representation

$$\delta(y) = (2\pi)^{-1} \int_{-\infty}^{\infty} \exp(iyx)\, dx \tag{7.35}$$

and then perform the sum

$$\sum_{N=0}^{\infty} P(N,\omega_k) \exp(-iNx) = \exp[-N_k\{1 - \exp(-ix)\}] \tag{7.36}$$

to obtain $p(B,t)$ in the form

$$p(B,t) = (2\pi)^{-1} \int_{-\infty}^{\infty} dx \exp\left\{ix(B_0 - B) - \sum_{k=0}^{\infty} N_k[1 - \exp(-i\omega_k x)]\right\}. \tag{7.37}$$

Now run the budget interval $\Delta\omega$ to infinitely small limits to obtain the continuous range representation

$$p(B,t) = (2\pi)^{-1} \int_{-\infty}^{\infty} dx \exp\left\{ix(B_0 - B) - \int_0^{\infty} \frac{dN(\omega,t)}{d\omega}[1 - \exp(-ix\omega)]\, d\omega\right\}, \tag{7.38}$$

where $dN(\omega,t)/d\omega$ is the average total number of requests arriving by time t in the amount requested interval ω to $\omega + d\omega$.

Equation (7.38) can be rewritten in the simpler form

$$p(B,t) = \pi^{-1} \int_0^{\infty} \cos\left[x(B_0 - B) - \int_0^{\infty} \frac{dN(\omega,t)}{d\omega} \sin\omega x\, d\omega\right] \\ \exp\left\{-\int_0^{\infty} \frac{dN(\omega,t)}{d\omega}(1 - \cos\omega x)\, d\omega\right\} dx, \tag{7.39}$$

which clearly shows that $p(B,t)$ is real.

It is not difficult (although it is tedious) to show from the exact form of $p(B,t)$ given by Eq. (7.38) that $p(B,t) = 0$ for $B > B_0$. Also, as the number of requests increases with time for a fixed budget B_0, there must eventually be a time, $t = \tau_1$, say, at which $p(B,t)$ is zero. For values of t greater than τ_1, $p(B,t)$ would then be negative. The correct inference is that $p(B,t)$ is to be set to zero for $t > \tau_1$ because there is then no money retained in the budget. As the original budget, B_0, is increased, all other factors being similar, the value of τ_1 at which budget exhaustion first occurs moves to larger values of time.

The cumulative probability, P, of *retaining* a budget of B or *larger* at time t is then given by

$$P(B, t) \equiv \int_B^{B_0} P(B', t)\, dB'$$

$$= \pi^{-1} \int_0^\infty y^{-1} \exp\left[-\int_0^\infty \frac{dN(t,\omega)}{d\omega}\{1 - \cos(\omega y)\}\, d\omega\right]$$

$$\times \left\{\sin\left[\int_0^\infty \frac{dN(t,\omega)}{d\omega} \sin(\omega y)\, d\omega\right]\right.$$

$$\left. - \sin\left[\int_0^\infty \frac{dN(t,\omega)}{d\omega} \sin(\omega y)\, d\omega - y(B_0 - B)\right]\right\} dy. \quad (7.40)$$

Now consider in more detail the distribution of arriving requests. The probability of N requests arriving in the dollar interval $\Delta\omega$ around a value ω_k is

$$P(N, \omega_k) = (N_k)^N \exp(-N_k)/N!, \quad (7.41)$$

where $N_k = \Delta\omega(\Delta N/\Delta\omega)_{\omega=\omega_k}$; and with $\Delta N/\Delta\omega$ the average number of requests in the dollar interval ω_k to $\omega_k + \Delta\omega$.

Then the probability of *not* receiving a request in the interval ω_k to $\omega_k + \Delta\omega_k$ is

$$P(0, \omega_k) = \exp(-N_k). \quad (7.42)$$

The probability of *not* receiving a request by time t in *any* dollar interval ω_0 to $\omega_0 + \Delta\omega$, *or higher*, is

$$P_0(t) = P(0, \omega_0)\, P_0(0, \omega_0 + \Delta\omega)\, P_0(0, \omega_0 + 2\Delta\omega)\ldots$$

$$= \exp\left\{-\sum_{\omega_k=\omega_0}^\infty N_k\right\}, \quad (7.43)$$

which, taking continuous limits, is

$$P_0(t) = \exp\left[-\int_{\omega_0}^\infty \frac{dN(t,\omega)}{d\omega}\, d\omega\right]. \quad (7.44)$$

The probability of *receiving* a request by time t at a dollar amount ω_0 *or higher* is then

$$P_R(t) = 1 - P_0(t). \quad (7.45)$$

Suppose, for example, that one knew that requests in each dollar interval ω to $\omega + \Delta\omega$ arrive at a constant rate with time so that

$$\frac{dN(t, \omega)}{d\omega} = R(\omega)t, \qquad (7.46)$$

where $R(\omega)$ is the rate of arrival of requests in the dollar interval ω to $\omega + \Delta\omega$. The total period of budget funding is T, so the total number of requests in ω to $\omega + \Delta\omega$ is $R(\omega)T$.

In the period $t = t$ to $t = T$ the probability of *not* receiving a request at a dollar amount ω_0 or *higher* is then

$$P_0(t, T) = \exp\left[-(T-t)\int_{\omega_0}^{\infty} R(\omega)\, d\omega\right]. \qquad (7.47)$$

Thus, in the time interval t to T the probability of *receiving* a request at a dollar amount ω_0 or *higher* is

$$P_R(t, T) = 1 - \exp\left[-(T-t)\int_{\omega_0}^{\infty} R(\omega)\, d\omega\right]. \qquad (7.48)$$

Correspondingly, in the time interval t to T, the probability of *receiving* a request at a dollar amount ω_0 or *lower* is

$$P_L(t, T) = 1 - \exp\left[-(T-t)\int_0^{\omega_0} R(\omega)\, d\omega\right]. \qquad (7.49)$$

Now the probability of a retaining budget B or *higher* at time t is $P(B, t)$ given by Eq. (7.40). Thus, if the budget B is to fund a request of dollar amount ω_0 or less at any time in the interval t_1 to T then we require that the minimum budget B be at least ω_0; that is, we require $B \geq \omega_0$ with

$$P(\omega_0, t_1) > P_L(t_1, T) \qquad \text{for } t_1 \leq t \leq T, \qquad (7.50)$$

where $P(\omega_0, t_1)$ is determined from Eq. (7.40) with $B = \omega_0$, $t = t_1$.

Note from Eq. (7.49) that as $t_1 \to T$, the probability of a request arriving at ω_0 or *lower* tends to zero, so that one need retain less and less of the budget. Spelling out inequality (7.50) in detail one requires, in general, that

$$\begin{aligned}
\pi^{-1} &\int_0^{\infty} y^{-1} \exp\left[-\int_0^{\infty} \frac{dN(t, \omega)}{d\omega}\{1 - \cos(\omega y)\}\, d\omega\right] \\
&\times \left\{\sin\left[\int_0^{\infty} \frac{dN(t, \omega)}{d\omega} \sin(\omega y)\, d\omega\right]\right. \\
&\left. - \sin\left[\int_0^{\infty} \frac{dN(t, \omega)}{d\omega} \sin(\omega y)\, d\omega - y(B_0 - \omega_0)\right]\right\} dy \\
&\geq 1 - \exp\left[-\int_0^{\omega_0}\left\{\frac{dN(T, \omega)}{d\omega} - \frac{dN(t, \omega)}{d\omega}\right\} d\omega\right].
\end{aligned} \qquad (7.51)$$

B. ANALYTIC EXAMPLE

An analytical illustration of the functional dependence can be given that is of use in seeing directly the dependence of results, as well as for gauging the accuracy of any numerical scheme used to perform the relevant integrals with various functional behaviors for $R(\omega)$.

As an example, let the distribution of rates of request, $R(\omega)$, vary in request amount as $\omega^{-3/2}$, from a low dollar amount of ω_L to a high dollar amount of Ω. Then with

$$\frac{dN}{d\omega} = tR(\omega) = tr\,\omega^{-3/2}, \tag{7.52}$$

where r is a scale constant, the rate of *all* requests, R_T, is given by

$$R_T = \int_{\omega_L}^{\Omega} R(\omega)\,d\omega = 2r[\omega_L^{-1/2} - \Omega^{-1/2}], \tag{7.53a}$$

while the total rate of dollar requesting is

$$\$_T = \int_{\omega_L}^{\Omega} \omega R(\omega)\,d\omega = 2r[\Omega^{1/2} - \omega_L^{1/2}]. \tag{7.53b}$$

The evaluation of the relevant integrals in Eq. (7.39) is then easily done, because

$$\int_0^\infty r\omega^{-3/2}(1 - \cos\omega y)\,d\omega = ry^{1/2}(2\pi)^{1/2}, \tag{7.54a}$$

$$\int_0^\infty r\omega^{-3/2}\sin\omega y\,d\omega = ry^{1/2}(2\pi)^{1/2}, \tag{7.54b}$$

so that Eq. (7.39) can be written

$$p(B, t) = \pi^{-1} \int_0^\infty \exp\{-tr(2\pi y)^{1/2}\}\cos\{y(B_0 - B) - tr(2\pi y)^{1/2}\}\,dy. \tag{7.55}$$

With the substitution $y = u^2$, Eq. (7.55) can be written

$$p(B, t) = 2\pi^{-1} \int_0^\infty u\exp(-au)\cos[bu^2 - au]\,du, \tag{7.56}$$

where $a = tr(2\pi)^{1/2}$ and $b = (B_0 - B)$. But the integral on the right-hand side of Eq. (7.56) is known (Gradshteyn and Ryzhik, 1965) to yield

$$\int_0^\infty u\exp(-au)\cos(bu^2 - au)\,du = \frac{a}{2}\left(\frac{\pi}{2b^3}\right)^{1/2}\exp\{-a^2/(2b)\}, \tag{7.57}$$

provided $b > 0$ (i.e., $B \leq B_0$).

Thus, Eq. (7.56) can be written exactly as

$$p(B, t) = tr\,(B_0 - B)^{-3/2}\exp\{-\pi t^2 r^2/(B_0 - B)\} \quad \text{for } B < B_0. \tag{7.58}$$

The cumulative probability of retaining a budget greater than or equal to B (but less than the original budget B_0) at time t is then

$$P(B, t) = \int_B^{B_0} p(B, t) \, dB = 2\pi^{-1/2} \int_{\xi_*}^{\infty} \exp(-\xi^2) \, d\xi \qquad (7.59)$$

where $\xi_* = \pi^{1/2} \, tr \, (B_0 - B)^{-1/2}$. Inequality (7.51) then can be put in the required form

$$2\pi^{-1/2} \int_{\xi_*}^{\infty} \exp(-\xi^2) \, d\xi \geq 1 - \exp[-2r(T - t)(\omega_L^{-1/2} - \omega_0^{-1/2})] \qquad (7.60)$$

with $\xi_* = t\pi^{1/2} r(B_0 - \omega_0)^{-1/2}$ and $B = \omega_0$.

The value of ω_0 at which the equality part of requirement (7.60) is satisfied is the highest value of ω_0 at time t that could be funded by the residual budget, and varies with time t; denote this value as $\omega_H(t)$. Because a value of $\omega_0 > \omega_H(t)$ at time t will both decrease the left-hand side of inequality (7.60) and increase the right-hand side, $\omega_H(t)$ is therefore given by

$$2\pi^{-1/2} \int_{\xi_H}^{\infty} \exp(-\xi^2) \, d\xi = 1 - \exp[-2r(T - t)(\omega_L^{-1/2} - \omega_H^{-1/2})] \qquad (7.61)$$

with $\xi_H = \pi^{1/2} rt \{B_0 - \omega_H(t)\}^{-1/2}$.

But Eq. (7.61) makes the presumption that the rate of funding of requests (dependent on r) after time t will be at the same rate as prior to time t. The left side of Eq. (7.61) yields the cumulative probability of a budget $\omega_H(t)$ being available, *given that funding requests for all earlier times were held at the rate* $2r(\omega_L^{-1/2} - \Omega^{-1/2})$, while the right-hand side is the probability of funding requests up to the amount ω_H for all times *later* than t at the same rate as prior to time t.

If one wishes to fund requests up to a *prescribed* amount, ω_0, after time t_1 then one must change the rate of funding requests for all times after t_1 (but less than T) to r_1 with

$$r_1 \leq -\frac{\ln\{1 - A(t_1)\}}{2(T - t_1)(\omega_L^{-1/2} - \omega_0^{-1/2})}, \qquad (7.62)$$

where

$$A(t_1) = 2\pi^{-1/2} \int_{\xi_*}^{\infty} \exp(-\xi^2) \, d\xi, \text{ with } \xi_* = t_1 \pi^{1/2} r(B_0 - \omega_0)^{-1/2}.$$

The total average number of requests funded is then

$$N_T = 2(\omega_L^{-1/2} - \Omega^{-1/2})[rt_1 + r_1(T - t_1)], \qquad (7.63a)$$

while the total amount funded is

$$\$ = 2(\Omega^{1/2} - \omega_L^{1/2})[rt_1 + r_1(T - t_1)]. \qquad (7.63b)$$

Alternatively, one could choose to consider funding requests at the rate of arrival in operation prior to time t_1, but then the probability that one will have retained a budget amount sufficient to fund a request up to the value ω_0 is diminished at all times in $T > t > t_1$.

It is somewhat simpler to phrase the problem slightly differently in terms of total number of requests funded, $N(t_1)$, (from $t = 0$ to time $t = t_1$), but retaining the same distributional form for amount requests ($\alpha\omega^{-3/2}$). Then the number $N(t_1)$ is given by replacement of rt_1 on the left-hand side of inequality (7.60) with $N(t_1)/[2(\omega_L^{-1/2} - \Omega^{-1/2})]$; while in place of $r(T - t_1)$ on the right-hand side one can write $N_R(T, t_1)/[2(\omega_L^{-1/2} - \Omega^{-1/2})]$, where $N_R(T, t_1)$ is the residual number of requests, at amounts less than ω_0, that could still be funded in the remaining time interval $T \geq t \geq t_1$, of the same distribution shape as prior to time $t = t_1$.

Inequality (7.51) can then be written in the more general form

$$2\pi^{-1/2} \int_\zeta^\infty \exp(-\xi^2)\, d\xi \geq 1 - \exp\left[-N_R \frac{\{1 - (\omega_L/\omega_0)^{1/2}\}}{\{1 - (\omega_L/\Omega)^{1/2}\}}\right], \quad (7.64)$$

where

$$\zeta = \frac{N(t_1)\omega_L^{1/2}\pi^{1/2}}{(B_0 - \omega_0)^{1/2}\{1 - (\omega_L/\Omega)^{1/2}\}}. \quad (7.65)$$

For instance, suppose that one would like to retain enough funds for at least one further project up to an amount ω_0 after time t_1. If so, then $N_R = 1$. The maximum number of funding requests N_0 up to time t_1 that can then be approved is given by

$$2\pi^{-1/2} \int_{\zeta_0}^\infty \exp(-\xi^2)\, d\xi = 1 - \exp\left[-\frac{\{1 - (\omega_L/\omega_0)^{1/2}\}}{\{1 - (\omega_L/\Omega)^{1/2}\}}\right], \quad (7.66)$$

where $\zeta_0 = \pi^{1/2} N_0 \omega_L^{1/2}(B_0 - \omega_0)^{-1/2}\{1 - (\omega_L/\Omega)^{1/2}\}^{-1}$.

If N_0 is too large, the left-hand side of Eq. (7.66) will be smaller than the right-hand side; and if $N_0 = 0$, the left-hand side is unity, greater than the right-hand side of Eq. (7.66). Hence, there is a value of N_0 for which at least one more request can be funded after time t_1 up to the amount ω_0.

There is, then, a close relationship between the residual number of requests, N_R, that could be funded at a dollar level of ω_0 or less after time t_1, versus the number of requests that have been funded, $N(t_1)$, until time t_1. Indeed, from inequality (7.61) one can write

$$N_R \leq -\{1 - (\omega_L/\Omega)^{1/2}\}\{1 - \omega_L/\omega_0)^{1/2}\}^{-1} \times \ln\left\{1 - 2\pi^{-1/2}\int_{\zeta_0}^\infty \exp(-\xi^2)\,d\xi\right\}.$$

(7.67)

If $N_R \leq 1$ then no further requests can be funded; while if it is *required* that a prescribed number, N_R, of projects be funded, then $N(t_1)$ is limited such that inequality (7.67) is obeyed.

While this analytical illustration has been focused on a particular functional form of requests ($\alpha \omega^{-3/2}$) in order to perform the integrals in closed form, the general sense of logic prevails with any choice of $N(t, \omega)$, although the relevant integrals would then normally need to be done numerically. The point here was to illustrate simply the different approaches to the problem of identification of residual requests for funding. This point is of use when more complex patterns of requests are evaluated, which would have to be done numerically.

C. NUMERICAL ILLUSTRATION

Because requests for funds arrive almost continuously at a corporation, it is normal procedure to evaluate each request (effectively instantaneously) for funding or rejection by whatever criteria the corporation has in force, possibly along the directions given in previous sections of this volume. The fundable requests are usually "binned" into categories of requested amounts and funding doled out (also done on an almost instantaneous basis, given both the high number of requests that arrive at corporate decision makers per day, and the fact that the individual approved requests are normally each a small fraction of the initially available budget, B_0).

To provide some rough rules of thumb, which enable "first pass" estimates to be made of likely funding capabilities for the remainder of the budget cycle, proceed as follows.

Note, from Eqs. (7.53a) and (7.53b), that the *average* request is for funding of

$$\$_T/R_T \equiv (\Omega \omega_L)^{1/2}. \tag{7.68}$$

Thus, with an initial budget of B_0, the total number, $\langle n \rangle$, of requests that can be funded at the average cost is

$$\langle n \rangle = B_0/(\Omega \omega_L)^{1/2}. \tag{7.69}$$

If requests arrive at a total rate, R_T, then when all requests are funded, on average the time, τ, to reach budget exhaustion is

$$\tau = \langle n \rangle/R_T \equiv (B_0/R_T)(\Omega \omega_L)^{-1/2}. \tag{7.70}$$

If the time, τ, exceeds the budget cycle time, T, then, on average, all is well and good. But if τ is less than the budget cycle time, T, then one must cut the rate at which requests are funded. The rate at which requests arrive for funding *higher* than the average cost per requests, $\omega_0 \equiv (\Omega \omega_L)^{1/2}$, is

$$R_{\text{HIGH}} = R_T (\omega_L/\Omega)^{1/4} \{1 + (\omega_L/\Omega)^{1/4}\}^{-1} \tag{7.71}$$

and the average cost of such higher requests is

$$\$_{\text{HIGH}}/R_{\text{HIGH}} = (\Omega \omega_0)^{1/2} \equiv \Omega^{3/4} \omega_L^{1/4} \equiv (\$_T/R_T) \times (\Omega/\omega_L)^{1/4}. \tag{7.72}$$

Thus, if one wanted to fund $N_H \{\equiv R_{HIGH}(T - t)\}$ requests after time t at or above the average costs then the average expenditure would be

$$E_{HIGH} = R_{HIGH}(T - t) \times (\$_{HIGH}/R_{HIGH}). \tag{7.73}$$

If one funded all requests prior to time t, then the total average expenditure until time t would be

$$E_{ALL} = R_T t (\Omega \omega_L)^{1/2}. \tag{7.74}$$

Because $E_{HIGH} + E_{ALL}$ must be less than the total initial budget B_0, one has

$$t\{R_T(\Omega \omega_L)^{1/2} - R_{HIGH}(\$_{HIGH}/R_{HIGH})\} \leq B_0 - R_{HIGH} T (\$_{HIGH}/R_{HIGH}), \tag{7.75}$$

which can be written more simply as

$$\frac{t}{T} \leq (\Omega/\omega_L)^{1/4} \left\{ \frac{B_0[1 + (\omega_L/\Omega)^{1/4}]}{R_T(\Omega \omega_L)^{1/2} T} - 1 \right\}. \tag{7.76}$$

Clearly, if the rate of funding projects $R_T(\Omega \omega_L)^{1/2}$ is large, so that over time T all the budget were to be spent, then there would be no time, t, at which one could shift to a strategy of funding "deserving" projects higher than the average, as per the requirements of inequality (7.76). To have $t \geq 0$ then requires a budget such that

$$B_0 \geq \frac{R_T T (\Omega \omega_L)^{1/2}}{\{1 + (\omega_L/\Omega)^{1/4}\}} \tag{7.77}$$

when it is possible to switch after time t to funding only projects above the average $(\Omega \omega_L)^{1/2}$ in cost. Otherwise, one immediately becomes selective from $t = 0$ onward. Similar rough estimates can, of course, be made for funding up to any preset amount, which is what the prior section was about.

To illustrate simply the rough rule of thumb given, consider a situation in which 100 requests per day arrive ($R_T = 100$), ranging from $\$10^2$ to $\$10^6$ ($\omega_L = 10^2$, $\Omega = 10^6$); and so the average request costs $\$(\Omega \omega_L)^{1/2} = \$10^4 \equiv \omega_0$. The rate of expenditure if all are funded is then $(\Omega \omega_L)^{1/2} R_T$ \$/day $\cong \$10^6$/day. The average high-end cost $\$_{HIGH}/R_{HIGH}$ is then $\$10^5$, while the average rate of high-end requests is

$$R_{HIGH} = (R_T/11) = 9/\text{day} \tag{7.78}$$

so that 90% of the funding goes to projects above the average in cost per project. Thus, funding *all* projects for a fraction, f, of the total budget period, T, incurs costs of

$$E_{ALL} = T f R_T (\Omega \omega_L)^{1/2} = T f \times 10^6 \text{ \$}, \tag{7.79}$$

while funding *only* the above average costs projects beyond time $t \, (\equiv fT)$ incurs costs of

$$E_{\text{HIGH}} = T(1-f) \, 0.9 \times 10^6 \text{ \$}. \tag{7.80}$$

Thus, the total outlay over time T would be $T \times 10^6 \{0.9 + f \, 0.1\}$ \$. For a total budget of \$280 MM spread over a working year of 300 days, $f = 1/3$. Thus, for the first 100 days of the budget cycle all projects can be funded, for a cost of \$100 MM, while for the remaining 200 days only those projects above the average cost of \$$10^4$ each are considered, for a total average cost of \$180 MM. Thus the *total* number of projects at, or above, average cost is about 2700 ($\equiv 1800 + 900$) while the number of projects funded below the average project cost of \$$10^4$ is about 9000, but they only cost an average of \$$10^3$ each, for a total of \$9 MM. Thus, \$271 MM of the \$280 MM goes to the high-cost projects—once the decision is made that only projects at above average cost will be funded beyond some time limit, and the time limit can itself be then determined once the total budget is given, together with the number of high-cost and low-cost projects. In performing this numerical illustration we again kept the cost distribution of projects proportional to $\omega^{-3/2}$, as in the more formal prior section. For other choices of cost distribution the calculations reported here can be done *de novo*, of course.

The point about this illustration is that it provides a simple example of how various corporate spending strategies can be satisfied once estimates of ranges and distributions of corporate project costs are available. Most often these values are available from historical information. In addition, one can always run "what if" scenarios to ascertain ranges and distributions of projects that will satisfy corporate mandates.

V. APPENDIX: APPROXIMATE PROBABILITY BEHAVIOR FOR B ≪ B₀

Consider a uniform *rate* of arrival of requests in each budget interval. One can then write

$$\frac{dN(\omega,t)}{d\omega} = t \frac{dR(\omega)}{d\omega}, \tag{7.81}$$

where $dR/d\omega$ is the rate of arrival of requests in ω to $\omega + d\omega$. Then

$$p(B,t) = (2\pi)^{-1} \int_{-\infty}^{\infty} dx \exp\left\{ ix(B_0 - B) - t \int_0^\infty \frac{dR(\omega)}{d\omega} [1 - \exp(-ix\omega)] \, d\omega \right\}. \tag{7.82}$$

APPENDIX: APPROXIMATE PROBABILITY BEHAVIOR FOR $B \ll B_0$

Note that for $t = 0$, the start of the budget cycle, one has

$$p(B, t = 0) = (2\pi)^{-1} \int_{-\infty}^{\infty} dx \exp\{ix(B_0 - B)\} = \delta(B_0 - B). \quad (7.83a)$$

Thus, the cumulative probability of having a residual budget *less* than an amount B is

$$P(B, t = 0) = \int_0^B p(u, t = 0) \, du = \begin{cases} 1 & \text{if } B > B_0; \\ 0 & \text{if } B < B_0; \end{cases} \quad (7.83b)$$

that is, one still retains the total budget since no requests have arrived. However, at any later time t in the budget cycle, the cumulative probability of having a residual budget of B or less is

$$P(B, t) = i(2\pi)^{-1} \int_{-\infty}^{\infty} dx \, x^{-1} \exp(ixB_0)\{1 - \exp(-ixB)\}$$

$$\exp\left\{-t \int_0^{\infty} \frac{dR}{d\omega}[1 - \exp(-ix\omega)] \, d\omega\right\}, \quad (7.84)$$

which diminishes as t increases.

Consider, for example, that requests arrive in two groups only, at requested amounts ω_1 and ω_2 and average rates r_1 and r_2, respectively. Then

$$\frac{dR}{d\omega} = r_1 \delta(\omega - \omega_1) + r_2 \delta(\omega - \omega_2), \quad (7.85)$$

which, when substituted into Eq. (7.84), yields

$$P(B, t) = i(2\pi)^{-1} \int_{-\infty}^{\infty} dx \, x^{-1} \exp(ixB_0)\{1 - \exp(-ixB)\}$$

$$\exp\{-t[r_1 + r_2 - r_1 \exp(-ix\omega_1) \quad (7.86)$$

$$- r_2 \exp(-ix\omega_2)]\}.$$

Now rewrite the variable x to u through $x = u/B$, when

$$P(B, t) = i(2\pi)^{-1} \int_{-\infty}^{\infty} du \, u^{-1} \exp(iuB_0/B) \{1 - \exp(-iu)\}$$

$$\exp\left[-t\left\{r_1 + r_2 - r_1 \exp\left(-iu\frac{\omega_1}{B}\right) - r_2 \exp\left(-iu\frac{\omega_2}{B}\right)\right\}\right]. \quad (7.87)$$

Let the residual budget $B \gg \omega_2$ (so that, automatically, $B \gg \omega_1$). Then expand the exponentials $\exp(-iu\omega_1/B)$ and $\exp(-iu\omega_2/B)$ to quadratic order in $1/B$ to obtain

$$P(B, t) = \frac{i}{2\pi} \int_{-\infty}^{\infty} du \, u^{-1} \exp(iuB_0/B) \{1 - \exp(-iu)\}$$

$$\exp\left\{[iu(r_1\omega_1 + r_2\omega_2)t/B] - \frac{t}{2}\frac{u^2}{B^2}(r_1\omega_1^2 + r_2\omega_2^2)\right\}. \quad (7.88)$$

Then, to the same degree of approximation, the probability density, $p(B, t)$, is analytically given through

$$p(B, t) = \begin{cases} (2\pi C^2 t)^{-1/2} \exp[-(B_0 - B - tA)^2 (4tC^2)^{-1}], & \text{for } B_0 > B \\ 0, & \text{for } B_0 \leq B \end{cases};$$

(7.89)

where

$$C^2 = \frac{1}{2}(r_1 \omega_1^2 + r_2 \omega_2^2) \equiv \frac{1}{2} \int_0^\infty \omega^2 \frac{dR}{d\omega} d\omega, \quad (7.90a)$$

$$A = \omega_1 r_1 + \omega_2 r_2 = \int_0^\infty \omega \frac{dR}{d\omega} d\omega. \quad (7.90b)$$

Clearly, as long as each request for funding is low compared to the total budget available, and as long as $B \ll B_0$, then the approximation given by Eq. (7.89) is accurate, with the use of the integral representations for all distributions of requests as given in Eqs. (7.90a) and (7.90b). Thus, one can generally write the probability of having a residual budget amount in the range B to $B + dB$ ($B_0 > B$) at time t as

$$p(B, t) = \frac{(2\pi C^2 t)^{-1/2}}{\exp[(B_0 - B - tA)^2 (4tC^2)^{-2}]} \quad (7.91)$$

with

$$C^2 = \frac{1}{2} \int_0^\infty \omega^2 \frac{dR}{d\omega} d\omega > 0, \quad (7.92a)$$

$$A = \int_0^\infty \omega \frac{dR}{d\omega} d\omega > 0. \quad (7.92b)$$

Note that $p(B, t)$ has a peak value at time $t = t_{\text{peak}}$ given by

$$t_{\text{peak}} = A^{-1}[\{(B_0 - B)^2 + C^4/A^2\}^{1/2} - C^2/A] \quad (7.93)$$

and, at $t = t_{\text{peak}}$, the peak value is

$$p(B, t_{\text{peak}}) = (2\pi C^2)^{1/2} \frac{A^{1/2}}{[\{(B_0 - B)^2 + C^4/A^2\}^{1/2} - C^2/A]^{1/2}} \quad (7.94)$$

$$\exp\left\{-\frac{A[(B_0 - B) + C^2/A - \{(B_0 - B)^2 + C^4/A^2\}^{1/2}]^2}{4C^2[\{(B_0 - B)^2 + C^4/A^2\}^{1/2} - C^2/A]}\right\}.$$

At large values of time, $t \gg (B_0 - B)/A$, the probability distribution tends toward

$$p(B, t) \cong C^{-1}(2\pi t)^{-1/2} \exp[-tA^2/4C^2] \equiv (B_0 - B)C^{-1}(2\pi t)^{-1/2} \exp(-t/\tau)$$

(7.95)

with $\tau = 4C^2/A^2$; while at small values of time, $t \ll (B_0 - B)/A$, the probability distribution tends toward

$$p(B, t) \cong C^{-1}(2\pi t)^{-1/2} \exp(-\tau_{\text{rise}}/t), \qquad (7.96)$$

where $\tau_{\text{rise}} = (B_0 - B)^2/4C^2$. Thus, it would seem that the probability of retaining a budget of B, or less, at time t would be

$$P(B, t) = \int_0^B p(x, t)\, dx, \qquad (7.97)$$

and then the sense of argument given in the body of the chapter can be repeated.

8

MAXIMIZING OIL FIELD PROFIT IN THE FACE OF UNCERTAINTY

I. INTRODUCTION

A decade and a half ago Nind (1981) provided a formula for estimating the number of wells, n, necessary to maximize the profit on an oil field. Nind's formula allows for the effects of continuous discounting, development costs per well, fixed development costs for the field, selling price of oil, and initial production rates as well as decline curves of production. The formula also takes into account the estimated recoverable oil volume.

Nind's formula is remarkably useful in providing estimates of likely present-day worth (PDW) net profit estimates as well as in providing an evaluation of likely number of wells needed to maximize cumulative profit. To date, two factors have restricted the application of Nind's formula to a wider variety of situations in oil field production.

On the geological and production sides, there is the question of the range of uncertainty of reservoir area, formation thickness, and production efficiency, each of which carries a degree of uncertainty. On the economic side there is the problem of uncertainties in selling price, development costs, and discount factor, each of which also provides some contribution to the uncertainties of total profit and optimum number of wells.

Clearly two questions are of major interest:

1. When allowance is made for geological, production, and economic uncertainties, what then is the likelihood of a positive PDW, and what is the range of uncertainty surrounding the optimum number of wells?

2. Which uncertainty factors are the most critical to address in order to minimize the uncertainty on maximum PDW and optimum number of

wells? That is, what is the relative importance of the different contributions to uncertainty in *PDW*?

Methods of providing answers to these two questions would go a long way toward making Nind's formula of greater practical utility in assessing production and development strategies in relation to anticipated total *PDW* and number of wells needed to maximize *PDW*.

The purpose of this chapter is to address these two concerns.

II. NIND'S FORMULA FOR PRESENT-DAY WORTH

Let the total recoverable oil reserves be R bbl; the initial production rate per well q_0 bbl/day; D the total field development costs in present-day dollars ($) independent of the number of wells; C the lifetime development costs, also in present-day dollars ($) per well; u the net oil value (selling price) in $/bbl. Let L be the total lifting costs (present-day value) per well, n the number of wells, and j the continuous annual discount rate. All costs and prices are after taxes, the lifetime costs D, C, and L are in fixed present-day dollars, and the selling price is in current dollars.

The problem, as stated by Nind (1981), is to supply a formula for determining the total present-day worth profit in terms of the number of wells, and to determine the maximum *PDW*, *PDW*$_{max}$, together with the optimum number of wells needed to provide the maximum *PDW*.

A. GENERAL ANALYSIS

Nind (1981) makes the three simplifying explicit assumptions that (1) the ultimate oil recovery is independent of well spacing; (2) the average well has a production rate declining exponentially with time with a scale constant b, so that at any time, t, after initial well production, the actual production rate is $q(t) = q_0 \exp(-bt)$; and (3) the economic limit may be taken to be zero to a good enough degree of approximation. Also, the implicit assumption is made that all wells are placed in production simultaneously at time $t = 0$.

Within the framework of these assumptions, the derivation of a *PDW* formula dependent on n, the number of wells, then proceeds as laid down by Nind (1981). The total production, p_0, per well is

$$p_0 = \int_0^\infty q(t)\, dt \equiv Q_0/b, \tag{8.1}$$

where $Q_0 = 365\, q_0$.

But the production is also taken to be R/n (Nind's first assumption). Hence, equating Eq. (8.1) and R/n leads to

$$b = Q_0 n / R. \tag{8.2}$$

At a selling price u after taxes, the constant value gross cash per well would be

$$\text{Cash} = \int_0^t q(t) \exp(-jt)\, u\, dt = \frac{uQ_0}{(j + Q_0 n/R)} \qquad (8.3)$$

so that, from n wells, one obtains a gross cash amount of

$$\text{Gross}_n = n \times \text{Cash} = \frac{nuQ_0 R}{(jR + nQ_0)}. \qquad (8.4)$$

From this gross, one must subtract the fixed development costs, D, as well as $n(L + C)$, the lifting costs and development costs per well, so that the profit ($) (present-day worth, PDW) is

$$PDW = \frac{nuQ_0 R}{(jR + nQ_0)} - n(L + C) - D, \qquad (8.5)$$

which is just Nind's (1981) formula.

For fixed values of $Q_0, u, R, j, L, C,$ and D it follows by simple differentiation with respect to n that Eq. (8.5) has a maximum at an optimum number of wells (ONW) given through

$$ONW = RQ_0^{-1}[\{jQ_0 u/(C + L)\}^{1/2} - j] \equiv jRQ_0^{-1}(z_m^{1/2} - 1), \qquad (8.6)$$

where $z_m = Q_0 u/\{j(C + L)\}$ and provided $ONW \geq 1$ for the total oil field. Equation (8.6) is given by Nind (1981). At the value of $n = ONW$, the maximum PDW is then given by

$$PDW_{\max} = R[u^{1/2} - \{j(C + L)/Q_0\}^{1/2}]^2 - D, \qquad (8.7a)$$

which can also be written in the form

$$PDW_{\max} = Q_0(Rj)^{-1}(C + L)(ONW)^2 - D. \qquad (8.7b)$$

Note from Eq. (8.7a) or (8.7b) that a positive $PDW_{\max}$ occurs if and only if one has

$$ONW \geq [DRj/Q_0(C + L)]^{1/2}, \qquad (8.8a)$$

while $ONW \geq 1$ occurs [from Eq. (8.6)] if and only if

$$z_m \geq 1 \quad \text{and } jQ_0 u/(C + L) > (j + Q_0/R)^2. \qquad (8.8b)$$

Also of interest are the minimum, n_-, and maximum n_+, number of wells for which $PDW \geq 0$; with too few wells the fixed field development costs will dominate over profitability, while at too many wells the lifting costs per well and development costs per well will outstrip profit.

It is not difficult to show from Eq. (8.5) that $PDW = 0$ on $n = n_\pm$ where

$$n_\pm = \frac{jR}{2Q_0}[z_m(1 - D/uR) - 1 \pm \{[1 - z_m(1 - D/uR)]^2 - 4z_m D/uR\}^{1/2}].$$

$$(8.8c)$$

It can also be shown, with somewhat more effort, that if PDW_{max} is positive, then $0 < n_- \leq ONW \leq n_+$; while if PDW_{max} is negative then n_- and n_+ are both pure imaginary; that is, there is no number of wells, n, at which $PDW = 0$ because then a PDW of zero would exceed the negative PDW_{max}. A field with $PDW_{max} < 0$ cannot be made profitable. However, if the fixed field development has already been undertaken, so that costs of D have been expended, one can minimize the total loss to the corporation by developing the field with ONW wells, so that PDW_{max} (<0) then represents the loss, which is a smaller total loss than D.

Note that on $D = 0$, then $n_- = 0$, $n_+ = jR(z_m - 1)/Q_0$. Also note that $n_- = n_+ = ONW$ whenever D is given through

$$[1 - z_m(1 - D/uR)]^2 = 4z_m D/uR \tag{8.9a}$$

and then $PDW_{max} = 0$. Fixed field development costs, D, must be kept lower than

$$D \leq jR(C + L)Q_0^{-1}(z_m^{1/2} - 1)^2 \equiv D_* \tag{8.9b}$$

in order to have a positive PDW.

For fixed field development costs, D, much less than the raw profit estimate uR, as is usually the case, it follows to order D/uR that

$$n_- \cong \frac{jR}{Q_0}\left(\frac{D}{uR}\right)\frac{z_m}{(z_m - 1)} \tag{8.10a}$$

$$n_+ \cong \frac{jR}{Q_0}\left[z_m - 1 - \left(\frac{D}{uR}\right)\frac{z_m^2}{(z_m - 1)}\right] \tag{8.10b}$$

so that the range of n over which a $PDW \geq 0$ can be achieved is narrowed compared to the situation for $D = 0$. The situations for $D < D_*$, $D = D_*$, and $D > D_*$ are sketched in Figure 8.1 for PDW versus well number, n.

B. NUMERICAL ILLUSTRATION

Nind (1981) provides a specific numerical illustration in which:

Average initial oil production/well ($Q_0/365$)	= 77 bbl/day
Development costs/well (C)	= $0.72 MM
Lifting costs/well (L)	= $0.09 MM
Oil value (u)	= $10.50/bbl
Discount rate (j)	= 0.1/yr (10%/yr)
Specific volume oil recoverability (s)	= 400 bbl/(acre ft)
Reservoir oil sand thickness (T)	= 20 ft
Recoverable oil ($R \equiv sT$)	= 8000 bbl/acre

For the numerical values reported by Nind (1981), the value of $z_m \cong 3.64$. Direct insertion of these values into Eq. (8.6) yields the optimum number

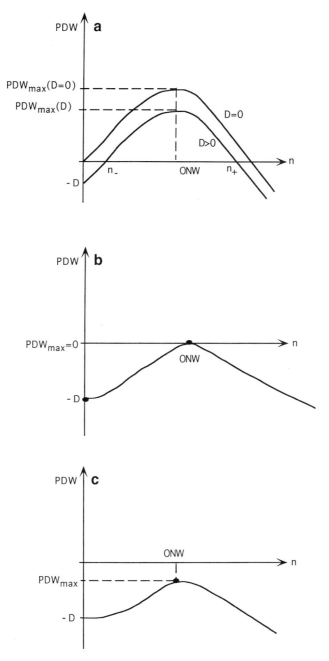

FIGURE 8.1 Sketch of the shape of *PDW* versus number of wells, n, for different field development costs, D, with $z_m > 1$: (a) $D < D_*$; (b) $D = D_*$; (c) $D > D_*$.

of wells/acre as

$$ONW = 26/1000 \text{ acres}, \tag{8.11a}$$

that is, 1 well about every 40 acres, while the maximum number of wells, n_+, at which $PDW = 0$ is 75/1000 acres for development cost $D = 0$. Then, from Eqs. (8.7),

$$PDW_{max} = (\$19.0 \text{ MM}/1000 \text{ acres} - D/1000 \text{ acres}). \tag{8.11b}$$

Nind (1981) did not specify a value for the fixed development costs, D, probably because they play no role in determining the optimum number of wells to maximize PDW. However, the fixed development costs do have a major role to play in determining profitability. If the development costs exceed about \$19,000/acre [Eq. (8.11b)] then, for the numerical values given, there is no possible way that field development can prove profitable and the project should be abandoned if the fixed development costs have not yet been expended. Note that a naive estimate of reserves, R, times selling price, u, would indicate a maximum value of \$84,000/acre if no cost at all had to be borne and if no inflation took place. The effect of both of these factors is to lower the PDW_{max} to less than 25% of the naive estimate (even ignoring fixed field costs, D).

Note also that the timescale for exponential production decline is then

$$\tau_{2/3} \cong 1/b = R/[Q_0 \; ONW], \tag{8.12}$$

which takes on the numerical value $\tau_{2/3} = 11$ yr so that $(1 - e^{-1})$ (i.e., about 2/3) of the reserves are drained in this period.

Several problems arise in attempting to use these estimated values in practical situations. First, over the course of an 11 year timescale the selling price of oil is highly variable (e.g., in 1984 West Texas Intermediate was selling for around \$30/bbl, in 1995 the price was about \$18/bbl), so that there is uncertainty on the price.

Second, the total continuous corporate discount rate of 10%/yr should also be allowed to vary so as to include changes in interest rates, changes in internal corporate discount factors (e.g., as corporate overheads are reduced or increased, as corporate personnel are reduced, or as internal costs are cut), and changes in the value of the dollar relative to other currencies.

Third, the so-called fixed field development costs, D, vary depending on continuing infrastructure needs, ongoing maintenance, and so on.

Fourth, the lifting costs per well, in addition to the development costs per well, vary with time, equipment age, the need for unanticipated acid-fracturing jobs, CO_2 flooding, changes in the oil-to-water fraction, etc.

Fifth, it is a rare event when the productive reservoir acreage is known precisely, and even rarer when the oil sand thickness is always precisely the same value. Thus, uncertainties also exist in terms of the total recoverability volume of hydrocarbons.

All of these uncertainties make it more difficult to be precise as to the number of wells necessary to maximize PDW or even to the maximum PDW itself. The problem is to address these concerns, which is the aim of the next section.

III. PROBABILITY AND RELATIVE IMPORTANCE

The total producible reserves, R, from a field of specific volumetric recoverability s, area A, and oil sand thickness T, are given by $R = sAT$ bbl. The specific volume recoverability is itself variable because the formation permeability varies around a mean value, with location of each well in the field, with the drainage area feeding a well site, and with changes in the oil-to-water ratio.

Interest centers on the probability of obtaining a positive PDW, the likely maximum PDW and its probability range; the ONW and its range of probable uncertainty; and an estimate of scale time, τ, for production decline and its range of probable values.

In addition, we need to determine the relative importance of each of the nine uncertainty factors (s, A, T, j, u, q_0, D, L, and C) in contributing to the uncertainties in $PDW_{\max}$, ONW, and τ, so that we have an idea of where to concentrate effort to narrow the range of uncertainty.

Because the total producible reserves, R, are estimated from

$$R = sAT, \qquad (8.13a)$$

while the total fixed field development costs are estimated from

$$D = A\Delta, \qquad (8.13b)$$

where Δ is the development cost per unit area, the PDW formula can then be written

$$PDW = \frac{nu365q_0 sAT}{(jsAT + 365q_0 n)} - n(L + C) - A\Delta, \qquad (8.13c)$$

while

$$PDW_{\max} = sAT[u^{1/2} - \{j(C + L)/365q_0 u\}^{1/2}]^2 - A\Delta \qquad (8.13d)$$

occurs at the optimum number of wells, ONW, given through

$$ONW = j(sAT)(365q_0)^{-1}[(365q_0 u/\{j(C + L)\})^{1/2} - 1]. \qquad (8.13e)$$

The timescale to about two-thirds oil depletion can also be written

$$\tau_{2/3} = \frac{sAT}{ONW(365q_0)}. \qquad (8.13f)$$

As each of the intrinsic variables, s, A, T, Δ, u, C, L, q_0, and j is allowed to vary within its range of uncertainty, a different value will be computed for n_-, n_+, PDW_{max}, $\tau_{2/3}$, and ONW. A series of Monte Carlo runs, with each of the intrinsic variables being selected at random from a probability distribution, will then produce a suite of values for n_-, n_+, PDW_{max}, $\tau_{2/3}$, and ONW, so that relative frequency of occurrence curves can be generated for each output.

However, of greater use are the *cumulative* probability of occurrence curves because they inform us about, for example, when a PDW greater than some amount (zero, say) is likely to occur under the ranges estimated for the intrinsic variables and their associated probability distributions.

In addition, not each of the intrinsic variables is equally important in contributing to the ranges of uncertainty of n_-, n_+, PDW_{max}, $\tau_{2/3}$, or ONW because of the different functional dependencies of the five output quantities on the intrinsic variables. The relative importance (RI) of each intrinsic variable and range of uncertainty in contributing to the uncertainty of n_-, n_+, PDW_{max}, $\tau_{2/3}$, or ONW needs to be addressed, and is measured here by the corresponding percentage contributions to the total variance of each of n_-, n_+, PDW_{max}, $\tau_{2/3}$, and ONW.

The probability distributions of each of the intrinsic variables, from which are drawn values for the Monte Carlo simulations, can be chosen to be uniformly distributed, triangular, Gaussian, or log-normal (with appropriate truncations if necessary), or can be based on a histogram distribution from a suite of samples. Thus, the relative importance of different distributions in contributing to the total uncertainty of the output quantities n_-, n_+, PDW_{max}, $\tau_{2/3}$, and ONW could also be investigated if required.

IV. NUMERICAL EXAMPLE

A. PROBABILITIES AND RANGES

To illustrate the dependence of PDW_{max}, ONW, $\tau_{2/3}$, n_-, and n_+ on the uncertainty in each of the nine intrinsic variables, a uniform distribution was chosen for each intrinsic variable with a 50% spread around a center value, as recorded in Table 8.1. Each variable is spread the same fractional amount around its center value so that, if each contributed equally to the uncertainty in the five output quantities, then the relative importance of the contribution to the variance would immediately show this fact.

As a template for comparison, note that if only the center values were to be chosen for each of the variables in Table 8.1, then one would estimate $uR = \$120$ MM, $PDW_{max} = \$25$ MM, $ONW = 27.4$, $\tau_{2/3} = 10$ yr, $n_- = 2$, $n_+ = 76$. Thus, while a maximum PDW of \$25 MM would be obtained with about 27 wells, some profit ($PDW > 0$) could be made with greater than about 2 wells and less than about 76 wells placed on the total oil field.

NUMERICAL EXAMPLE

TABLE 8.1 Ranges of Parameters for Oil Field Development

Factor	Minimum	Center	Maximum
Discount, j (%/yr)	5	10	15
Costs per well, C ($MM)	0.5	1.0	1.5
Lifting costs per well, L ($MM)	0.05	0.10	0.15
Initial production rate, q_0 (bbl/day)	40	80	120
Selling price, u ($/bbl)	7.5	15	22.5
Field development costs, Δ, per acre ($)	2500	5000	7500
Specific recoverability, s (bbl/acre ft)	200	400	600
Oil sand thickness, T (ft)	10	20	30
Field acreage (acres)	500	1000	1500

However, due to the uncertainties in each of the intrinsic variables the problem is that the trustworthiness of these estimates is not yet known.

For each of the intrinsic variables, a Monte Carlo simulation was run on a small PC with several thousand iterations, with variables selected at random from uniform distributions limited by the minimum and maximum ranges given in Table 8.1. The total computer time was 6 min.

For each of the output variables PDW_{max}, ONW, $\tau_{2/3}$, n_-, and n_+, it was then possible to provide cumulative probability of occurrence curves illustrating the expected ranges of values. It was also possible to provide similar plots for total recoverable oil reserves or, indeed, for any other appropriate quantity.

1. Recoverable Reserves Estimates

Figure 8.2a displays the cumulative probability for the expected total recoverable oil reserves from the field. Note that there is a 10% chance (P_{10}) of obtaining less than 3.5 MMbbl, a 90% chance (P_{90}) of less than 13.5 MMbbl, a 50% chance (P_{50}) of less than 7 MMbbl, and about a two-thirds chance (P_{68}) of less than about 9 MMbbl. P_{68} is highlighted in this discussion because it provides an approximate description of the mean value. Two measures of uncertainty are often used to quantify the volatility of estimates of reserves. The volatility measure, ν, is here defined by

$$\nu = (P_{90} - P_{10})/P_{68},$$

which, for the present illustration, takes on the value $\nu = 1.1$. A high value of volatility ($\nu \gg 1$) indicates considerable uncertainty on the mean estimated reserves, while a low value ($\nu \ll 1$) indicates a statistically sharp determination. In the present case, the volatility of 1.1 indicates about a 50% uncertainty in the P_{68} value, so that one records the estimate of $9^{+4.5}_{-5.5}$ MMbbl, indicating directly the uncertainty at the 10 and 90% cumulative probability ranges.

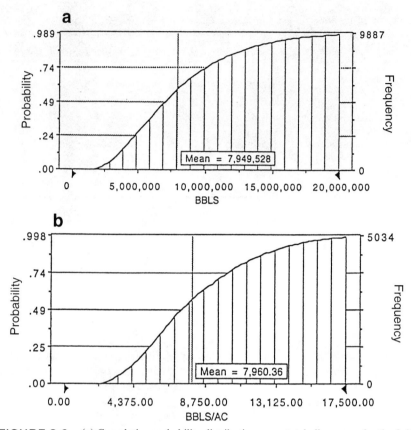

FIGURE 8.2 (a) Cumulative probability distribution verus total oil reserves for the field allowing for the uncertainties given in Table 8.1. (b) Cumulative probability distribution versus oil reserves per acre using Table 8.1 uncertainties.

A second measure is the ratio of standard error, σ, to mean value, E, which, for the curve of Figure 8.2a is $\sigma/E = 4.25/8 \cong 0.5$, indicating that the mean value could range up to about 50% on either side of E. In the case of Figure 8.2a the mean is 8 MMbbl, occurring at about P_{60}, so one records the estimate 8 ± 4 MMbbl to indicate the uncertainty. Volatility is used here as the measure of uncertainty throughout the remainder of this numerical example section.

As well as an uncertainty on the total recoverable reserves, there is also uncertainty on the recoverable amount per acre due to reservoir thickness and specific volume productivity being variable. For the values in Table 8.1, Figure 8.2b records the cumulative probability of obtaining recoverable reserves/acre.

NUMERICAL EXAMPLE

Note that P_{68} occurs at about 9500 bbl/acre, while P_{10} yields 4300 bbl/acre and P_{90} occurs at about 13,300 bbl/acre. In this case the volatility is $\nu = 1.05$, very close to that for the total field. One records the estimate of 9500^{+3800}_{-5200} bbl/acre as the likely range of recoverable oil per acre.

2. Optimum Number of Wells

Figure 8.3a displays the cumulative probability distribution for *ONW*. Note that there is a 50% chance (P_{50}) that fewer than 21 wells are needed, a 90% chance (P_{90}) that fewer than 47 wells is appropriate, a two-thirds chance (P_{68}) of needing fewer than 25 wells, and only a 10% chance that fewer than 13 wells will be needed. Thus, the volatility on well number is $\nu_{\text{wells}} = 1.4$, suggesting a considerably broader range of

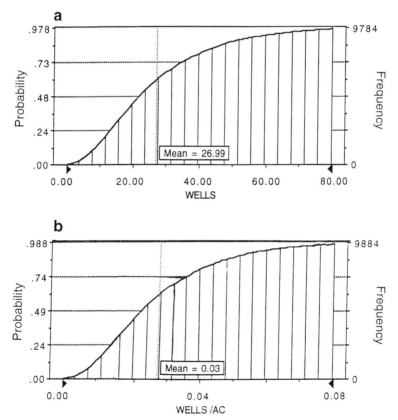

FIGURE 8.3 (a) Cumulative probability distribution for the optimum number of wells, *ONW*, for the total field using Table 8.1 uncertainties. (b) Cumulative probability distribution for the optimum number of wells per acre using Table 8.1 uncertainties.

uncertainty than for the producible oil reserve estimates. Thus one writes that assessment as $ONW = 25^{+22}_{-12}$ to give some idea of the range to be expected.

The uncertainty in the ONW/acre is also of interest and is recorded in Figure 8.3b through the cumulative probability distribution. Here one has P_{90} occurring on $ONW/1000$ acres $= 56$; P_{68} occurring on $ONW/1000$ acres $= 32$; P_{50} occurring on $ONW/1000$ acres $= 24$; P_{10} occurring on $ONW/1000$ acres $= 10$; so that the volatility of the P_{68} estimate is $\nu \cong 1.5$, yielding the assessment of ONW as $32^{+24}_{-22}/1000$ acres. This assessment is of broader range, and with a marginally higher volatility, than the ONW for the total oil field due to the fact that the range of acreage is only from 500 to 1500, so that there is a 50% chance of needing fewer than 1000 acres (with the uniform probability assumption on acreage), and a low-acreage field requires proportionately fewer wells.

3. Maximum Present-Day Worth

Perhaps of greatest interest to a corporation is the maximum expected worth of a field and, in particular, the probability of *not* making a positive return, so that corporate risk tolerance factors can be applied to the decision to develop a particular oil field, and can then be used to evaluate working interest, if any, that should be taken in the development project.

Shown in Figure 8.4a is the cumulative probability of estimated PDW_{max}, as the nine intrinsic variables randomly vary. Reading from Figure 8.4a, one has $P_{90} = \$67$ MM, $P_{68} = \$32$ MM, $P_{50} = \$19$ MM, and $P_{10} \cong \$0$ MM. Thus, there is only a 10% chance that development of the oil field will *not* be profitable ($PDW < 0$), while the volatility is $\nu = 2.1$, indicating a fairly broad range of uncertainty on the profit to be expected, recorded as $\$32^{+35}_{-32}$ MM at the 90% probability level. Of interest also is the PDW/acre, displayed in Figure 8.4b as though there were zero development costs per acre ($\Delta = 0$). Reading from Figure 8.4b one has P_{10} on $(-\$0.28\text{ M} - \$\Delta)$; P_{90} on $(\$65.2\text{ M} - \$\Delta)$, and P_{68} on $(\$37\text{ M} - \$\Delta)$. In this case there is a volatility of about $\nu = 65/(32 - \Delta)$ with Δ in units of \$M. Thus, in order to be 90% sure of a profit, the development costs have to be kept to less than \$65.2 M/acre, while to be 68% sure of a profit, it is sufficient to keep development costs at less than about \$37 M/acre.

Hence, one can use the estimates of PDW_{max}/acre, even without knowledge of the development costs, Δ, to figure out what maximum level of development costs is permissible in order for the field to have a chance of turning a profit.

4. Field Lifetime

Also of interest is the time, $\tau_{2/3}$, to deplete the field to $1 - e^{-1}$ (i.e., about two-thirds) of total reserves when the reserves, costs, etc., are also uncertain, because such estimates provide some indication of the time that

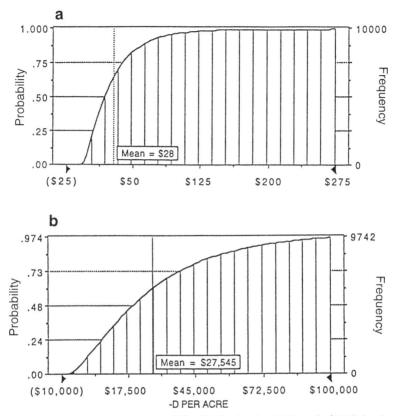

FIGURE 8.4 (a) Cumulative probability distribution for PDW_{max} (in $MM) for the oil field using Table 8.1 uncertainties. (b) Cumulative probability distribution for PDW_{max} per acre using Table 8.1 uncertainties, but not including field development costs, Δ, per acre.

one should be prepared to spend maintaining aging equipment, doing site remediation, or planning for new projects as the old field slowly dies.

For the parameter ranges of Table 8.1, Monte Carlo simulations provide the cumulative probability plot of $\tau_{2/3}$ given in Figure 8.5. Here one has P_{10} at 2.1 yr, P_{50} at 10.3 yr, P_{68} at 14.0 yr, and P_{90} at 27.7 yr, for a volatility on the P_{68} lifetime estimate of $\nu = 1.8$, yielding the estimate $\tau_{2/3} = 14^{+13.7}_{-11.9}$ yr.

Thus, there is a good chance that one should plan the field production over a decade or more; it is unlikely that the field production will take only a few years, because the logistics of drilling the wells and bringing them on line normally precludes the ability to drain the field by two-thirds in less than a few years. (While this result is formally counter to the underlying assumption of instantaneous/simultaneous drilling and initial production, it serves the purpose of pointing out that even when one can come close

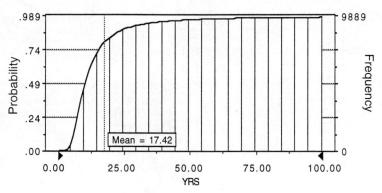

FIGURE 8.5 Cumulative probability distribution for the total field lifetime, $\tau_{2/3}$, using Table 8.1 uncertainties.

to honoring the underlying assumption it will still take a decade or more to drain the field.)

5. Minimum and Maximum Number of Wells

While PDW_{max} and ONW provide the optimal values (and corresponding cumulative probabilities) for maximizing profit, it can happen that these optima are not possible to achieve. For example, perhaps because of federal or state regulations one cannot have greater than a maximum number of wells per acre, or perhaps in an offshore environment a shipping lane has to be kept open, or perhaps encroachment on an environmentally sensitive area is forbidden. For these and a variety of other reasons, it is often the case that the number of wells is less than ONW. In such cases the question is this: What is the minimum number of wells, n_-, necessary to produce a $PDW > 0$?

Equally, oftentimes a government or corporation may be desperate for cash flow or production, so it requests that a very large number of wells be drilled once a field is discovered. In this sort of situation the question is this: What is the maximum number of wells, n_+, that can be drilled and still return a $PDW > 0$?

For fixed values of the nine intrinsic parameters the analytic formulas for n_- and n_+, given by Eq. (8.6), can be used to provide appropriate constraints yielding positive PDW values. However, because the parameters vary randomly, both n_- and n_+ will also vary, permitting only cumulative probability estimates for each to be made. For the parameter ranges of Table 8.1, Figures 8.6a and 8.6b provide the corresponding cumulative probability plots for n_- and n_+, respectively. Reading from Figure 8.6a, P_{10} is at 0.33, P_{50} at 1.27, P_{68} at 1.95, P_{90} at 3.97, indicating a high volatility to n_- of $v_- = 1.87$, so that a $PDW \geq 0$ is likely (68% chance) to be achieved at the 90% probability interval for $n_- \geq 2^{+2}_{-1}$ wells.

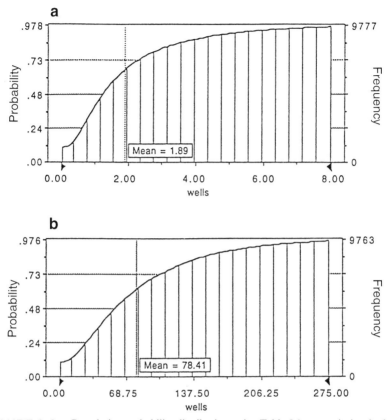

FIGURE 8.6 Cumulative probability distribution, using Table 8.1 uncertainties, for both (a) the minimum, n_-, and (b) the maximum, n_+, number of wells to ensure $PDW > 0$.

Equally, reading from Figure 8.6b, P_{10} is at 17.85, P_{50} at 77.4, P_{68} at 115, and P_{90} at 216, for a high volatility of $\nu_+ = 1.72$, so that a $PDW \geq 0$ is likely (68% chance) to be achieved at the 90% probability interval for $n_+ \leq 115^{+101}_{-93}$.

Thus, there is an extremely broad range for the number of wells (both minimum and maximum) that might permit a profit to be made. One can, then, use this information when negotiating positions of involvement with field-controlling agencies.

B. SENSITIVITY AND RELATIVE IMPORTANCE

Apart from the ability to identify the uncertainty of values relating to reserves, PDW, lifetime, etc., there is also a question of concern as to which intrinsic variables are providing the largest contributions to the uncertainty (i.e., the relative importance). For example, if initial selling price, u, were to be providing, say, 70% of the uncertainty to PDW_{max}, then it makes sense to spend considerable effort to arrange to sell oil at the highest price to minimize the effect of the broad range of uncertainty of u. Of interest, then, is a determination, for given ranges of variations, of the nine intrinsic factors that control uncertainties in output variables. In this way one can determine where to concentrate efforts to minimize the range of uncertainty of the intrinsic variable so that the broadest swings of variation in output variables are reduced.

Using the ranges of variables given in Table 8.1, the relative importance (in %) is illustrated on the variability of outputs, as determined by the percentage contribution of the range of uncertainty of each intrinsic factor to the variance.

1. Recoverable Reserves

The relative importance (RI) charts of Figures 8.7a and 8.7b for oil per acre and total oil, respectively, indicate the dependence of the uncertainty (variance) on the range of each intrinsic variable. Because the numerical example was chosen with the same *fractional* range of uncertainty of each intrinsic variable around its center value, and because the values for each intrinsic variable are drawn from uniform populations, one anticipates, from Eq. (8.13a), that field acreage, A, specific productivity, s, and oil sand thickness, T, should each provide RI contributions of 33⅓% to the uncertainty in total recoverable oil reserves; and that s and T should each provide $RI = 50$% to the oil per acre. Note from Figure 8.7a that, to within 0.5%, the Monte Carlo computations indicate that the mathematical expectations are being honored by the numerical values, while Figure 8.7b would indicate that, to within about 1%, similar agreement is reached.

Thus, planning for improvements in recoverable reserves should be split equally between attempts to delineate better acreage and oil sand thickness, and between attempts to improve the range of specific productivity per well. At this stage there is no preference for one factor over the other two. Ranges of uncertainty that are specific to data from a particular oil field could result in a clearer preference.

2. Optimum Numbers of Wells

Figure 8.8 indicates the RI (%) values of factors influencing the range for the optimum number of wells, ONW, which maximizes PDW_{max}. Note here that initial selling price, u, has about a 30% impact on ONW, while costs per well, C, specific productivity, s, and oil sand thickness, T, are roughly next equally important at about 23% each. In this situation, if the range of the initial selling price can be narrowed or negotiated (futures trading) then effort should be put into doing so. However, if it is felt that there is little capability of influencing the selling price, then roughly equal

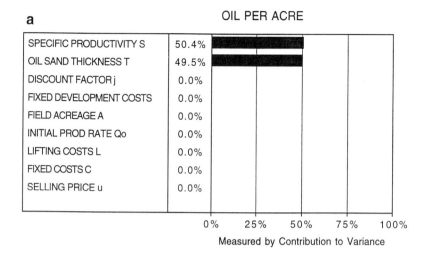

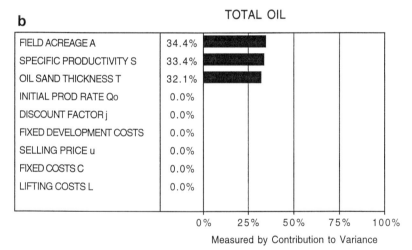

FIGURE 8.7 Relative importance (%) of factors contributing to the variance (a) for amount of oil per acre; (b) for the total estimate of oil from the field.

effort should be put into improving the uncertainty on costs per well, oil sand thickness, and specific productivity per average well; there is no reason, for the example given, to prefer to expend effort on any one of these variables at the expenses of the others.

3. Maximum Present-Day Worth

Figure 8.9 provides the *RI* (%) values for the ranges of input variables of Table 8.1. Note now the increased predominance of initial selling price with an *RI* of 44%, with initial production, Q_0, discount factor, j, specific

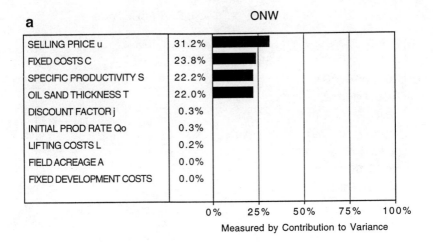

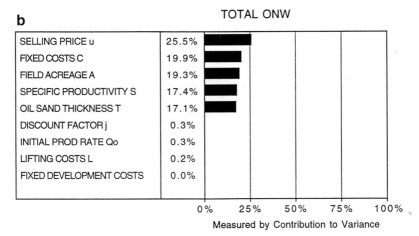

FIGURE 8.8 Relative importance (%) of factors contributing to the variance in (a) *ONW*/acre; (b) total *ONW* for the field.

productivity, s, oil sand thickness, T, and fixed well costs, C, each having an *RI* of around $10 \pm 2\%$, well below the *RI* for selling price. The uncertainties in field acreage, A, fixed development costs, D, and lifting costs, L, can be ignored at this level.

Thus the effort needed is clear (based on the ranges of uncertainty provided): concentrate dominantly on narrowing the range of uncertainty of initial selling price if possible. If such is thought not to be possible then work roughly equally at narrowing the ranges of uncertainty on Q_0, j, s, T, and C; but do not yet expend the same level of effort to narrowing the

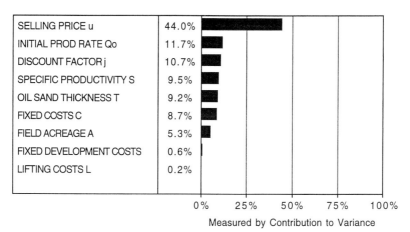

FIGURE 8.9 Relative importance (%) of factors contributing to the variance in the maximum PDW_{max}.

uncertainties on A, D, or L, which have much less influence on improving both the upside potential and downside variability of PDW_{max}.

4. Field Lifetime

The *RI* contributions to the field lifetime, $\tau_{2/3}$, are given in the chart of Figure 8.10. Note here the slight dominance of initial selling price, u, and initial yearly production rate, Q_0, both of which contribute almost 35% of the variance, with fixed costs, C, at about 30%, and negligible contributions

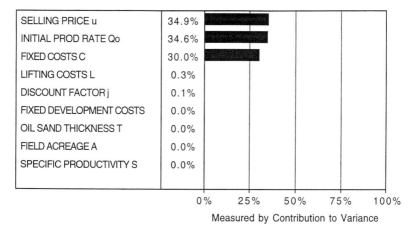

FIGURE 8.10 Relative importance (%) of factors contributing to the variance in the field lifetime, $\tau_{2/3}$.

from other factors. Again the concentration of effort should be on improving the range of selling price to narrow the range of uncertainty on lifetime but, equally, improvements in uncertainty on initial production rate, Q_0, should be subject to similar effort. Improvements in fixed costs, C, for each well should be given almost as much effort, and is likely to be more under the control of the field development plan than selling price.

5. Minimum and Maximum Number of Wells

The number of wells needed (both minimum and maximum) in order to turn a profit ($PDW \geq 0$) is clearly critical in deciding whether to involve a corporation in the field development. Thus, the ability to determine where to place effort to improve uncertainty on the values of n_- and n_+ is crucial. For example, as shown in the previous section, the volatilities of n_- and n_+ are high (of order two) so that there is an uncertainty of nearly a factor of two on the maximum number of wells—an already high value of 115^{+101}_{-93}. Clearly, if the upside number (216) for n_+ can be lowered, then costs can be kept low and profitability rises.

Figure 8.11a provides the RI chart of factors influencing the uncertainty in n_-, while Figure 8.11b provides the same information for n_+. To be noted from Figure 8.11a is that uncertainty in field acreage is a dominant contributor ($RI \sim 27\%$) to uncertainty in n_-, with discount factor, j, and fixed development costs, D, roughly next equal at $RI \sim 20\%$; initial selling price, u, and initial production rate, Q_0, are each slightly lower in importance, at $RI \sim 15\%$, with all other factors being negligible.

Thus, uncertainty in n_- can be helped by better field acreage delineation, and by narrowing (if possible) uncertainty on fixed development costs and discount factor, with a slightly lesser effort paid to improving uncertainty on selling price and production rate. Because n_- is already absolutely small ($n_- = 2^{+2}_{-1}$ for the numerical values used) it is probably not significant to use the improvement factors as relevant to doing a better job. The point here is that any number of wells, n, in $n_- \leq n \leq n_+$ will provide a positive PDW. Because n_+ is already in the range 115^{+101}_{-93}, and because ONW is in the range 25^{+22}_{-12}, both of which are considerably in excess of n_- even when the 90% cumulative probability ranges are allowed for, it follows that doing a better job at improving the range of uncertainty of n_- is really not significant.

However, the same argument cannot be used for n_+, which, at the P_{68} level, is a factor of 4.6 higher than the P_{68} value for ONW. Clearly one would very much like to narrow the range of uncertainty of n_+. Indeed the P_{10} value for n_+ is only 22, while the P_{90} is 216, so that costs could vary by a factor 10, depending on the range of n_+, making for a high degree of uncertainty on achievable profit.

Shown in the chart of Figure 8.11b are the dominant players, in terms of RI, contributing to the uncertainty in n_+. Note that selling price, u, takes 30% of the uncertainty, with fixed costs per well at about 21% (as might

a

FIELD ACREAGE A	35.4%
FIXED DEVELOPMENT COSTS	22.5%
DISCOUNT FACTOR j	17.2%
INITIAL PROD RATE Qo	12.1%
FIXED COSTS C	4.9%
SELLING PRICE u	4.2%
SPECIFIC PRODUCTIVITY S	2.4%
OIL SAND THICKNESS T	1.3%
LIFTING COSTS L	0.0%

b

SELLING PRICE u	34.5%
FIXED COSTS C	22.2%
SPECIFIC PRODUCTIVITY S	13.0%
OIL SAND THICKNESS T	12.5%
FIELD ACREAGE A	10.5%
DISCOUNT FACTOR j	3.6%
INITIAL PROD RATE Qo	3.3%
LIFTING COSTS L	0.2%
FIXED DEVELOPMENT COSTS	0.1%

Measured by Contribution to Variance

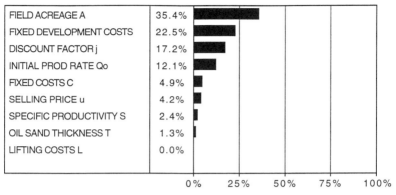
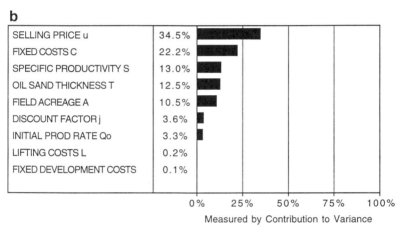

FIGURE 8.11 Relative importance (%) of factors contributing to the variance in (a) n_- and (b) n_+.

have been anticipated given the range of uncertainty of n_+), while factors to do with reserves (namely, oil sand thickness, specific productivity, and field acreage) each have RI values around $13 \pm 2\%$, with all other factors of minor importance.

Thus, again, initial selling price is the dominant factor where effort should be placed to narrow the range of uncertainty, but the contributions of fixed costs per well are not negligible ($RI \cong 21\%$) and should be kept as low as possible, while the reserves need to be evaluated better in terms of recoverability and acre-feet available.

A summary chart of priorities for each category can then be provided as shown in Table 8.2. From the perspective of a corporation, the highest priority to improve the uncertainties on $PDW_{\max}$, ONW, lifetime, τ, and to minimize the uncertainty on the maximum number of wells, n_+, is the

TABLE 8.2   Priorities for Improvement of Uncertainty[a]

| | Output variables | | | | |
Input variables	PDW_{max}	ONW	τ	n_-	n_+
Selling price (u)	1	1	1	4	1
Oil sand thickness (T)	5	4	—	8	3
Field acreage (A)	7	—	—	1	5
Specific productivity (s)	4	3	—	7	4
Fixed costs (C)	6	2	3	6	2
Initial production rate (Q_0)	2	5	2	5	6
Fixed development costs (D)	8	—	—	3	8
Discount factor (j)	3	7	—	2	7
Lifting costs (L)	9	6	4	9	9

[a] 1, dominant; 9, least influential; Dash Means No Dependence.

initial selling price uncertainty. All other factors are of less significance. The uncertainties in initial production and fixed costs both have the highest number of priority class 2 assessments in Table 8.2 for PDW_{max}, ONW, $\tau_{2/3}$, and n_+, suggesting that each should be given attention.

At the other end of the prioritization, lifting costs per well have very little impact at all and do not need any significant work done to improve the ranges of their uncertainty relative to all other factors.

V. DISCUSSION AND CONCLUSIONS

One of the more important aspects in oil field development is to assess the likely present-day worth and number of wells needed to achieve maximum profit.

The example reported here illustrates how to evaluate that goal in the light of uncertainties on selling price, discount factor, field reserves, producibility, acreage, oil sand thickness, fixed development costs, well development costs, and lifting costs.

While the specific numerical example used uniform population probabilities and consistent ranges for the nine intrinsic variables as a simple illustrative procedure, from which Monte Carlo parameter selections were made, there is no fundamental difficulty in using other population probabilities (such as Gaussian, triangular, log-normal, etc.) and ranges as demonstrated earlier in the case of exploration projects.

The points of the illustration were twofold: (1) to illuminate how the ranges of uncertainty can be taken into account when estimating probable profitability of an oil field development decision; (2) to illuminate which factors are causing the largest contributions to uncertainties in present-day worth, field lifetime, and optimum number of wells, so that decisions can be made about where to focus effort to improve matters if desired. Thus, the *absolute* uncertainty (as measured by the P_{90}, P_{68}, and P_{10} ranges) provides information on *when* improvements are needed, while the relative importance provides information on *where* improvement is needed should it be deemed appropriate to narrow uncertainty.

The assessment of oil field well spacing examined here uses Nind's (1981) formula for *PDW*. This formula makes the three explicit assumptions that (1) ultimate oil recovery is independent of well spacing; (2) the average well has a production rate declining exponentially in time; and (3) the economic limit is approximated well by zero. Nind's formula also makes the implicit assumption that all wells go into production at the same time.

These assumptions provide a simple formula for assessing *PDW* in terms of the number of wells. More complex models of production can be invoked (rather than just a simple exponential as reported here); equally, more complex models of well development with time, or of economic threshold limits, etc., can also be included. While such factors are undoubtedly of importance in specific applications to particular oil field development situations, the point is that they provide only technical changes to the general logic procedure outlined here, which is best served by the simple model given.

The advantage to having available a method for assessing not only the range of uncertainty in potential worth of oil field development, but also for assessing which factors have the largest relative importance in influencing uncertainties of PDW_{max}, ONW, τ, n_-, and n_+, is that one can assess relative risk of particular oil fields, and one can assess what needs to be done to narrow the range of uncertainty if necessary. It is these two intertwined components that determine the utility of the procedure given to the evaluation of oil fields.

9

THE VALUE OF ADDED INFORMATION: CATEGORIES OF WORTH

I. INTRODUCTION

One of the classic arguments used to determine if it is worthwhile to pay to acquire more data in a hydrocarbon exploration opportunity is essentially as follows. If the expected value (EV) of the opportunity is E_0 before acquiring further information, and the expected value is estimated to increase to E_1 after acquiring the information, then the maximum amount, A, that one should spend to acquire the information (the so-called *value of added information*) is the difference:

$$A = E_1 - E_0. \qquad (9.1)$$

If it is anticipated that $E_1 < E_0$ then there is usually a negative worth attached to acquiring the additional data and so one does not proceed. An alternative way of expressing this concern is to consider the simple decision tree diagram of Figure 2.1a, where the success branch has a probability, p_s, gains, G, and costs, $-C$, and the failure branch has probability p_f ($\equiv 1 - p_s$) and costs $-C$. For this diagram the expected value is $E_0 = p_s G - C$. The maximum value of *perfect information* (which provides information only about p_s) is then given by estimating the average costs of failure at $p_f C$, so that it is worthwhile spending up to $A_{max} = p_f C$ in order to acquire the perfect data, which will eliminate the failure branch.[1] The same argument can be applied to Eq. (9.1). If no further gains are anticipated as a consequence of acquiring perfect information, then the maximum value of E_1 is $p_s G$ so that $E_1 - E_0 = p_f C$ and hence $A_{max} =$

[1] Also, one has to know what perfect data to acquire, which will eliminate the failure branch!

p_fC—provided the information only resolves the chance of success and not the uncertainty with respect to p_s, G, or C. Yet this argument on the value of added information is not the only viewpoint that can be adopted to evaluate the worth of acquiring new data (perfect or imperfect) from a corporate perspective. As an illustration consider, for instance, some different rationales for acquiring more seismic information:

Rationale 1. Acquiring more seismic data may better define a drill location, There does not need to be any change in expected value of the opportunity, but a better understanding can be achieved of what is likely to be encountered by the well-bore. Without the additional seismic information a poor drill location may have been chosen due to the coarser scale assumptions of a geological model.

Rationale 2. Acquiring more seismic data may provide greater certainty on trap boundaries, size, and reservoir thickness variations. In this case, changes could either lower or increase the *EV*, but in either case can make *EV more certain.*

The *EV* could *increase* because the better defined reservoir boundaries may end up being larger than the assessments made in doing the *EV* evaluation prior to determining whether more seismic data needed to be acquired. Historically, the preponderance of observations makes this larger boundary size an unlikely occurrence, because initial evaluations are usually much more optimistically biased (possibly due to the coarser grid of initial data) that turns out to be the case. However, once in a while, better defined trap boundaries do turn out to be larger than initially estimated—a positive aspect when it happens, and one that it is often assumed *will* occur as a justification for acquiring more seismic data.

The *EV* could *decrease* because more sharply defined trap boundaries define better a smaller, but more accurately delineated, field size. This situation is often considered a positive aspect in equivalent value because there is less uncertainty on the reservoir, even though the *EV* is lowered.

The *EV* may remain the *same* as it was prior to acquiring the seismic data. But a clearer definition of the trap size can reduce the range of uncertainty of the *EV*, so this result is considered of positive worth.

Rationale 3. Acquiring more seismic data may provide more certainty (confidence) on the assigned chances of source, seal, reservoir, and/or structure. There may be an increase, decrease, or no change in the expected value, but the results are considered positive because of the increased certainty in the numbers.

Rationale 4. Acquiring more seismic data may end up neither improving the *EV* of a particular opportunity nor decreasing its uncertainty, but the data can be used at later times to address other problems not involved in the particular opportunity under consideration. For instance, the presence of a stratigraphic trap on the flanks of a structural trap may be suggested by a later geologic model of depositional environment changes, and that new model was not available when one was attempting to evaluate the original structural trap. While no gain may result from applying the acquired seismic data to the structural trap, reuse of the data can help evaluate the potential EV for the satellite stratigraphic trap. This aspect is also a positive contribution.

Rationale 5. Acquiring more seismic data may identify new opportunities that were totally unanticipated—a serendipity factor. There is usually little, if any, change in *EV* prior to the acquisition of the seismic data, but the identification of new, potentially lucrative, unanticipated opportunities makes the seismic acquisition a positive attribute.

To be sure, some of these aspects of the value of acquiring more seismic data have been addressed elsewhere; in particular, Withers (1992) has provided an excellent review of the increase in knowledge of reservoir geophysics for source, seal, structure, and trap as a consequence of the addition of more seismic information. Withers's paper is strongly recommended reading.

From this seismic acquisition illustration, it would appear that the value of added information has three essential, but independent, components:

1. To increase the expected value without increasing the uncertainty on the *EV* beyond the point where the difference between the updated *EV* and the prior *EV* is smaller than the initial uncertainty spread (otherwise, little is accomplished)
2. Either to decrease or leave constant the *EV*, but to decrease sufficiently the uncertainty associated with the new *EV* that the cost of the added information is considered of positive worth
3. To do little to nothing to the *EV* or its uncertainty for a particular opportunity, but to be of use either in later opportunities, in generating new opportunities, or in identifying better alternatives because of completely unanticipated results arising from the data acquisition

The following sections consider *quantitative descriptors* for these three components in turn.

II. INCREASED EXPECTED VALUE AND DECREASED UNCERTAINTY

A. QUANTITATIVE ANALYSIS

Suppose that, before acquiring any further information, an exploration opportunity has a cost, C_0, potential gains, G_0, and a success probability, p_0 (and failure probability $p_{f0} = 1 - p_0$), yielding an expected value

$E_0 = p_0 G_0 - C_0$, and a variance σ_0^2, of the expected value at $\sigma_0^2 \equiv p_0(1-p_0)G_0^2$. If an amount A is paid to acquire additional information, the values are changed to a success probability of p_1 (and $p_{f1} = 1 - p_1$), gains of G_1, and direct costs of C_1, respectively, yielding a new expected value of $E_1 = p_1 G_1 - C_1$, with corresponding variance $\sigma_1^2 = p_1(1-p_1)G_1^2$. The twin overriding requirements of (1) increased EV and (2) decreased uncertainty then become

$$A \equiv E_1 - E_0 = p_1 G_1 - C_1 - (p_0 G_0 - C_0) \geq 0 \qquad (9.2a)$$

and $\sigma_1^2 \leq \sigma_0^2$ (less uncertainty on E_1 than on E_0), which can be written

$$p_1(1-p_1)G_1^2 \leq p_0(1-p_0)G_0^2. \qquad (9.2b)$$

When inequality (9.2b) is in force, the subordinate requirement that (3) the increase in *EV* exceed the standard error range prior to acquiring the new data then demands

$$(E_1 - E_0)^2 \geq p_0(1-p_0)G_0^2. \qquad (9.2c)$$

Consider first the two overridge requirements (9.2a) and (9.2b). If inequality (9.2b) (smaller variance) is to be obeyed, then

$$G_1 = \lambda_1 G_0 [(p_0 p_{f0})/(p_1 p_{f1})]^{1/2}, \qquad (9.3)$$

where $\lambda_1 \leq 1$.

Use expression (9.3) in inequality (9.2a) to write

$$p_1 G_1 \geq E_0 + C_1 \qquad (9.4a)$$

in the form

$$(p_1/p_{f1})^{1/2} \geq (E_0 + C_1)(p_0 p_{f0})^{-1/2} \lambda_1^{-1} G_0^{-1}, \qquad (9.4b)$$

which can be rewritten as

$$p_1 = \lambda_2 \lambda_1^2 (E_0 + C_1)^2 / [\sigma_0^2 \lambda_1^2 + (E_0 + C_1)^2], \qquad (9.5)$$

where $\lambda_2 \geq 1$. Because $p_1 \leq 1$, one also has the requirement $\lambda_2 \leq \lambda_1^{-2}[\sigma_0^2 \lambda_1^2 + (E_0 + C_1)^2]/(E_0 + C_1)^2 \equiv \lambda_*$. The twin requirements $\lambda_* \geq \lambda_2 \geq 1$ are automatically satisfied when $\lambda_1 \leq 1$. Then, using Eq. (9.5) for p_1 in Eq. (9.3), one has

$$G_1 = G_0(p_0 p_{f0})^{1/2}[\sigma_0^2 \lambda_1^2 + (E_0 + C_1)^2]\lambda_2^{-1/2}(E_0 + C_1)^{-1}$$
$$\times [\sigma_0^2 \lambda_1^2 + (E_0 + C_1)^2 - \lambda_2 \lambda_1^2(E_0 + C_1)^2]^{-1/2}. \qquad (9.6)$$

Now, because $\lambda_* \geq \lambda_2 \geq 1$, it follows that requirement (9.6) can be written

$$\lambda_1^2[(E_0 + C_1)^2 - \sigma_0^2] \leq (E_0 + C_1)^2. \qquad (9.7)$$

If $\sigma_0^2 > (E_0 + C_1)^2$ then inequality (9.7) is automatically valid because the left-hand side is negative while the right-hand side is positive. If

$\sigma_0^2 < (E_0 + C_1)^2$ (the most common situation in exploration opportunities), then inequality (9.7) is valid if and only if

$$\lambda_1^2 \leq 1, \text{ which is automatically true.} \tag{9.8}$$

The general conclusion, then, is that (1) only if $1 \leq \lambda_2 \leq \lambda_*$ and $1 \geq \lambda_1$ [for $\sigma_0^2 < (E_0 + C_1)^2$] can one maintain the twin override requirements of $E_1 \geq E_0$ and $\sigma_1 \leq \sigma_0$; and (2) for $\sigma_0^2 > (E_0 + C_1)^2$ then any $\lambda_1 \leq 1$, and $1 \leq \lambda_2 \leq \lambda_*$, must both be obeyed if $E_1 \geq E_0$ and $\sigma_1 \leq \sigma_0$. And when the override requirements are satisfied, then p_1 is given by Eq. (9.5) and G_1 by Eq. (9.6). Because there are ranges of values allowed for λ_1 and λ_2, it follows that p_1 and G_1 likewise have ranges of values that will satisfy both $E_1 \geq E_0$ and $\sigma_1 \leq \sigma_0$. The *worst* situation that can occur is when the mean value is not improved (and so $E_1 = E_0$) and the uncertainty is not lowered (and so $\sigma_1 = \sigma_0$). Both of these equalities are satisfied with $\lambda_1 = 1 = \lambda_2$. Any larger values of λ_1^{-1} and λ_2 will either increase E_1 in excess of E_0, or diminish σ_1 below σ_0, or both.

When the override requirements are in force, one can then consider possible satisfaction of the subordinate requirement (9.2c), which can be written

$$E_1 \geq E_0 + \sigma_0. \tag{9.9}$$

Hence, if it is required that $\sigma_1 \leq \sigma_0$ and also that inequality (9.9) be in force simultaneously, then a similar argument to that just employed yields the requirements

$$p_1 = \Lambda_2 \Lambda_1^2 (E_0 + \sigma_0 + C_1) / [\sigma_0^2 \Lambda_1^2 + (E_0 + \sigma_0 + C_1)^2] \tag{9.10a}$$

and

$$G_1 = G_0 (p_0 p_{f0})^{1/2} [\sigma_0^2 \Lambda_1^2 + (E_0 + \sigma_0 + C_1)^2] \Lambda_2^{-1/2} (E_0 + \sigma_0 + C_1)^{-1} \\ \times [\sigma_0^2 \Lambda_1^2 + (E_0 + \sigma_0 + C_1)^2 - \Lambda_2 \Lambda_1^2 (E_0 + \sigma_0 + C_1)^2]^{-1/2}, \tag{9.10b}$$

with

$$1 \leq \Lambda_2 \leq \Lambda_1^{-2} [\sigma_0^2 \Lambda_1^2 + (E_0 + \sigma_0 + C_1)^2] / (E_0 + \sigma_0 + C_1)^2 \equiv \Lambda_{2*} \tag{9.11a}$$

and

$$1 \geq \Lambda_1; \tag{9.11b}$$

otherwise, one cannot have both $E_1 \geq E_0 + \sigma_0$ and $\sigma_1 \leq \sigma_0$ satisfied simultaneously. Thus the subordinate condition (9.2c) places a greater restriction on the range of variability than does the override condition (9.2a). The worst case is when $E_1 = E_0 + \sigma_0$ and $\sigma_1 = \sigma_0$, which occurs when $\Lambda_1 = \Lambda_2 = 1$, corresponding to

$$p_{f1}/p_1 = \{\sigma_0^2 / (E_0 + \sigma_0 + C_1)^2\}, \tag{9.12a}$$

which is a stronger requirement (higher p_1 required) on the expected success probability, p_1, than the override requirement $E_1 \geq E_0$, which only needs a larger minimum (lower p_1 required) of

$$p_{f1}/p_1 = \sigma_0^2/(E_0 + C_1)^2. \tag{9.12b}$$

Because λ_1, λ_2, and Λ_2 are arbitrary within their limiting ranges, estimating the worth of acquiring additional data to satisfy the constraints $E_1 \geq E_0 + \sigma_0$ and $\sigma_1 \leq \sigma_0$ can be expressed in one of two ways: (1) Either one specifies the anticipated improvements from the data, tantamount to specifying the values of p_1, G_1, and C_1, and so one computes the expected worth $\langle A \rangle$ of the data from

$$\langle A \rangle = E_1 - E_0 \tag{9.13a}$$

and, at the same time, one computes the uncertainty variance in the worth, $\langle \delta A^2 \rangle$, relative to E_0 as a fixed base of comparison, from

$$\langle \delta A^2 \rangle \equiv \langle A^2 \rangle - \langle A \rangle^2 \equiv p_1(1 - p_1)G_1^2; \tag{9.13b}$$

or (2) one allows for uncertainty in p_1, G_1, and C_1, and so performs a series of Monte Carlo calculations to determine cumulative probable worth distributions from the intrinsic uncertainties in p_1, G_1, and C_1.

In either event both the expected worth and its uncertainty are then determined. Suppose the additional data to be acquired are *perfect*, in the sense that such data resolve absolutely the opportunity in a statistically sharp manner, to the point at which there is no longer any probability of failure (i.e., $p_1 = 1$), but without changing gains or costs so that $G_1 = G_0$, $C_1 = C_0$. Then one has $E_1 = G_0 - C_0$ and $\sigma_1 = 0$, so that the requirement $\sigma_1 < \sigma_0$ is automatically obeyed and the subordinate requirement $E_1 \geq E_0 + \sigma_0$ reduces to

$$G_0 p_{f0}\{1 - (p_0/p_{f0})^{1/2}\} \geq C_0. \tag{9.14}$$

For $p_0 < p_{f0}$ (the usual situation in hydrocarbon exploration opportunities where success probability estimates are usually much less than failure probability estimates) inequality (9.14) can be satisfied, but for opportunities initially estimated to be likely successful (i.e., with $p_0 > p_{f0}$), inequality (9.14) cannot be satisfied. In this second situation (i.e., when $p_0 \geq 50\%$) it is not possible to both *decrease* uncertainty and *increase* expected value enough ($E_1 \geq E_0 + \sigma_0$) to satisfy the rationale for acquiring more information, although the less restrictive requirement $E_1 > E_0$ is satisfied.

In terms of return for dollar invested, the original opportunity evaluation called for an investment of C_0 to produce an expected value $E_0 = p_0 G_0 - C_0$, for an average return per dollar invested of

$$R_0 = p_0 G_0/C_0 - 1, \tag{9.15a}$$

which is positive when $p_0 G_0 \geq C_0$. If one invests an amount A in acquiring new data, thereby changing gains to G_1, direct costs to C_1 (and so total costs to $C_1 + A$), and success probability to p_1, then the new average return per total dollar invested is

$$R_1 = p_1 G_1/(C_1 + A) - 1, \quad (9.15b)$$

which is positive when

$$p_1 G_1 - C_1 \geq A, \quad (9.16a)$$

which requires $p_1 G_1 \geq C_1$ in order to be satisfied at all. The average return per investment dollar is increased by the data acquisition when $R_1 \geq R_0$, which condition can be written

$$A \leq \left(\frac{p_1 G_1}{p_0 G_0}\right) C_0 - C_1. \quad (9.16b)$$

For instance, if gains do not change ($G_1 = G_0$) and direct costs remain the same ($C_1 = C_0$), and if the data acquired provided perfect information ($p_1 = 1$) then

$$A \leq C_0 p_{f0}/p_0 \equiv A_{\max}. \quad (9.16c)$$

But if one indeed attempted to expend the maximum worth on acquiring new data then, because it is also required that $p_1 G_1 - C_1 \geq A$, it follows that with $p_1 = 1$, $G_1 = G_0$, $C_1 = C_0$, and $A = A_{\max} = C_0 p_{f0}/p_0$, then $p_1 G_1 - C_1$ exceeds $A_{\max}$ when $G_0 \geq C_0 p_0^{-1}$, which keeps $E_0 \geq 0$. If $G_0 \leq C_0 p_0^{-1}$ one cannot expend the *maximum* amount, $A_{\max}$, on new perfect data (retaining $p_1 = 1$, $G_1 = G_0$, $C_1 = C_0$) in order to convert a losing proposition ($E_0 < 0$ and $R_0 < 0$) into a profitable opportunity ($E_1 > 0$ and $R_1 > 0$). To obtain $R_1 > 0$ then requires

$$A \leq G_0 - C_0, \quad (9.17a)$$

which yields a positive value for A as long as $G_0 > C_0$, and A approaches the limit $A_{\max}$ from below as G_0 increases from C_0 toward $C_0 p_0^{-1}$.

If one chooses to limit A to no more than $p_{f0} C_0$ (so that $A = \varepsilon p_{f0} C_0$ with $\varepsilon \leq 1$), and if $G_1 = G_0$, $C_1 = C_0$, then R_1 will not automatically exceed R_0, but only do so when

$$p_1 \geq p_0(1 + p_{f0}\varepsilon). \quad (9.17b)$$

In this case one does not pay enough ($A \leq p_{f0} C_0$) to acquire enough perfect data to *force* p_1 to unity independently of all other considerations although, serendipitously, it can happen that p_1 will be unity.

B. NUMERICAL ILLUSTRATION

Suppose an exploration opportunity has been assessed as having $p_0 = 0.25$, $G_0 = \$3$ MM, and estimated costs of $C_0 = \$0.5$ MM. Then one has an expected value of $E_0 = \$0.25$ MM, and a standard error on E_0 of $\sigma_0 =$

$3^{3/2}/4$ MM ≡ $1.3 MM so that there is considerable uncertainty in the estimated expected value ($\sigma_0/E_0 \cong 5.2$). Analysis of available data suggests that a seismic survey, sampled at 2 ms, might be able to isolate better leakage regions from the reservoir and so increase the geologic success chance to $p_1 \cong 0.3$ (a 20% improvement in the success chance). However, drilling costs are not anticipated to improve ($C_1 = C_0$), and it is not clear whether expected gains, G_1, will be larger or smaller than the gains, G_0 (≡$3 MM), assessed prior to the new seismic survey.

If the expected value is to increase as required, then the relation between amount A (in $MM) to expend on the survey and the new gains, G_1 (in $MM), is given by Eq. (9.2a) as

$$A = 0.3G_1 - 0.5 - (0.25 \times 3 - 0.5),$$

that is,

$$A = 0.3G_1 - 0.75, \tag{9.18a}$$

which requires that $G_1 \geq $2.5 MM in order to be satisfied at all.

If the uncertainty on the new expected value E_1 is to be less than on the assessment of the opportunity prior to authorizing the seismic survey, then inequality (9.2b) requires

$$G_1 \leq G_0[p_0(1-p_0)/\{p_1(1-p_1)\}]^{1/2} = \$2.83 \text{ MM}. \tag{9.18b}$$

Thus, the twin requirements of increased expected value *and* lower variance force the anticipated gains after the survey to lie in the range

$$\$2.5 \text{ MM} \leq G_1 \leq \$2.83 \text{ MM}. \tag{9.18c}$$

If G_1 is less than $2.5 MM (a reduction of $0.5 MM from the presurvey assessment), then it is impossible to recover the cost of the survey from the estimated increase in success probability; whereas if $G_1 \geq $2.83 MM then while the costs of the survey can be recovered *on average*, the uncertainty on the exploration opportunity has increased ($\sigma_1 > \sigma_0$) relative to the uncertainty prior to the survey.[2] To honor both higher *EV* and lower uncertainty means that the maximum amount one should spend on the survey occurs when $G_1 = $2.83 MM, and then

$$A_{\max} = \$0.1 \text{ MM}. \tag{9.19}$$

The *total* costs of the opportunity are then $C_1 + A_{\max} = $0.6 MM, whereas $p_1G_1 = $0.85 MM, so the new expected value, *including survey costs*, is $E_1 = p_1G_1 - C_1 - A_{\max} = $0.25 MM, precisely the same as prior to the survey and the uncertainty, σ_1, is then the same as $\sigma_0 = $1.3 MM.

Thus, only if the survey costs less than $0.1 MM is it worthwhile authorizing under the joint requirements of higher *EV* and lower variance. In

---

[2] One has to ensure that lower variance is, indeed, a rational objective, or else one can end up in the strange situation of preferring *not* to acquire data that could increase the estimate of gain!

addition, the maximum survey costs of $0.1 MM depend critically on the assessed chance of success increasing to 0.3 from 0.25. Other estimates of the increase in success chance change the maximum one should pay for the survey.

If one wanted to ensure that the new expected value was sufficiently high ($\geq E_0 + \sigma_0$) after the survey compared to the assessment before the survey, then inequality (9.9) requires

$$A \leq p_1 G_1 - C_1 - (p_0 G_0 - C_0 + \sigma_0). \tag{9.20a}$$

With $p_1 = 0.3$ and $C_1 = C_0$, one has

$$A \leq 0.3 G_1 - 0.75 - 1.3 \equiv 0.3 G_1 - 2.03. \tag{9.20b}$$

In this case one requires a minimum value of $G_1 \geq \$2.03/0.3$ MM $\equiv \$6.77$ MM in order to have any value at all ($A \geq 0$) for performing the survey. But, because one also wishes to ensure a lower variance after the survey, which requires $G_1 \leq \$2.83$ MM, both of the constraints *cannot* be satisfied simultaneously with the estimated increase in success probability only as high as 0.3 (from the presurvey estimate of 0.25).

If one also allows p_1 to be unknown, then the requirements $E_1 \geq E_0 + \sigma_0$ [inequality (9.9a)] and $\sigma_1 < \sigma_0$ [inequality (9.2b)] yield the constraints (for $C_1 = C_0$)

$$0 \leq A \leq p_1 G_1 - 2.03 \tag{9.21a}$$

and

$$p_1(1 - p_1) G_1^2 \leq 2.69. \tag{9.21b}$$

Because one requires $p_1 G_1 \geq 2.03$ [from inequality (9.21a)] in order to have any hope at all that the survey is worthwhile, it follows from inequality (9.21b) that one then requires

$$(2.03)^2 \leq (p_1 G_1)^2 \leq 2.69 p_1/(1 - p_1). \tag{9.21c}$$

The outer pair of inequalities (9.21c) require that $p_1 \geq 0.6$ in order to be satisfied at all, and then one requires $G_1 \geq \$(2.03/p_1)$ MM. At a *minimum* value for p_1 of 0.6, the *maximum* G_1 required is $\$3.38$ MM, while at the *minimum* G_1 of $\$2.03$ MM, one requires $p_1 = 1$, that is, perfect information must result from the survey.

Thus, prior to authorizing the survey the technical assessment of requiring an improvement in the success probability to greater than 0.6 (from 0.25) has to be in force, a difficult thing to arrange practically.

The alternative viewpoint is to ask whether a survey at some preordained cost is worthwhile doing. For instance, if $0.2 MM is to be spent on a survey, what increases in gains and success chance are required of the survey? In this case one would work inequalities (9.21a) and (9.21b) slightly differently. With $A = \$0.2$ MM, inequality (9.21a) requires $p_1 G_1 \geq 2.23$. Then, combining this inequality with the lower variance statement of inequality (9.21b), it follows that

$$(2.23)^2 \leq (p_1 G_1)^2 \leq 2.69 p_1/(1 - p_1). \tag{9.22}$$

Thus, a minimum improvement is required in success probability to $p_1 \geq 0.65$ if \$0.2 MM is to be spend on a survey.

III. CONSTANT OR DECREASED EXPECTED VALUE AND DECREASED UNCERTAINTY

A. QUANTITATIVE ANALYSIS

If the expenditure of an amount A on acquiring new data either lowers or does not change the EV, but does decrease the uncertainty in the EV, then the classical measure of expected return per dollar invested would indicate that one has not improved the opportunity. Yet, a more precise determination (lower variance) of a smaller expected return may be preferable to a less well-determined (higher variance) estimate of a higher expected value for an opportunity, because the uncertainty is reduced. In such a case one needs a different measure of worth for acquiring more data.

Four measures are readily available: (1) volatility-weighted expected value, (2) bias-weighted expected value, (3) volatility-weighted gain per investment dollar, (4) bias-weighted gain per investment dollar. Consider each in turn.

The volatility, ν, is customarily defined by

$$\nu = \frac{1}{2}\{\exp(\mu) - \exp(-\mu)\} = \sinh(\mu), \tag{9.23}$$

where $\mu^2 = \ln(1 + \sigma^2/E^2)$, where σ^2 is the variance ($\sigma^2 = p_s p_f G^2$) and E is the mean value ($E = p_s G - C$). This definition of volatility is precise if the cumulative probability of the opportunity having a value less than or equal to V (with mean E) is exactly log-normally distributed.

One measure of volatility-weighted expected value is

$$M_1 = E/(1 + \nu^2)^{1/2}, \tag{9.24}$$

which, for a precisely log-normally distributed situation, corresponds to the 50% cumulative probability value.

Bias-weighting for a precisely log-normally distributed value is customarily defined by[3]

$$\nu_+/\nu_- = [P(84) - P(50)]/[P(50) - P(16)], \tag{9.25}$$

where $P(84)$, $P(50)$, and $P(16)$ are the cumulative probability values at 84, 50, and 16%, respectively. For an exactly log-normal distribution

---

[3] In the same notation the volatility, ν, is given by $\nu = [P(84) - P(16)]/P(50)$.

one has
$$v_+/v_- = \exp(\mu). \tag{9.26}$$
Thus, a bias-weighted measure of expected value is
$$M_2 = (v_+/v_-)E, \tag{9.27a}$$
which provides a measure of the expectation of upside potential relative to downside risk.

Equally, a volatility-weighted measure of gain per investment dollar is provided by
$$M_3 = \{(p_s G - C)/C\}/(1 + v^2)^{1/2}, \tag{9.27b}$$
while a correspondingly bias-weighted measure is
$$M_4 = \{(p_s G - C)/C\}(v_+/v_-). \tag{9.28}$$
One can obviously generalize to define a measure that is both volatility and bias-weighted through
$$M_5 = \{(p_s G - C)/C\}(v_+/v_-)/(1 + v^2)^{1/2}. \tag{9.29}$$
The point to make about these four (or five) measures is that they allow for lowering of expected value with data acquisition while at the same time decreasing uncertainty.

In a general sense, for any of the measures M_1–M_5, the worth of the data acquisition, A, is expressed through the difference
$$(M_{i,1} - M_{i,0}) \geq 0 \quad \text{for } i = 1\text{–}5, \tag{9.30}$$
where $M_{i,0}$, $M_{i,1}$ are the measures pre- and postacquisition, respectively.

For instance, consider that both $\sigma_1^2/E_1^2 \ll 1$ and $\sigma_0^2/E_0^2 \ll 1$, so that the respective volatilities, v, can be written through
$$v_0 \equiv \sinh(\mu_0) \cong \mu_0 \cong \sigma_0^2/E_0^2 \tag{9.31a}$$
and
$$v_1 \cong \sigma_1^2/E_1^2, \tag{9.31b}$$
so that
$$M_1 = E\left(1 - \frac{1}{2}\sigma^2/E^2\right). \tag{9.31c}$$
Hence,
$$A \equiv M_{1,1} - M_{1,0} \cong (E_1 - E_0) - \frac{1}{2}\left(\frac{\sigma_1^2}{E_1} - \frac{\sigma_0^2}{E_0}\right). \tag{9.32}$$
With $E_1 - \varepsilon_1 - A$, where $\varepsilon_1 = p_1 G_1 - C_1$, Eq. (9.32) can be written
$$2A = (\varepsilon_1 - E_0) - \frac{1}{2}\left(\frac{\sigma_1^2}{\varepsilon_1 - A} - \frac{\sigma_0^2}{E_0}\right), \tag{9.33a}$$

which can be put in the quadratic form

$$2X^2 - X\left[\varepsilon_1 + E_0 - \frac{1}{2}\sigma_0^2/E_0\right] - \frac{1}{2}\sigma_1^2 = 0, \qquad (9.33b)$$

where $A = \varepsilon_1 - X$. Equation (9.33b) has the relevant solution

$$A = \varepsilon_1 - X \cong \frac{1}{2}\left\{\varepsilon_1 - E_0 + (E_0 + \varepsilon_1)^{-1}\left[-\sigma_1^2 + \frac{1}{2}\sigma_0^2(1 + \varepsilon_1/E_0)\right]\right\}, \qquad (9.34)$$

showing that if $\varepsilon_1 \geq E_0$ and $\sigma_1^2 \leq \frac{1}{2}\sigma_0^2 (1 + \varepsilon_1/E_0)$, then there is worth in acquiring new data. Even if $0 < \varepsilon_1 < E_0$ there is *still* worth in acquiring new data ($A > 0$) as long as

$$\sigma_0^2 \geq 2[\sigma_1^2 + (E_0^2 - \varepsilon_1^2)](1 + \varepsilon_1/E_0)^{-1} \equiv \sigma_*^2; \qquad (9.35)$$

that is, as long as there is sufficient uncertainty ($\sigma_0 > \sigma_1$) on the preacquisition opportunity estimates, even though the postacquisition estimate expresses a lower expected value ε_1 (or E_1) than prior to data acquisition.

Similar arguments can be made for each of the measures of worth even when the variances may be large compared to the mean values, although solution of the relevant equations for the worth, A, of the data acquisition is not then so easy because of the nonlinearity. The case of low variance displayed here simplifies the analysis without losing any of the sense of the general argument.

To illustrate how to use the measures, consider than an amount A is to be spent on acquiring new data, but that the new expected value is such that $E_1 < E_0$ with, however, a lowering of uncertainty, that is, $\sigma_1 < \sigma_0$. We examine in detail the consequences using the M_3 measure.

Prior to acquisition, M_3 takes on the value

$$M_{3,0} = \{(p_0 G_0 - C_0)/C_0\}(1 + v_0^2)^{-1/2} \qquad (9.36a)$$

with

$$v_0 = \frac{1}{2}\{\exp(\mu_0) - \exp(-\mu_0)\} \qquad (9.36b)$$

and

$$\mu_0^2 = \ln\{1 + p_0(1 - p_0)G_0^2/(p_0 G_0 - C_0)^2\}. \qquad (9.36c)$$

After acquisition the corresponding value is

$$M_{3,1} = \{(p_1 G_1 - (C_1 + A)\}/(C_1 + A)](1 + v_1^2)^{-1/2} \qquad (9.37a)$$

with

$$v_1 = \frac{1}{2}\{\exp(\mu_1) - \exp(-\mu_1)\} \qquad (9.37b)$$

and

$$\mu_1^2 = \ln[1 + p_1(1 - p_1)G_1^2/(p_1G_1 - C_1 - A)^2]. \tag{9.37c}$$

The requirement for satisfaction of the rationale for data acquisition is that

$$M_{3,1} - M_{3,0} \geq 0, \tag{9.38}$$

with $E_1 \leq E_0$, $\sigma_1 \leq \sigma_0$, and $\sigma_1/E_1 < \sigma_0/E_0$, so that the value of E_1 ($\equiv p_{s1}G_1 - C_1 - A$) is better defined than is E_0. The worth, A, of data acquisition is then determined by the range of values of A satisfying Eq. (9.34), which is a nonlinear equation for A. The question is this: How is the worth A of the data expressed in terms of parameters of the opportunity:

From the requirement $E_1 \leq E_0$ one can already set the limit

$$A \geq (p_1G_1 - C_1) - (p_0G_0 - C_0). \tag{9.39}$$

Notice that this requirement implies that the expected value $(p_1G_1 - C_1)$ of the opportunity (in the absence of any cost for acquiring the new data) should exceed the original expected value, E_0. There is also the requirement $A \leq p_1G_1 - C_1$ in order to keep $E_1 \geq 0$.

From the requirement $\sigma_1 \leq \sigma_0$ one can set the constraint

$$p_1(1 - p_1)G_1^2 \leq p_0(1 - p_0)G_0^2, \tag{9.40a}$$

implying, for instance, that the estimated gains, G_1, after acquiring the data would be less than G_0 prior to the data acquisition for the same success probability pre- and postacquisition.

From the requirement $\sigma_1/E_1 < \sigma_0/E_0$ one can write the constraint

$$A \leq p_1G_1 - C_1 - E_0\sigma_1/\sigma_0, \tag{9.40b}$$

which is automatically satisfied when both $\sigma_1 < \sigma_0$ and when inequality (9.39) is obeyed.

Then, the requirement that $M_{3,1} \geq M_{3,0}$ (a lower expected value per cost after acquisition but a higher volatility-weighted expected value per cost) can be written in the form

$$A \leq \varepsilon_1 - \sigma_1/[\exp\{u(A)^2\} - 1]^{1/2}, \tag{9.41a}$$

where

$$u(A) = \ln[\Lambda + (\Lambda^2 - 1)^{1/2}], \tag{9.41b}$$

with

$$\Lambda = \{(\varepsilon_1 - A)/E_0\}\cosh\mu_0 \geq 1, \tag{9.41c}$$

where $\varepsilon_1 = p_1G_1 - C_1$, and with $\mu_0^2 = \ln(1 + \sigma_0^2/E_0^2)$.

To obtain the limiting value of A, use the parametric representation

$$A = \varepsilon_1 - \{E_0\Lambda/\cosh(\mu_0)\} \tag{9.42a}$$

to write

$$\Lambda \geq \left(\frac{\sigma_1}{E_0}\right)\cosh(\mu_0)[\exp[(\ln\{\Lambda + (\Lambda^2 - 1)^{1/2}\})^2] - 1]^{-1/2} \quad (9.42b)$$

and $\Lambda \geq 1$. Hence, it is certain that

$$A \leq A_{\max} = \varepsilon_1 - E_0/\cosh(\mu_0). \quad (9.43)$$

If the data acquisition cost, A, is only a small fraction of the expected value, ε_1 ($\equiv p_1 G_1 - C_1$) then a useful method for estimating A with any of the measures, M_i, is as follows. Write

$$M_{i,1}(A) - M_{i,0} \geq 0, \quad (9.44a)$$

$$\cong M_{i,1}(0) - M_{i,0} + A\left(\frac{\partial M_{i,1}(A)}{\partial A}\bigg|_{A=0}\right) \geq 0. \quad (9.44b)$$

Then by rearrangement of Eq. (9.44b) one has

$$A \lesssim [M_{i,1}(0) - M_{i,0}]\left[-\frac{\partial M_{i,1}(A)}{\partial A}\bigg|_{A=0}\right]^{-1}, \quad (9.45)$$

where $M_{i,1}(0)$ is $M_{i,1}(A)$ evaluated on $A = 0$.

One does have to check (1) that $A \ll \varepsilon_1$ and (2) that $A \geq 0$ if the acquisition is to be worthwhile, but as a practical matter the approximation is useful.

B. NUMERICAL ILLUSTRATION

Consider the same parameters for an exploration opportunity as in the previous numerical illustration, namely, $p_0 = 0.25$, $G_0 = \$3$ MM; $C_0 = \$0.5$ MM, $E_0 = \$0.25$ MM, $\sigma_0 = \$1.3$ MM. Also again assume that the direct drilling costs will not change after new data acquisition so that $C_1 = C_0$ and, at the moment, let the updated estimate of gains, G_1, and the updated success probability, p_1, be unspecified, so that $E_1 = p_1 G_1 - C_0 - A$, $\sigma_1^2 = p_1(1 - p_1)G_1^2$.

The volatility, ν_0, of the opportunity prior to data acquisition is given through $\nu_0 = 2.32$, so that the initial M_3 measure is

$$M_{3,0} = 0.5\{1 + (2.32)^2\}^{-1/2} = 0.198. \quad (9.46)$$

The corresponding $M_{3,1}$ measure, after paying an amount A for the additional data, is

$$M_{3,1}(A) = [(p_1 G_1 - 1/2 - A)/(A + 1/2)](1 + \nu_1^2)^{-1/2} \quad (9.47)$$

with

$$\nu_1 = \sinh\mu_1, \text{ and } \mu_1^2 = \ln\left\{1 + \frac{p_1(1 - p_1)G_1^2}{(p_1 G_1 - A - 1/2)^2}\right\}.$$

The difference between $M_{3,1}(A)$ and $M_{3,0}$ gives the maximum amount one should pay for extra data under the M_3 measure control, that is, those A values for which

$$M_{3,1}(A) - M_{3,0} \geq 0. \tag{9.48}$$

Use the Taylor series expansion of $M_{3,1}(A)$ around $A = 0$ with inequality (9.48) to obtain

$$A \lesssim [2p_1 G_1 - 1 - 0.198 \cosh\mu_*] \\ \times [4p_1 G_1 + \mu_*^{-1} \sinh\mu_* \exp(-\mu_*^2)(p_1 G_1 - 1/2)^2]^{-1}, \tag{9.49}$$

where μ_* is just μ, evaluated on $A = 0$.

Inequality (9.49) requires that

$$p_1 G_1 \geq 0.5 + 0.099 \cosh\mu_* \geq 0.599$$

in order to be satisfied at all, while the requirement $0 \leq E_1 \leq E_0$ requires

$$p_1 G_1 - 0.75 \leq A \leq p_1 G_1 - 0.5. \tag{9.50}$$

Thus, the override requirement to make it worthwhile spending any money at all to acquire new data is

$$p_1 G_1 \geq 0.599. \tag{9.51}$$

The absolute maximum of the right-hand side of inequality (9.49) is 0.5, so that under *no* conditions should one spend more than $0.5 MM to acquire more data while still attempting to preserve the constraints $E_1 < E_0$ and $\sigma_1 < \sigma_0$. If it is anticipated that any extra data would increase the success probability to $p_1 = 0.8$ but would also lower assessed gains to $G_1 = \$0.75$ MM, then $\mu_* = 1.76$, so that, from inequality (9.49), it follows that $A < 0$. Thus, while both $E_1 < E_0$ and $\sigma_1 < \sigma_0$, in this case it is not worthwhile spending any money to acquire further data under the M_3 measure.

The point here is that each measure of worth of spending money will provide some criterion of the usefulness of acquiring information under particular corporate philosophies, which is why more than one measure is available when it is anticipated that $E_1 < E_0$ and $\sigma_1 < \sigma_0$. Each corporation acts and reacts differently to its perceived needs and desires.

IV. UNANTICIPATED BENEFITS OF DATA ACQUISITION

Occasionally, data acquisition is undertaken at some cost, CD, which ends up having little to no influence on the opportunity, in that neither the EV nor the variance of the opportunity is significantly improved by the acquisition. But when the data are analyzed, a new opportunity is identified that was completely unanticipated. In that case the data obviously have some worth.

The simplest measure of worth of the data is just the difference between the expected value, E_1, of the newly identified opportunity and the value (zero) that would have prevailed without the data, that is,

$$A = E_1. \tag{9.52}$$

Equally, if any of the measures M_1–M_5 are used as identifiers of worth, then each M value is set to zero in the absence of the data, so that the worth of the previously acquired data is determined by the values of $M_{i,1}(A)$ such that

$$A = M_{i,1}(A) \geq 0, \tag{9.53}$$

which, for values of A that are small compared to the expected value $\varepsilon_1(\equiv p_1 G_1 - C_1)$, can be written

$$A \lesssim M_{i,1}(0)/[-(\partial M_{i,1}(A)/\partial A)_{A=0}]. \tag{9.54}$$

Thus, when an unanticipated opportunity arises as a consequence of data acquisition, an amount A of the total data acquisition cost, CD, can be written off against the value of the opportunity. An ideal circumstance is one in which $A \geq CD$ so that all costs of acquisition are recovered and the serendipitously discovered opportunity then also yields a profit. A less ideal, but still acceptable, circumstance is one in which only a fraction of the total acquisition costs can be written off ($A < CD$) because the effective loss to the corporation is thereby reduced. Thus, it may then be worthwhile to develop and produce the unanticipated opportunity, because the corporation will lose less money by doing so than it would by ignoring the unanticipated opportunity.

10

COUNTING SUCCESSES AND BIDDING STATISTICS ANALYSES

I. INTRODUCTION

Historical precedent is one of the major components that goes into scientific and economic projections of the worth of an exploration opportunity. Oil field statistics in terms of number found and volume distributions, production statistics in terms of pump rates with time and depletion estimates to ultimate recoverable reserves, economic statistics in terms of selling price fluctuations with time, ratios of gains to costs with time, etc., are all used in attempts to assess different aspects of the likely potential of an exploration opportunity.

In addition, petrophysical information (permeability versus porosity rules for individual facies) and organic carbon geochemical information (kinetic descriptions of kerogen burning to different hydrocarbon fractions) are also used in attempts to control the geologic risk.

Thus, the quantitative models developed for the scientific and economic components of an exploration risk assessment are beholden to a variety of input assumptions, to input parameters and their ranges of uncertainty, and to the quality, quantity, and frequency distribution of data used as control information in the models. Because of uncertainities in each of the components, any outputs from both the scientific models and from the economic models carry uncertainty. As has been seen in previous chapters, risk uncertainty and the relative importance of individual model factors in contributing to uncertainty play significant roles in determining whether the exploration opportunity should be undertaken and what probable range of working interest a corporation should consider. Yet the ranges of risk uncertainty are even more problematic than has so far been considered because of data analysis difficulties with historical precedents. Three case studies are presented here to illustrate some of the problems that have to be guarded against. First we consider the results for a corporation that appears to be improving its ratio of successful wells when in fact precisely the opposite is occurring. We next consider the results of Gulf of Mexico Lease Sale 147 for the 10 blocks on which the highest bids were recorded, illustrating the problems with different corporate assessments and different corporate strategies. Third we consider *all* of the results for Gulf of Mexico Lease Sale 157, illustrating that there is an enormous variation (at a very statistically significant level) in the way individual corporations perceive the worth of blocks.

Finally, a brief discussion of bidding in relation to estimated block worth is given, illustrating some of the complexities involved in deciding on a bid value.

II. CORPORATE SUCCESSES AND FAILURES: SIMPSON'S PARADOX

Corporations are paying considerable attention to their profitability in the present era of high costs and low selling prices for product, and are continually exhorting their divisions to improve the success ratio of exploration wells. Overall corporate performance is often summarized annually by collecting all data from each division and then providing an average success ratio each year for the wells drilled by the corporation. Such an analysis is, at best, misleading and, at worst, can be in complete contradiction with what is actually happening to the corporation.

Consider the Majil Oil Company[1] overall results for 1994 and 1995 as shown in Table 10.1. A rapid inspection of Table 10.1 would appear to indicate a major improvement in finding success, from 29 to 41%, which would seem to suggest a vibrantly healthy company. But the seeds of ruin are, in fact, planted in the company.

Consider in closer detail how the drilling activity has changed in the period 1994 through 1995. The Majil Oil Company is divided into two major divisions: the Existing Properties Division and New Properties Division, with the Existing Properties Division subdivided into an Exploitation Group and a Producing Properties Group, and with the New Properties Division subdivided into a Producing Basins Group and a New Basins Group as sketched in Table 10.2. Each of the four groups is functionally separate, with the New Properties Division subgroups being responsible

[1] Both the real name of the corporation and the actual statistics have been changed to maintain confidentiality but the point to be made is validly depicted.

TABLE 10.1 Corporate Summary

Year	Company total
1994	
Wells	100
Successes	29
Rate	29%
1995	
Wells	100
Successes	41
Rate	41%

for new exploration on new acreage in either known producing basins or in virgin basins, while the Existing Properties Division subgroups are responsible for enhancing existing acreage production.

Detailed data, which went into composing the corporate summary statement of Table 10.1, are given in Table 10.3 for both the Existing Properties Division and the New Properties Division over the 1994–1995 period. Perusal of Table 10.3 shows that *neither* division improved their success rates; both had the same success rates in 1995 as they had in 1994. The only change was the level of drilling activity per division, with the Existing Properties Division increasing its drilling activity from 30 wells in 1994 to 70 wells in 1995, and the New Properties Division decreased its drilling activity from 70 wells in 1994 to 30 wells in 1995.

Also available were the detailed data for each of the two groups of each division, and these data are recorded in Table 10.4 for the Existing

TABLE 10.2 Corporation Divisions

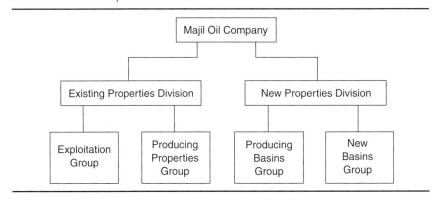

TABLE 10.3 Divisional Information

Year	Existing Properties Division	New Properties Division	Company total
1994			
Wells	30	70	100
Successes	15	14	29
Rate	50%	20%	29%
1995			
Wells	70	30	100
Successes	35	6	41
Rate	50%	20%	41%

Properties Division subgroups, and in Table 10.5 for the New Properties Division subgroups.

The detailed data of Table 10.4 show that *each* of the two subgroups had serious declines in performance between 1994 and 1995, so that these groups could not, apparently, be responsible for the dramatic improvement in success rate listed in Table 10.1 at the corporate summary level. But, when attention was therefore turned to the detailed data from the New Properties Division subgroups (Table 10.5), those subgroups also showed decreases in success rates between 1994 and 1995. Thus, all four of the groups show declining success rates, but the divisions show constant success rates—and the corporation as a whole shows a dramatic improvement in performance as measured by the success rate. This sort of contradictory phenomenon from data is known as *Simpson's paradox*, which is defined as the reversal of the direction of a comparison or association when data from several subsets are combined to form a single set.

TABLE 10.4 Group Information (Existing Properties Division)

Year	Exploitation Group	Producing Properties Group	Existing Properties Division
1994			
Wells	10	20	30
Successes	7	8	15
Rate	70%	40%	50%
1995			
Wells	40	30	70
Successes	25	10	35
Rate	63%	33%	50%

TABLE 10.5 Group Information (New Properties Division)

Year	Producing Basins Group	New Basins Group	New Properties Division
1994			
Wells	30	40	70
Successes	9	5	14
Rate	30%	13%	20%
1995			
Wells	20	10	30
Successes	5	1	6
Rate	25%	10%	20%

For the Majil Oil Company statistics presented above, the key to the apparent paradox is the clear shift from high-risk wells to low-risk wells. But, from a 1-year perspective, management still has to contend with the question of whether an improvement has occurred in 1995 relative to 1994. Is it that the data describe correctly a true improvement in corporate performance and only an apparent decline in performance of each group, or vice versa? In point of fact, the change in performance is an early warning sign of a long-term trend of declining performance. The groups will only be able to maintain the shift to lower risk wells for a while; eventually the division performances will reflect the group strategy, and then the corporate summary performance will be seen to decline precipitously. The long-term outlook for the Majil Oil Company is serious, even though the occasional shifting in categories will continue so that apparent improvements in results will be reported.

It is fortunate that the constant level of activity in the 1994 and 1995 years, and the functional separation of the exploration drilling categories, permitted a sharp focus to be made rapidly of the key to the problem. Groups were not drilling wells in the higher risk categories and were substituting wells in lower risk categories in attempts to meet success rate goals set at the corporate and division levels. This shift to a more risk averse strategy would have been less obvious in a geographically separated company, or in one with a variable level of drilling activity on a year-to-year basis, or in one without functional separation of exploration categories.

The lesson to be learned from this example is that Simpson's paradox has to be guarded against in assessing the fiscal health of a corporation. Directions of change can occur, particularly with rates, which are inconsistent between detailed and compiled data. Part of the problem is that corporations routinely set objectives on success rates and strive to maintain or improve such rates at the expense of other objectives, which may be of greater ultimate benefit to the corporation. Accordingly, Simpson's paradox is pervasively invading corporations on an almost routine basis. It becomes more difficult to disentangle the statistics to ascertain what is really happening and masks the long-term decline in performance of the corporation.

While the particular illustration here focused on success rates for drilling wells, it should be obvious that any time groups of data of any sort are combined and then compared and contrasted, one always faces the likelihood that the results of the comparison will fall within the bailiwick of Simpson's paradox. An old Chinese proverb remarks that "One should be careful of what one wishes to obtain, for the wish may come true." In the case of detailed and compiled data comparisons a restatement is "Beware of what one asks groups to achieve, for they very likely will do just that!"

III. BID ANALYSIS AND INFERRED CORPORATE STRATEGIES FOR LEASE SALE 147[2]

The U.S. Minerals Management Service (MMS) held Gulf of Mexico Lease Sale 147 on 30 March 1994. The sale attracted considerable attention because of the sub-salt play concept, demonstrated to be commercial through the Phillips' Mahogany prospect (AAPG Explorer, 1994).

The general idea of laterally extensive impermeable salt sheets acting as seals for underlying hydrocarbon accumulations has been around for a number of years, with a summary of physical conditions likely to be found under salt sheets provided in O'Brien and Lerche (1994).

The interest created by the sub-salt play concept at Lease Sale 147 led to interesting bidding ranges for different blocks and also to a significant number of bids overall, as well as to varied patterns of bidding per block.

The purpose of this section is to analyze the bid patterns of the 10 most expensive blocks in order to see what corporate strategies and/or distributions of bids by block were involved in the sale. Raw bid data provide the basis for computing statistical facts concerning the bidding behavior. Of concern in this section are several questions:

1. Overall bid statistical distribution
2. The extent to which bid distributions for individual blocks reflect the overall bid distribution pattern

---

[2] The work reported in Sections III and IV of Chapter 10, together with all interpretations, inferences, and figures, does not represent the official position of Texaco Inc., or of any of its employees, including J. A. MacKay. No attribution to Texaco Inc., or any of its employees (including J. A. MacKay), shall be made of work in Sections III and IV of Chapter 10. Responsibility for these sections rests solely with Ian Lerche. A version of Section III has been previously published in the *Oil and Gas Journal* with Ian Lerche as sole author; and a version of Section IV is in preparation for journal publication, again with Ian Lerche as sole author.

3. Bid distribution patterns by corporate participation
4. What is learned from the various behaviors.

Answers to these questions may be of some importance for influencing bid amounts and strategies in future Gulf of Mexico lease sales involving subsalt sheet plays—particularly when the success rate for finding probabilities for hydrocabon accumulations increases in the future for prospects and blocks leased at Lease Sale 147.

This general appreciation of strategies and learning curves may then be a factor to incorporate in the future.

A. BID DISTRIBUTIONS

1. Overall Distributions

For the 10 most expensive blocks on which bids ere made (Table 10.6), the bids were first pooled together irrespective of who made which bid or on which block. The pooled bids were then organized in order of lowest to highest and normalized so that, instead of plotting the bid frequency

TABLE 10.6 Lease Sale Bid Values by Blocks and Corporate Groups

Bidder	Block: Ship Shoal South 337 amount ($10^6)
Anadarko	40.00
Amoco/Phillips/Anadarko	6.25
Omni/Fina/EP Operating	2.95
Chieftan/Murphy/Mobil	2.50
Fina/Oxy/Total, etc.	0.80
Kerr McGee	0.666
Vastar/Nomenco/LL&E	0.377
Conoco/Union	0.242
Enron	0.240
	9 bids

Bidder	Block: Vermilion South 375 amount ($10^6)
Phillips/Anadarko	7.3
Amoco/Murphy/Mobil	3.75
Samedan/Energy Dev.	1.65
Kerr McGee/Chevron	0.801
British-Borneo/Cool	0.531
Vastar/LL&E	0.376
Hunt	0.276
	7 bids

(*continues*)

TABLE 10.6 (*continued*)

Bidder	Block: East Cameron South 357 amount (10^6)
Anadarko	8.2
Marathon/Texaco	1.1
Murphy/Mobil	0.76
Energy Dev./Chevron/LL&E	0.667
Vastar	0.237
Chieftan/Omni/EP Operating	0.175
	6 bids

Bidder	Block: Vermilion South 295 amount (10^6)
Amoco	4.137
Phillips/Anadarko	2.500
Murphy/Mobil	0.412
Hunt	0.407
Sun OPLP	0.302
	5 bids

Bidder	Block: East Cameron South 358 amount (10^6)
Anadarko	10.7
Marathon/Texaco	2.57
Energy Dev./LL&E	0.385
Amoco/Murphy/Mobil	0.305
	4 bids

Bidder	Block: Vermilion South 376 amount (10^6)
Phillips/Anadarko	5.8
Amoco/Murphy/Mobil	0.934
Samedan/Energy Dev.	0.382
Hunt	0.276
	4 bids

Bidder	Block: Vermilion South 307 amount (10^6)
Amoco	5.5
Sun OPLP	0.434
Phillips/Anadarko	0.277
	3 bids

Bidder	Block: Vermilion South 307 amount (10^6)
Phillips/Anadarko	7.2
Amoco/Murphy/Mobil	3.75
	2 bids

Block: Eugene Island South 346; 1 bid: Phillips/Anadarko @ 10×10^6
Block: Eugene Island South 345; 1 bid: Phillips/Anadarko @ 7.6×10^6

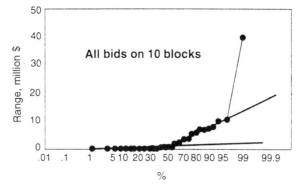

FIGURE 10.1 Cumulative bid distribution for all bids on the 10 most expensively acquired blocks at Gulf of Mexico Lease Sale 147.

over all bids, a cumulative probability of bid value versus bid made was plotted (Figure 10.1). Two factors stand out:

1. There is a break in slope at a bid value of $2 million, with a linear slope below the break point and a second linear slope above the break.
2. There is an isolated bid (Anadarko's $40 million bid on Ship Shoal 337) that is not consistent with the linear trend above $2 million in Figure 10.1.

Indeed, if the anomalous bid at $40 million bid is removed and the remaining bids replotted, again on a cumulative probability plot, the result (Figure 10.2) clearly shows the break of slope around $2 million; two straight-line fits on a cumulative probability axis match the total data field of bids remarkably well.

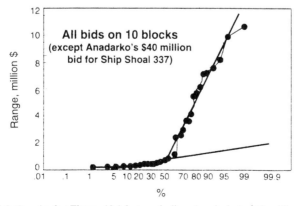

FIGURE 10.2 As for Figure 10.1 but excluding Anadarko's $40 million bid for Ship Shoal 337.

Independent of any variations deducible from bids on individual blocks, three conclusions can be drawn from the overall bid distribution:

1. The low bid group (≤$2 million) on its own seems to be roughly log-normally distributed (Figure 10.3a).
2. The high bid group (≥$2 million) on its own also seems to be roughly log-normally distributed (Figure 10.3b).
3. Anadarko's $40 million bid on Ship Shoal 337 is anomalous relative to the distribution of all other high bids (≥$2 million).

Perusal of the bid data (Table 10.6) also indicates that individual companies, or company groups bidding as a unit, sometimes are in the high bid group and sometimes the low bid group. Thus, Anadarko and Phillips bid $277,000 on Vermilion South Addition Block 307 (a bid in the low group)

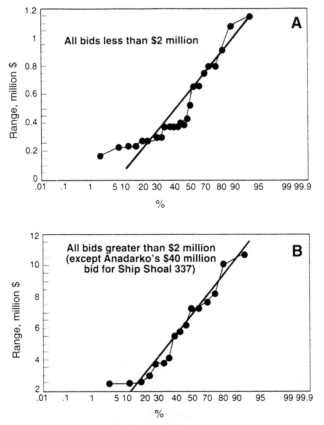

FIGURE 10.3 (a) As for Figure 10.1 but only for bids less than the break point of $2 million. (B) As for Figure 10.1 but only for bids in excess of $2 million, and excluding Anadarko's $40 million bid for Ship Shoal 337.

but also bid $7.6 million on Eugene Island South Addition Block 345 (a bid in the high group). So individual corporate strategies are not geared to being solely in the high or low bid groups.

What is remarkable is the sharp division at about $2 million. A total of 25 bids lie below and 17 bids lie above $2 million, making for a significant statistical framework. The perception is that either corporations did very little work on particular blocks and so bid on the low end, or did significant homework prior to the lease sale and so bid on the high end; or that not much money was available in a given corporation to provide anything except a minimal bid.

The pattern can be likened to walking through a casino and dropping a quarter in a slot machine with a random, distant chance of hitting the jackpot versus playing a no-stakes-limit progressive poker game—there are both low rollers and high rollers.

2. Bid Distributions by Blocks

Of the bid values for the 10 most expensive blocks listed in Table 10.6, four blocks have 5 or more bids and so can be analyzed for further information. These blocks are as follows:

1. Ship Shoal South 337 (9 bids)
2. Vermilion South 375 (7 bids)
3. East Cameron South 357 (6 bids)
4. Vermilion South 295 (5 bids)

The remaining six blocks each have 4 or fewer bids and so do not provide measures on a block-by-block basis from which to draw statistically significant conclusions.

Figure 10.4a shows the bid range for Ship Shoal South 337 including Anadarko's anomalous $40 million bid. When the Anadarko bid is removed and the remaining data are replotted (Figure 10.4b), a mix of the low bid and high bid groups is present, suggesting that minimal or close to minimal bids were offered (roughly log-normally distributed below $2 million) with little to no hope or chance of winning the block, while the high-end bids sit on a rough log-normal curve above $2 million, suggesting that significant effort was put into deciding on a bid.

The same grouping of patterns of behavior is apparent in the remaining three blocks that had enough bids to provide a statistically meaningful exercise—although the number of bids per block is progressively less. A similar bidding pattern is seen for Vermilion South 375 with seven bids (Figure 10.4c), East Cameron 357 with six bids (Figure 10.4d)—although only one member of the high end bid group is present here—and Vermilion South 295 with five bids (Figure 10.4e).

In every case the pattern of bidding range is the same, separating those bidders who had done significant homework on the blocks from those who

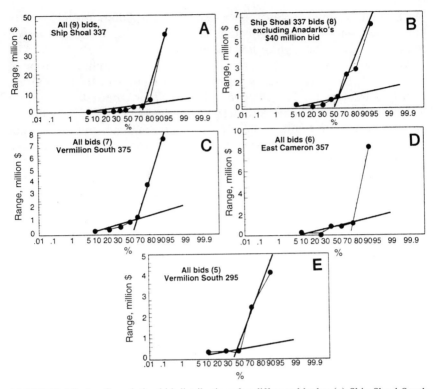

FIGURE 10.4 Cumulative bid distributions for different blocks: (a) Ship Shoal South 337, 9 bids; (b) Ship Shoal South 337 (excluding Anadarko's $40 million bid), 8 bids; (c) Vermilion South 375, 7 bids; (d) East Cameron South 357, 6 bids; (e) Vermilion South 295, 5 bids.

had not (or separating those who had indeed done all of their homework but had arrived at radically different interpretations of the likelihood of a block being hydrocarbon bearing).

There is, apparently, no consistent pattern of variation of bids with respect to physical location in the Gulf of Mexico—perhaps because MMS preselected the blocks to hone in on the sub-salt play in and around the Phillips' Mahogany prospect. One would then expect similar bidding patterns by block, as is consistent with the observations.

3. Bids by Corporations

Two major bidding groups bid on enough of the 10 blocks given in Table 10.6 that statistically significant inferences can be drawn. Amoco on its own or with partners bid on 8 of the 10 blocks and was the high bidder on its own on 2 of the blocks.

Anadarko on its own or with Phillips—and in the case of Ship Shoal South 337 both on its own and together with Phillips and Amoco—bid on all 10 blocks and was the successful bidder on 8 blocks. A question of interest is to examine the bid distribution of both of these major groups. Two groupings were formed from the data of Table 10.6:

1. Amoco on its own or with partners.
2. Phillips and Anadarko together and, in addition, Anadarko on its own.

The Amoco group's bidding range (Figure 10.5a) is approximately lognormally distributed, suggesting that a roughly fixed pool of money was

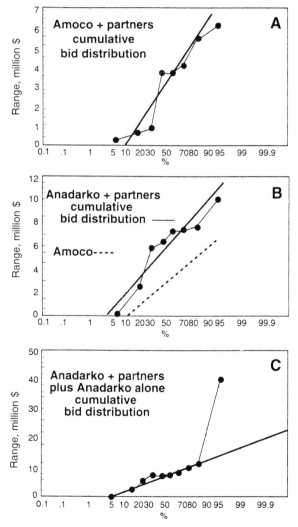

FIGURE 10.5 Corporate bid distributions: (a) Amoco plus partners; (b) Anadarko plus partners; (c) Anadarko plus partners, and Anadarko alone, combined.

available with a log-normal division of the money into individual bids, a very common occurrence in bidding strategies.

Thus, the Amoco group followed some corporate strategy rules relative to an expected bid *expenditure* limit, as opposed to a bid *exposure* limit. Therefore, the Amoco group appears to have made some estimate of the likely average total amount it would have to spend if their most likely scenario of successes prevailed, in contrast to an exposure scenario if they were to have been successful on all blocks on which they bid.

A similar bid distribution for the grouping of Phillips plus Anadarko (Figure 10.5b) shows a behavior pattern similar to that of Amoco. For contrast, plotted in Figure 10.5b is a dashed line that is the Amoco group indicating that, while roughly parallel to the Phillips plus Anadarko group and so, presumably, following a somewhat similar corporate expenditure strategy, nevertheless the Amoco group's bids are not as optimistic overall as those of Phillips plus Anadarko.

This optimism on the part of Phillips plus Anadarko can result from many causes: a truly more optimistic scientific outlook on the play, a mandated corporate higher risk factor, a larger pool of money with which to bid, economic risk assessment factors that calculate higher expenditure thresholds for break-even, and so on. The optimism may also reflect the well-known fact that when multiple corporations band together to make a combined bid, such bids tend to be more conservative than the most optimistic corporation of the band would wish and less conservative than the most conservative corporation of the band—with a bias toward conservative rather than optimistic bidding.

Of even greater interest is to put together the combined statistics of corporate bidding by Anadarko on its own and with partners (Figure 10.5c). Two factors clearly stand out:

1. The pattern of development of all bids is approximately log-normally distributed with the exception of Anadarko's $40 million bid for Ship Shoal South 337.
2. The $40 million bid represents either a switch in corporate strategy by Anadarko or is a likely result of individual control overriding a basic corporate strategy.

This point can be made more strongly because Anadarko had all of its own bidding distribution information plus that of its dominant partner, Phillips, *prior* to the lease sale. On that information alone, Anadarko could have chosen to bid about $16 million with a 99% chance of success based on the data of Figure 10.5c, or about $20 million with a 99.99% chance of success; a bid of $40 million is then well outside any standard error on corporate strategy as depicted through Figure 10.5c.

Consider the following:

1. Both Amoco *and* Phillips were joint partners with Anadarko on their combined $6.25 million bid for block Ship Shoal South 337.

2. Anadarko was privy to the joint bidding statistics (before the sale) of Phillips and itself on eight blocks (and also knew what its own internal bidding strategy would have been if taken in isolation) and so could infer Phillips' internal bidding strategy.

3. Anadarko, together with partners or on its own, was already exposing $65.927 million on the other nine blocks of Table 10.6 for an average of about $7.3 million/block.

Therefore, it would seem that Anadarko's bid for Ship Shoal South 337 does not fit any of the statistical controls deducible before and after the sale, either with respect to perceived for inferred spatial distribution bids on different blocks, overall bid probabilities, or internal and/or with partner bidding arrangements and statistics. Nor does the bid fit any other pattern available, suggesting either a switch in corporate strategy or an individual overriding control of the internal corporate exploration economic assessment process.

B. SUMMARY

Analysis of the bid distributions for the 10 most expensively acquired blocks at the March 1994 Gulf of Mexico Lease Sale 147 indicates that the global distribution of bids splits into two groups at a bid of around $2 million.

Both the low bid group—25 bids—and the high bid group—17 bids—follow approximately log-normal distributions with different slopes. On an individual block basis, four blocks had sufficient numbers of bids (≥ 5) to indicate a statistical "mix" of low and high group bids, but no preference with respect to physical location of the blocks is apparent.

On a corporate basis, two bidding groups—Amoco on its own or with partners, and Anadarko on its own or with partners—each made a sufficient number of bids to indicate that both groups followed a very similar bidding pattern, both of which are approximately log-normally distributed, but the Amoco group was systematically conservative relative to the Anadarko group.

For the Anadarko group, the bidding pattern follows a conventional bid expenditure behavior with the exception of an isolated bid of $40 million made by Anadarko on its own for Ship Shoal South Addition Block 337. That bid is much in excess of any extrapolation of the bid pattern of Anadarko on its own or of Anadarko with partners, suggesting that either a shift in corporate strategy took place, individual preference overrode corporate exploration assessment procedures, or that the quality of Anadarko's interpretation for that block was uniquely different than that of Anadarko and its partners (Amoco and Phillips) for the same block. A bid of $20 million would have had a 99.9% chance (or greater) of success.

IV. BLOCK STATISTICS FOR LEASE SALE 157 AND BID RATIOS

A. OVERVIEW

One of the largest lease sales (in terms of number of blocks offered and number of bids made) in recent memory was the Gulf of Mexico Offshore Lease Sale 157, so that a statistical analysis of the bid data from that sale offers a unique opportunity to obtain some statistically representative measures of corporate patterns of behavior, which were not as easily obtained from the smaller Lease Sale 147 discussed in the previous section.

B. BID RATIOS

For instance, one of the questions that often arises in corporations concerns the amount that one should bid in relation to competitors when one does not know, ahead of the lease sale, what the competitors will do. In such cases, past histories of bids at other lease sales are often used as a means of assessing bids to make. There is a generally prevalent statement, commonly thought to be appropriate, that the ratio of the next lower bid to the highest bid is around 0.7, and companies often act on that premise to figure out how much premium they should add to their bids to ensure a 90% chance of successfully securing a block. For all of the blocks at Lease Sale 157, on which there were two or more bids, the ratio of second highest to highest bid was computed. Results are displayed in Figure 10.6. It is clear that there is, in fact, no preferred value for the ratio; the distribution

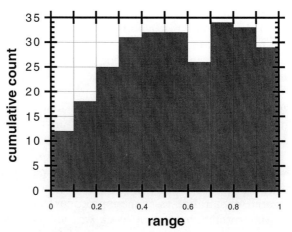

FIGURE 10.6 Distribution (not cumulative) of second highest bid to highest bid for bids from Gulf of Mexico Lease Sale 157 drawn at intervals of 0.1.

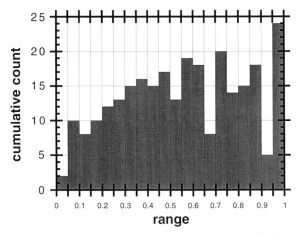

FIGURE 10.7 As for Figure 10.6 but drawn on the finer scale interval of 0.05.

is almost uniform above a ratio value of 0.2. It might be argued that the histogram ratio intervals of 0.1 are too broad and that a narrower range would show more structure. To obviate any such questions, ratio intervals of 0.05 were also used for plotting purposes, with the result shown in Figure 10.7, reinforcing the point that there is a lack of clustering around any particular value of the ratio.

C. BID DISTRIBUTIONS

1. All Bids

All bids on all blocks were superimposed on a single cumulative percentage diagram (Fig. 10.8). The absolute dynamic range is large, from a lowest bid of about 4×10^3 to a high of just over 10^7. Note that between about 10 and 90% cumulative percentages the bids form an overall arcuate structure but, just as in the previous example of Lease Sale 147, the total bids can once more be fit by two straight lines, above and below about a bid of 2.5×10^5, as sketched on Figure 10.9, and presumably reflecting senses of purpose similar to those of the corporations at Lease Sale 147.

2. Bids by Blocks

Because of the large number of blocks offered at Lease Sale 157, the number of bids per block on some blocks was higher than for Lease Sale 147, so that the ability to draw statistically significant conclusions is enhanced. One block, South Marsh Island 261 drew 10 bids, two blocks (Eugene Island 65 and Main Pass 280) drew 8 bids each, West Cameron 492 received 7 bids, while 6 bids were recorded for each of Mississippi

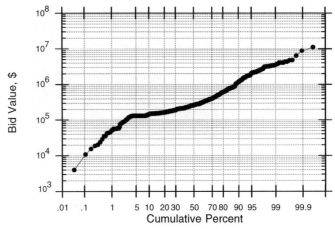

FIGURE 10.8 Cumulative distribution of all bids for Gulf of Mexico Lease Sale 157.

Canyon 728 and 772, and for South Timbalier 59 and Viosca Ridge 179. All other blocks received 5 or fewer bids so that it is not possible to draw any statistically significant conclusions for each such block. Shown in the different parts of Figure 10.10 are the bid values drawn on cumulative probability paper for each of the blocks with 6 or more bids. Also drawn on each part are dashed straight lines through the data, showing how close each bid distribution is to a cumulative log-normal result between 10 and 90% probability. There are to be sure, minor fluctuations of the data away from the straight lines but, by and large, one has a fairly good representation of approximately log-normal bidding on each block.

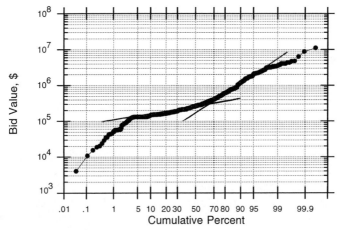

FIGURE 10.9 As for Figure 10.8 but with two superimposed straight lines showing a "break" at a bid of around $250,000.

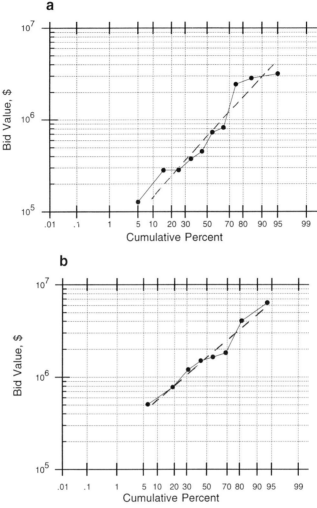

FIGURE 10.10 Cumulative bid distributions for different blocks: (a) South Marsh Island 261, 10 bids; (b) Eugene Island 65, 8 bids; (c) Main Pass 280, 8 bids; (d) West Cameron 429, 7 bids; (e) Mississippi Canyon 728, 6 bids; (f) Mississippi Canyon 772, 6 bids; (g) South Timbalier 59, 6 bids; (h) Viosca Ridge 179, 6 bids.

3. Bids by Corporations

Because of the large number of blocks offered at Lease Sale 157 the absolute number of bids was large and so was the number of permutations and combinations of corporations bidding singly, in pairs, or as parts of large consortiums. It would be difficult to analyze all such combinations for strategic patterns in a short space, and probably not a particularly rewarding exercise overall, but a couple of interesting factors stand out.

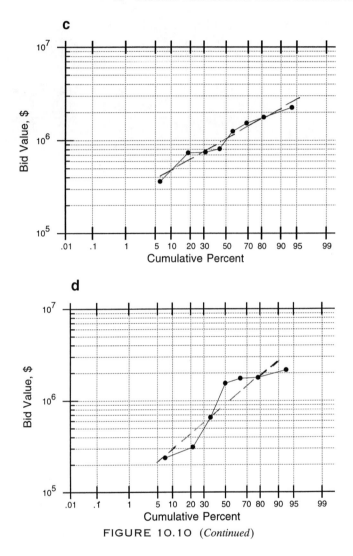

FIGURE 10.10 (*Continued*)

For example, Figures 10.11a–d show that Elf Exploration, alone or with partners, bid in the range 10^5 to 10^6 except when bidding in combination with BP Exploration (Figure 10.11d), when bids in excess of 10^6 occur, suggesting a more aggressive bidding pattern from that particular corporate combination. Equally, when BP Exploration bid on its own (Figure 10.11e) all of its bids, except one, were less than 10^6, but when BP Exploration was allied with BHP Petroleum (Figure 10.11f) or with Elf Exploration (Figure10.11d), the alliance was more aggressive, with significant numbers of bids in excess of 10^6.

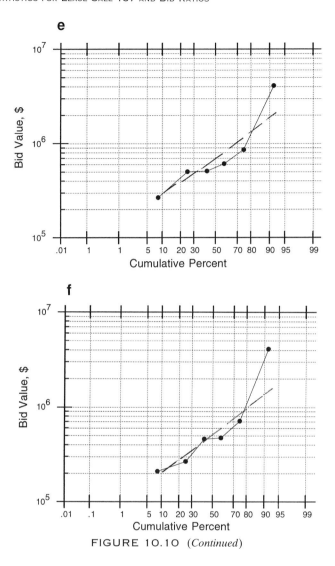

FIGURE 10.10 (*Continued*)

Of the 19 companies (either alone or with partners) that had 10 or more bids on blocks, 16 made bids above 10^6 with very different distributions, as shown in Figures 10.12a–s. There is, obviously, a mix of bids for each bidding group above and below the break point of 2.5×10^5 deduced from the curves of Figure 10.9, but there does not seem to be any other particular rationale that can be deducible from the 19 bidding groups.

However, when ranked by the depth of water in which the acreage bids were sited, the bids take on a different character. For water depths of less than 200 m there were 463 high bids, from 200 to 900 m a total of 93 high

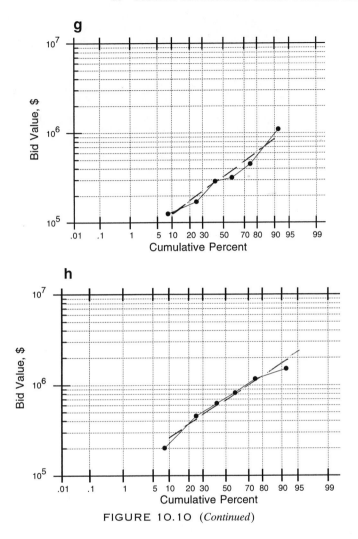

FIGURE 10.10 (*Continued*)

bids, and for water depths in excess of 900 m a total of 368 high bids. Thus, the low corporate bids tend to reflect the shallow water acreage with the high bids being more for the deep water offshore (Table 10.7), which accounts for why Zilkha Energy, with no very high bid at all, captured the largest acreage of more than 500,000 acres —mostly shallower than 400 ft of water—while Texaco captured the largest acreage (more than 350,000 acres) of very deep water (more than 900 m).

The major point of difference that emerges from this case history relative to the prior case history is that the regional differences of acreage in shallow water versus deep water are prevalent; all other patterns appear similar.

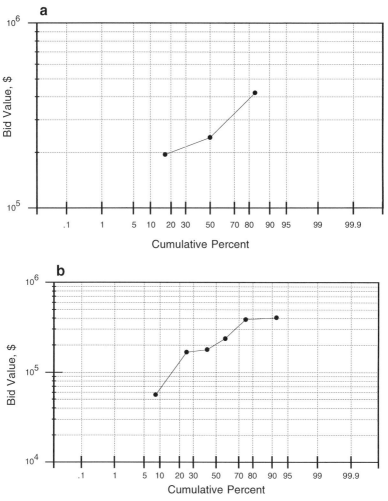

FIGURE 10.11 Cumulative bid distributions by corporations: (a) Elf Exploration alone; (b) Elf Exploration plus Vastar plus Chevron; (c) Elf Exploration plus Enserch Corporation; (d) Elf Exploration plus BP Exploration; (e) BP Exploration alone; (f) BP Exploration plus BHP Petroleum.

V. WHAT IS A BID WORTH?

Once the scientific and economic evaluations of an exploration opportunity have been made including costs, optimum working interest, potential gains, success probability, corporate risk tolerance, and portfolio balancing aspects, an estimate of the expected value and its uncertainty are then available. The costs of acquisition of opportunities and the bidding strategy

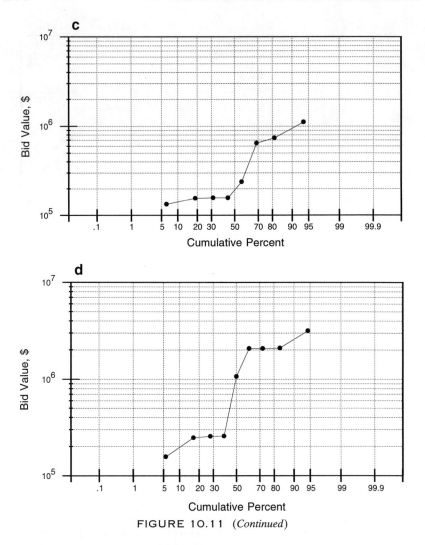

FIGURE 10.11 (*Continued*)

necessary to obtain some prospects against other competing corporations now need to be included in the exploration assessment.

Any successful bid, B, will *lower* the value of an exploration opportunity by the amount of the bid. On the one hand, a bid of zero in the face of competitors will surely be unsuccessful, while on the other hand too high a bid will certainly ensure successful acquisition of an opportunity but will also lower the expected value to a negative amount. The problem is to assess the amount to bid in the face of competition, but in such a way as to end up with a positive worth for the opportunity *and* to have a good chance of acquiring the opportunity. Part of the difficulty of resolving this problem uniquely is due to the fact that different corporations evaluate

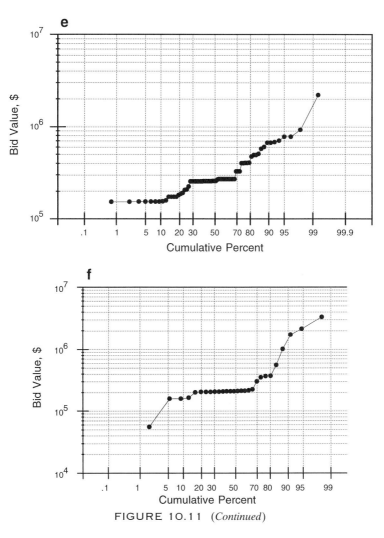

FIGURE 10.11 (*Continued*)

differently a given exploration opportunity and so have different perceptions of the value and of the uncertainty of that value. In turn, this variation in value implies that each corporation has a different idea of how much to bid on an opportunity, which accounts for some of the variation in the statistical patterns evident in Lease Sales 147 and 157 for the Gulf of Mexico.

Part of the remaining variation in bid values is due to the different risk tolerance values each corporation mandates and to the total bidding budget each corporation has available.

Methods for handling each of these uncertainties have been developed and short précises of each are presented here, with more detailed technical developments available elsewhere (Lerche, 1991).

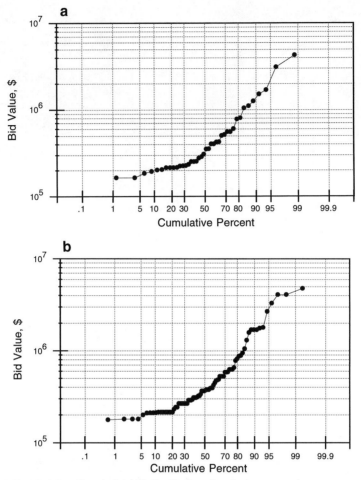

FIGURE 10.12 Cumulative bid distributions for single corporations, or consortia of corporations, making 14 or more bids: (a) BHP Petroleum; (b) Texaco plus Chevron; (c) Chevron; (d) Amoco; (e) Marathon Oil; (f) Enserch Corporation; (g) Conoco; (h) Walter Oil and Gas; (i) UMC Petroleum plus Barrett; (j) Chieftan International plus Carin Energy plus Enserch; (k) Callon Petroleum plus Murphy E&P; (l) Enron Oil and Gas; (m) Vastar; (n) Texaco E&P; (o) LL&E; (p) Shell Offshore; (q) Exxon; (r) Zilkha Energy; (s) Mobil Oil plus Phillips Petroleum.

A. BID PROBABILITIES AND WORTH UNCERTAINTIES

A particular equivalent cash bid value B will, if successful, lower the total expected cash flow to the corporation such that the net constant-value *worth* W of the exploration opportunity to the corporation decreases to

$$W = E_1(V) - B, \quad (10.1)$$

where $E_1(V)$ is the expected value.

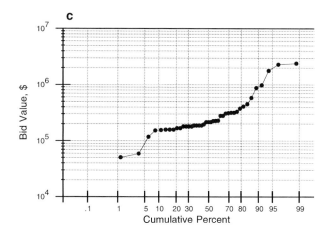

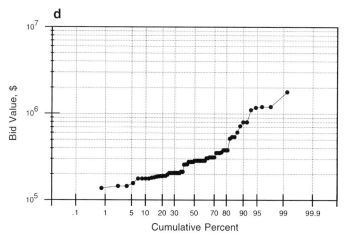

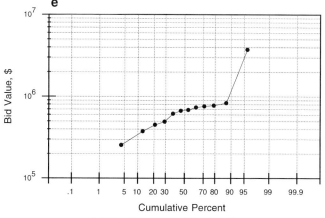

FIGURE 10.12 (*Continued*)

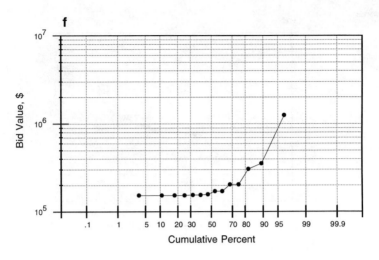

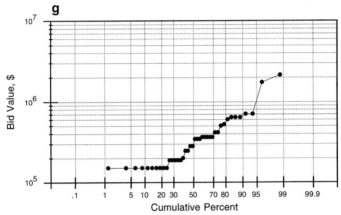

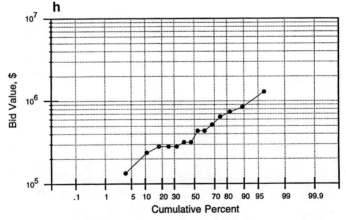

FIGURE 10.12 (*Continued*)

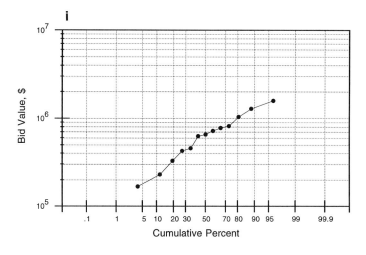

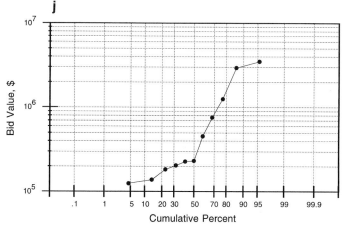

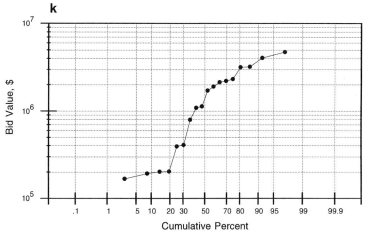

FIGURE 10.12 (*Continued*)

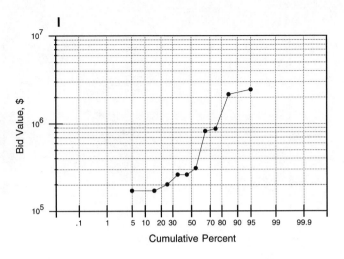
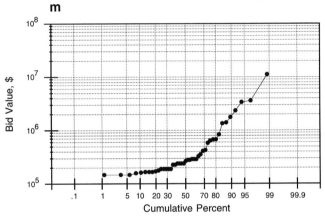
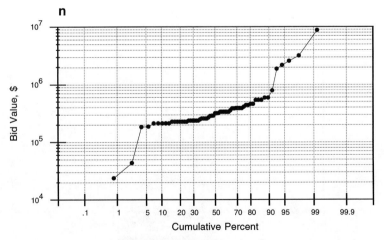

FIGURE 10.12 (*Continued*)

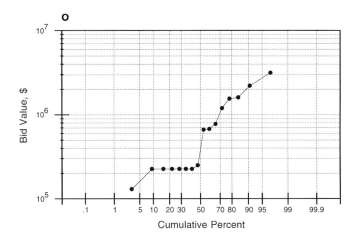

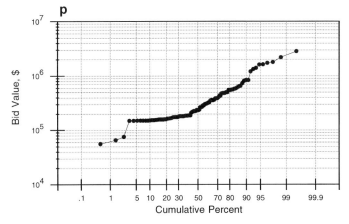

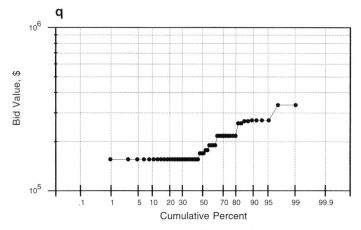

FIGURE 10.12 (*Continued*)

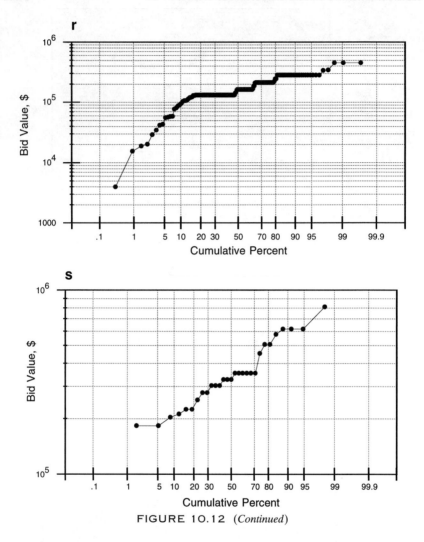

FIGURE 10.12 (*Continued*)

A bid such that $W < 0$ immediately implies an anticipated loss to the corporation on average. But the problem is now compounded by the bid value itself. The difficulty stems from the fact that if the bid is truly astronomical, then there is 100% chance of being awarded the rights but the corporation loses money (sometimes a necessary evil in order to demonstrate ability, capability, and willingness to develop an exploration presence in a country). If the bid is too low, the authorities granting exploration rights will not award an exploration lease and so potential profit to the corporation is lost. Somewhere between these two extremes lies a bid that optimizes the worth of the exploration

TABLE 10.7 Acreage Acquired (in MM Acres)

Company	Water depth			Total acreage
	Shallow (0–200 m)	Medium (201–900 m)	Deep (>901 m)	
Zilkha	0.50	—	0.01	0.51
Texaco	0.07	0.05	0.35	0.47
Chevron	0.12	0.05	0.23	0.4
BP	—	0.04	0.36	0.4
Amoco	0.05	0.01	0.21	0.27
Shell	0.11	0.05	0.10	0.26
BHP	0.01	0.06	0.19	0.26
Exxon	—	—	0.18	0.18
Conoco	0.002	0.01	0.17	0.18
Vastar	0.12	0.02	0.04	0.18

Note: All entries rounded to nearest significant two figures.

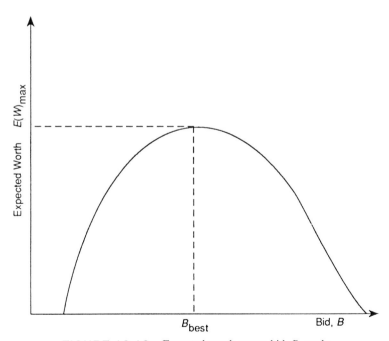

FIGURE 10.13 Expected worth versus bid, B, made.

opportunity. But that bid cannot maximize the absolute worth to the corporation (which occurs at a bid of $B = 0$). The corporation must settle for less than the maximum ideal total cash flow in order to have any total cash flow at all.

The expected worth $E(W)$ to the corporation is then obtained by multiplying the worth at a particular bid B by the probability $P(B)$ of the bid being successful at the value B:

$$E(W) = [E_1(V) - B]P(B). \tag{10.2}$$

The general shape of $E(W)$ as a function of bid value B must be approximately as shown in Figure 10.13: $E(W)$ drops to zero at the lowest acceptable bid value $B_{\min}$, and again drops to zero at $B = E_1(V)$. In between, $E(W)$ has a maximum. The question is how to determine the bid value B_{best} that produces a maximum expected worth $E(W)_{\max}$. The answer to this question clearly depends on the modeled behavior for successful bidding probability $P(B)$.

For instance, if it is assumed that the minimum acceptable bid is zero ($B_{\min} = 0$), then any single corporation bidding higher than zero is sure to acquire the prospect if all other corporations bid zero. Thus, the probability $P(B)$ must tend to zero as $B \to 0$. A simple rule is $P(B) \propto B^q (q > 0)$ as $B \to 0$. Equally, $P(B)$ must tend to unity as the bid tends to infinity (although the expected worth will then be large and negative).

Thus, $P(B) \to 1$ as $B \to \infty$. A simple rule encompassing both end members is

$$P(B) = \tanh[(B/B_*)^q], \tag{10.3}$$

where B_* is a scaling value such that for $B \le B_*$, $p(B) \approx (B/B_*)^q$; while for $B \ge B_*$, $P(B) \approx 1 - 2\exp[-(B/B_*)^q]$.

Alternatives for $P(B)$ abound. Thus, the simple power law

$$P(B) = \begin{cases} (B/B_*)^q & \text{for } B < B_* \\ 1 & \text{for } B > B_* \end{cases} \tag{10.4}$$

is one such option;

$$P(B) = \text{erf}(B/B_*) \tag{10.5}$$

is another; and so on. Empirical analyses of past bids by corporations in open domain lease sales are often used to construct approximate behaviors of $P(B)$ to be used at the next lease sale in a similar setting (see previous two subsections of this chapter).

1. Expected Worth

The simple power law, Eq. (10.4), will suffice here for illustrative purposes. Then

$$E(W) = [E_1(V) - B](B/B_*)^q. \tag{10.6}$$

Maximizing $E(W)$ with respect to the bid B [obtained by finding the value of B at which $\partial E(W)/\partial B = 0$] one has

$$B_{\text{best}} = \frac{q}{(q+1)} E_1(V) < E_1(V) \qquad (10.7a)$$

and, at this value of B, the maximum of $E(W)$ is given by

$$E(W)_{\max} = E_1(V) \left[\frac{q^q}{(1+q)^{q+1}}\right] \left[\frac{E_1(V)}{B_*}\right]^q. \qquad (10.7b)$$

For a linear power increase ($q = 1$)

$$E(W)_{\max}/E_1(V) = \frac{1}{4}\left[\frac{E_1(V)}{B_*}\right], \qquad (10.8a)$$

while for a quadratic increase ($q = 2$)

$$E(W)_{\max}/E_1(V) = \frac{4}{27}\left[\frac{E_1(V)}{B_*}\right]^2. \qquad (10.8b)$$

To maximize $E(W)_{\max}$, it is then appropriate, with the simple power law probability, to have the expected value of $E_1(V)$ as close to the bid scale value B_* as possible, because then $E_1(V)/B_* = 1$ and the worth of the project is just

$$E(W)_{\max} = E_1(V) \left[\frac{q^q}{(1+q)^{q+1}}\right]. \qquad (10.9)$$

Small fractional powers for q are better than values of q exceeding unity for optimizing $E(W)_{\max}$. Thus, if estimates of q and B_* can be obtained from prior information, then projects with $E_1(V)$ closest to B_* will provide higher expected worth to the corporation.

The ratio of optimal bid B_{best} to maximum expected worth $E(W)_{\max}$ gives an estimate of the fractional return to be anticipated, so that bid index BI can be used to rank competing exploration opportunities with

$$B_I = E(W)_{\max}/B_{\text{best}}. \qquad (10.10)$$

A high BI implies a significant expected return relative to bid value expended.

2. Uncertainties in Expected Worth

Just as with the expected value $E_1(V)$ of an exploration opportunity, an uncertainty is associated with the expected worth of a project including the bid value. One has

$$E_2(W) = [E_2(V) - 2E_1(V)B + B^2] P(B) \qquad (10.11)$$

so that the mean square uncertainty, δW^2, on the expected worth is given through

$$\delta W^2 = E_2(W) - E_1(W)^2 \\ = [E_2(V) - E_1(V)^2 + B^2]\, P(B)\,[1 - P(B)]. \tag{10.12}$$

Note that δW is zero on $P = 0$ and $P = 1$ as expected, but that between these two values, δW is positive. The uncertainty on $E_1(W)$ is $\pm \delta W$, reflecting the combined uncertainty of the exploration opportunity [customarily measured by the variance $\sigma^2 \equiv E_2(V) - E_1(V)^2$] and the bidding success uncertainty.

Two interesting facets of the worth uncertainty are examined here: (1) the value of δW^2 when the expected worth is at its maximum value and (2) the possibility of having both expected worth and δW^2 at maximum values simultaneously. The calculations are again carried out for a power law dependence of the bidding success probability, $P(B)$ [$\equiv (B/B_*)^q$] for ease of illustration, although the calculations can equally be done with any chosen form for bidding success probability.

a. Uncertainty at Maximum Expected Worth

When the best bid value, B_{best}, is directly inserted into Eq. (10.12), and under the requirement

$$E_1(V) < B_*(1 + q^{-1}) \tag{10.13}$$

in order to keep $B_{\text{best}} < B_*$, which is the maximum permitted bid, then

$$\delta W^2 = (\sigma^2 + B_{\text{best}}^2)\, P(B_{\text{best}})\,[1 - P(B_{\text{best}})]$$

that is,

$$= [\sigma^2 + E_1(V)^2 q^2/(q+1)^2]\, P(B_{\text{best}})\,[1 - P(B_{\text{best}})] \tag{10.14}$$

A limit can be put on δW^2 because $P(1 - P)$ cannot exceed one-quarter, so that from Eq. (10.14) one has

$$\delta W^2 \leq \frac{1}{4}[\sigma^2 + E_1(V)^2 q^2/(q+1)^2]. \tag{10.15}$$

One also has

$$E(W)_{\max} = E_1(V)\,(q+1)^{-1}\, P(B_{\text{best}}), \tag{10.16a}$$

which can be cast in the form

$$E(W)_{\max} = q^{-1} B_{\text{best}}\, P(B_{\text{best}}) \leq B_{\text{best}}/q. \tag{10.16b}$$

Now if $E(W)_{\max}^2 \leq \delta W^2$ then the uncertainty on the expected worth can lead to a negative worth within one standard deviation from the expected value [i.e., $E(W)_{\max} - \delta W$ is negative]. Such an event will occur whenever

$$P(B_{\text{best}}) \leq (\sigma^2 + B_{\text{best}}^2)/[\sigma^2 + B_{\text{best}}^2(1 + q^{-2})]. \tag{10.17}$$

If the standard error uncertainty, σ, on the exploration opportunity is large compared to the best bid, B_{best}, ($\sigma \gg B_{\text{best}}$) then $E_1(W) - \delta W$ will be negative for

$$P(B_{\text{best}}) \leq 1 - (B_{\text{best}}/\sigma)^2 q^{-2}. \tag{10.18}$$

Thus, there is then a high probability of a negative worth within one standard error of the mean worth. If $B_{\text{best}} \gg \sigma$, then $E(W) - \delta W$ will be negative when

$$P(B_{\text{best}}) \lesssim q^2/(1 + q^2), \tag{10.19}$$

which, for $q \leq 1$ is less than 50% probability, but for $q \gg 1$ yields $1 - q^{-2}$, very close to unity. Hence, even when $E_1(V)$ is positive, there is a large uncertainty on the expected worth, $E(W)$, to the point that within one standard error of the expected value there is a high probability of a negative worth when bidding is at the value B_{best}, which maximizes $E(W)$.

If one wishes to retain $E_1(W) - \delta W$ at a positive value, then it is necessary to choose a bid value B such that

$$P(B) \geq (\sigma^2 + B^2)/[\sigma^2 + B^2 + \{E_1(V) - B\}^2]. \tag{10.20}$$

b. Maximum Uncertainty on Worth

Because the variance, δW^2, on expected worth is zero at $P = 0$ and $P = 1$, it has a maximum at some bid value in $0 \leq B \leq B_*$. It is useful to know what the maximum variance is so that one can assess the worst case economic hazard of obtaining a negative worth.

From Eq. (10.12), the maximum value of δW^2 occurs on $B = B_0$, where B_0 is given through

$$P(B_0) = (B_0/B_*)^q = \frac{\frac{1}{2}[B_0^2(2 + q) + q\sigma^2]}{[B_0^2(1 + q) + q\sigma^2]} \tag{10.21}$$

and then

$$\delta W_{\text{max}}^2 = \frac{\frac{q}{4}(\sigma^2 + B_0^2)^2 [B_0^2(2 + q) + q\sigma^2]}{[B_0^2(1 + q) + q\sigma^2]^2}. \tag{10.22}$$

Note that if the uncertainty, σ, in the exploration opportunity is large compared to the bid value, B_0 ($\sigma \gg B_0$) then Eq. (10.21) returns the values

$$P(B_0) \cong 1/2 \quad \text{and} \quad B_0 \cong B_* 2^{-1/q}, \tag{10.23a}$$

while, similarly, Eq. (10.22) provides

$$\delta W_{\text{max}}^2 \cong \sigma^2/4. \tag{10.23b}$$

Alternatively, if $\sigma \ll B_0$, then Eq. (10.21) returns the values

$$P(B_0) \cong (1 + q/2)/(1 + q) \quad \text{and} \quad B_0 \cong \frac{B_*(1 + q/2)^{1/q}}{(1 + q)^{1/q}}, \tag{10.24a}$$

while Eq. (10.22) then provides

$$\delta W_{\max}^2 \cong B_0^2 \frac{q}{4} \frac{(2+q)}{(1+q)^2}. \tag{10.24b}$$

It is also of interest to ask if it is possible for an optimal bid to be made (so that the expected worth is optimized), while at the same time having maximum uncertainty on the expected worth. If such an eventuality is desired, then one requires $B_{\text{best}} = B_0$. But, because $B_{\text{best}} = qE_1(V)/(q+1)$ [Eq. (10.7a)], direct insertion of this value for B_{best} into Eq. (10.21) sets up a requirement on the scale value B_* in terms of parameters of the exploration opportunity as

$$B_* = B_{\text{best}}\{2[B_{\text{best}}^2(1+q) + q\sigma^2]/[B_{\text{best}}^2(2+q) + q\sigma^2]\}^{1/q}. \tag{10.25}$$

Then it is indeed possible to have maximum expected worth and also maximum uncertainty on that worth.

The two values $E_1(W)$ and δW^2 are sufficient to permit an assessment to be made of a bid, B, which will probably return a positive worth, by regarding the actual worth W as drawn from a Gaussian distribution with mean value, $E_1(W)$, and variance, δW^2, each of which depends on the bid, B. Thus, one would write the differential probability $p(W)\,dW$ of obtaining a worth in the range W to $W + dW$ as

$$p(W)\,dW = (2\pi|\delta W|)^{-1/2} \exp[-(W - E_1(W)^2/2\delta W^2)]\,dW \tag{10.26}$$

and the probability, $P_+(B)$, of obtaining a positive value of worth as

$$P_+(B) = \int_0^\infty p(W)\,dW. \tag{10.27}$$

As the bid value, B, varies so do $E_1(W)$ and δW^2, and so too does $P_+(B)$. Hence, one can decide on a bid range such that there is (1) a known probability of successfully bidding on the exploration opportunity *and* (2) a probability that the worth of the opportunity will then be positive. Sketches of the behavior of $p(W)$ as $E_1(W)$ and δW vary are given in Figure 10.14.

B. NUMERICAL ILLUSTRATIONS

1. Expected Worth versus Expected Value

For $E_1(V) \equiv B_*$ and for different values of the power index, q, Table 10.8 gives ratios of optimal bid to expected value, $E_1(V)$, and of the expected worth of the opportunity as a fraction of the expected value, together with the estimated success probability.

Effectively, the higher the power of q, the slower the success probability increases as the bid, B, increases, so there needs to be a greater fraction of the expected value expended to increase the chance of being successful;

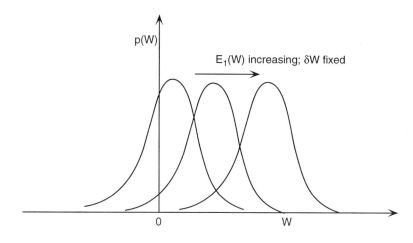

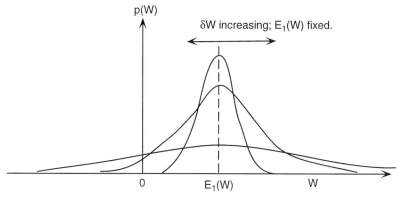

FIGURE 10.14 Schematic variation of $p(W)$ with W as (a) the expected value $E_1(W)$ increases for fixed variance δW^2 and (b) the variance, δW^2, increases for fixed expected value, $E_1(W)$.

TABLE 10.8   Bid Worth and Success Probability for Variable q

q	Optimal bid $B_{best}/E_1(V) \equiv q/(q+1)$	Opportunity worth $E(W)_{max}/E_1(V)$	Success probability at $B = B_{best}$
0.1	0.09	0.7	0.79
0.5	0.33	0.4	0.58
1.0	0.50	0.25	0.5
1.5	0.6	0.18	0.46
2.0	0.67	0.13	0.45
2.5	0.71	0.11	0.43
3	0.75	0.10	0.42
5	0.83	0.07	0.39

TABLE 10.9 Uncertainty on Expected Worth for Various q

(a) $B_0 \ll \sigma q^{1/2}$

q	$P(B_0)$	B_0/B_*	$\delta W^2_{max}/\sigma^2$
0.1	0.5	10^{-4}	0.25
0.5	0.5	0.25	0.25
1.0	0.5	0.5	0.25
2.0	0.5	0.7	0.25

(b) $B_0 \gg \sigma q^{1/2}$

q	$P(B_0)$	B_0/B_*	$\delta W^2_{max}/B_0^2$
0.1	0.95	0.6	0.04
0.5	0.83	0.69	0.14
1.0	0.75	0.75	0.19
2.0	0.67	0.82	0.22

the expected worth then decreases for increasing q, suggesting that one should keep q as low as possible to increase the worth.

2. Uncertainty of Expected Worth

Table 10.9 provides the values of $P(B_0)$, B_0, and δW^2_{max} for various q, using either a bid, B_0, much larger than σ, the uncertainty on expected value, or a bid, B_0, much smaller than σ. Except for very small values of $q(\leq 0.2)$, the majority of cases in Table 10.9 indicate that a bid value, B_0, of around 0.3–0.8 of B_* will yield a success probability of greater than about 50%, irrespective of whether B_0 is larger or small compared to σ, and will also yield an uncertainty δW_{max} that is around 50% of B_0 (or σ), depending on whether $\sigma \gtrless B_0$. Thus the uncertainty, δW_{max}, is relatively insensitive to the bid value, and is fairly close to its maximum under all except extreme conditions.

11

Economic Model Uncertainties

I. INTRODUCTION

The scientific exploration assessment of geologic conditions leading to estimates of the hydrocarbon finding probability, of the volume in place, and of the fractions and types of hydrocarbons (oil, gas, condensate) likely to be in an undrilled reservoir has undergone major development during the last decade or so. This development is due in no small part to advancement of quantitative procedures for evaluating assumptions in both geologic concepts, model uncertainties, and in considering the extent to which scattered, uncertain data control evolutionary behaviors of the models. Thus, a strong understanding is now available of scientific uncertainty, scientific risk, and appropriate strategies to follow to minimize such factors.

On the exploration economic side the conversion of potential hydrocarbon volumes to potential economic worth is fraught with problems similar to those occurring on the geotechnical side. One has to develop an estimate of how an exploration opportunity will be produced and, at the same time, convert the production on an ongoing basis to selling price of product so that one can estimate the total profit (in fixed-year dollars) over the estimated life of the opportunity. Then one has an estimate of total gains to be achieved from the opportunity. Over the same period of time one also has to make an estimate of costs associated with exploration, field development, and production (again, all in fixed-year dollars) so that one has an assessment of the total costs of the opportunity. It is the combined estimates of production modeling and product sales modeling that provide an economic gain estimate and a total cost estimate.

268   11   ECONOMIC MODEL UNCERTAINTIES

Just as with any other area of quantitative investigation, the estimates made of gains and costs contain uncertainties due to model assumptions, model parameters, and the quality, quantity, and sampling distributions of any data used as controls on model behaviors. One also has to evaluate whether such economic model uncertainties are causing more of the total uncertainty in the expected value of the opportunity than the geotechnical model uncertainties, for then the prioritization of further effort to narrow the range of total uncertainty on the opportunity can be placed within the economic aspects or the geotechnical aspects, as appropriate.

Here the exploration economic model aspects are illustrated using very simple production and product sales models. More complex models can be constructed to examine more realistic representations of production and product sales behaviors, but such models also contain even more assumptions and more parameters than the simple illustrative models to be used here. Uncertainties due to such extra factors would also need to be examined for their influence on gain and cost estimates. Although not a difficult thing to do, such a more germane evaluation would merely introduce technical points that complicate, without clarifying, and the logic procedure would be the same as with the simple illustrative models given here anyway.

The economic evaluation is first considered in two separate parts, which are then combined to provide an overall economic value for an opportunity.

II. A SIMPLE PRODUCTION MODEL

To illustrate the procedure, the production curve used to illustrate salient and dominant features is given in Figure 11.1, in which production is taken

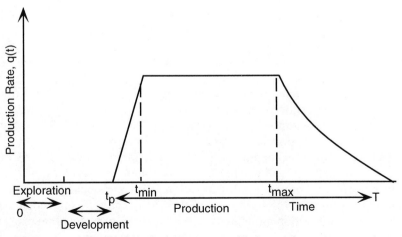

FIGURE 11.1   Simplified production rate curve with time used to assess economic worth of a project.

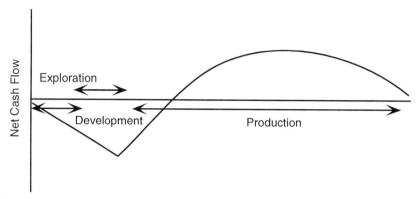

FIGURE 11.2 Net cash flow with time during exploration, development, and production phases of a project using the simplified production rate curve of Figure 11.1.

to start at time t_p after exploration commencement (which occurs at time $t = 0$). Production rate then increases linearly from zero with time until time t_{min}, production rate is then at its maximum value q_{max} from $t = t_{min}$ until $t = t_{max}$. Over the same time period the net cash flow will follow the general shape of the curve given by the solid line in Figure 11.2.

For times greater than t_{max}, but less than the economic lifetime T of the total project (production lifetime: $T - t_p$), the production rate is taken to drop exponentially with time.

Although other behaviors could be chosen to represent production rates, the form used is of relatively good fit, in a broad sense, to many real-life production curves and, in the sense of using the production curve to assess impact on exploration economic evaluation, is more than adequate. Expressed quantitatively one can write the planned production rate $q(t)$ in the form

$$q(t) = \begin{cases} 0, & \text{for } 0 \leq t \leq t_p; \quad (11.1a) \\ q_{max}(t - t_p)/(t_{min} - t_p), & \text{for } t_p \leq t \leq t_{min}; \quad (11.1b) \\ q_{max} & \text{for } t_{min} \leq t \leq t_{max}; \quad (11.1c) \\ q_{max} \exp[-\lambda(t - t_{max})], & \text{for } t_{max} \leq t \leq T, \quad (11.1d) \end{cases}$$

where the decline constant λ is related to the total estimated recoverable reserves R by

$$R = \int_{t_p}^{T} dt\, q(t)$$

$$\equiv \frac{1}{2} q_{max}(t_{min} - t_p) + q_{max}(t_{max} - t_{min}) \quad (11.2)$$

$$+ q_{max}[1 - \exp\{-\lambda(T - t_{max})\}]/\lambda.$$

From estimates of the reserves R, the maximum production rate q_{max}, and associated times (t_p, t_{min}, t_{max}, and T) and with the assumption $\lambda(T - t_{max}) \gg 1$, an estimate of the decline constant λ is then provided by

$$\lambda = \left[R/q_{max} + \frac{1}{2}(t_{min} + t_p) - t_{max} \right]^{-1}. \quad (11.3)$$

The problem is to provide estimates of q_{max}, t_p, t_{min}, t_{max}, and T for a given estimate of the recoverable reserves R that will maximize the total gains to the corporation. This problem is considered in the second part of the economic evaluation.

With the planned production model given by Eq. (11.1) the next problem is to develop an appropriate economic model.

III. A SIMPLE ECONOMIC MODEL

To provide a consistent basis for evaluation of the exploration project deal with constant-value dollars at time $t = 0$, that exploration commences.

A. REVENUE

Over the total life ($0 \leq t \leq T$) of the exploration product, it follows that the constant-value present worth total revenue (*CPWR*) is given by

$$CPWR = \int_0^T q(t)S(t) \exp[-(i + min)t] \, dt, \quad (11.4a)$$

where i is the rate of inflation; min is the minimum acceptable rate of return, which includes the discount factor, escalation, and any other pertinent factors; and $S(t)$ is the selling price of product (dollars/STB) at the instant of time t [i.e., $S(t)$ is measured in *current* dollars/STB].

B. PRODUCTION COSTS

In addition to revenue there are, of course, costs associated with production that are usually split into fixed production costs at each instant of time plus a cost per STB produced, which also varies with time. In terms of total constant present worth production costs (*CPWPC*) over the life of the exploration project one can write

$$CPWPC = \int_0^T [q(t)cb(t) + fp(t)] \exp(-it) \exp(-\Psi t) \, dt, \quad (11.4b)$$

where $cb(t)$ represents the current dollar cost/STB of production and $fp(t)$ the fixed costs (current dollars). The factor $\exp(-it)$ represents the conversion to constant worth dollars at the start of exploration ($t = 0$), after the factor $\exp[-\Psi t]$ has allowed for escalation in the costs of doing business above and beyond the general inflation factor. The fixed production costs, $fp(t)$, will be zero in $0 \le t \le t_p$ because production rate occurs only for $t \ge t_p$.

Three further costs need to be factored into the exploration economic model. They are exploration costs, development costs, and depreciation recovery on exploration and development costs.

C. EXPLORATION COSTS

At the start of exploration a rate of expending cash is required in order to obtain a discovery. This flow of cash is conveniently partitioned into pure exploration costs (whether one finds hydrocarbons or not) and development costs (after one has found hydrocarbons). The former operates in a time interval $0 \le t \le t_{D1}$, while development costs operate in $t_{D1} \le t \le T$. (The fact that a well is producing does not necessarily mean that development of a field ceases. The end of development time, t_{D2}, is often the same as the total economic lifetime T.) For an ongoing rate of exploration expenditure $\dot{E}(t)$, which can itself vary with time, the total constant-value present worth exploration expenditure (*CPWEE*) is given by

$$CPWEE = \int_0^T \dot{E}(t) \exp[-(i + x)t]\, dt, \tag{11.4c}$$

where $\dot{E}(t) = 0$ in $t > t_{D1}$, and where the factor $\exp(-xt)$ measures the hidden costs of doing exploration over and above those accounted for by the general rate of inflation, measured through $\exp(it)$.

D. DEVELOPMENT COSTS

At time t the total amount expended in exploration costs is

$$E(t) = \int_0^t \dot{E}(t)\, dt. \tag{11.5a}$$

Likewise the total investment for development is

$$D(t) = \int_0^t \dot{D}(t)\, dt, \tag{11.5b}$$

where $\dot{D}(t)$ is the ongoing rate of development expenditure.

Thus the total constant present worth of development expenditure (*CPWDE*) is given by

$$CPWDE = \int_0^T \dot{D}(t) \exp[-(i + y)t]\, dt, \tag{11.6}$$

where $\dot{D}(t) = 0$ in $t < t_{D1}$, and where the factor $\exp(-yt)$ measures the fractional escalation of costs to be charged to the development phase of the project over and above charges accounted for by the general rate of inflation, $\exp(-it)$.

E. DEPRECIATION RECOVERY

In most situations the development and exploration costs can be depreciated on an ongoing basis at some fractional rate, and so be used to lower the taxes charged against producing wells. For a fractional depreciation rate $FD(t)$, which may itself vary with time, and with an investment in exploration $E(t)$, and development costs $D(t)$ at time t, the total constant present-worth depreciation relief against taxes ($CPWDR$) is given by

$$CPWDR = \int_0^T [\dot{D}(t) + \dot{E}(t)]FD(t) \exp[-(i+z)t]\, dt, \qquad (11.7)$$

where $\exp(-zt)$ measures the costs of doing business in order to maintain a depreciation relief position over and above charges accounted for by the general rate of inflation.

F. NET CASH FLOW

With the individual components of the cash flow picture defined as above one can now use fiscal arguments to obtain a total net cash flow present worth in constant dollars (CF). For simplicity in illustration we make the assumptions:

1. The constant-value present worth royalty ($CPWROY$) is a fixed fraction f of the constant-value present worth revenue.
2. The constant-value present worth tax ($CPWTAX$) is a fixed fraction τ of $\{CPWR - CPWROY - CPWPC - CPWDR\}$.

It then follows that the total cash flow CF to the corporation (in constant-value present-worth dollars) throughout the life of the exploration project is given by

$$CF = [CPWR - CPWROY - CPWPC - CPWTAX - CPWEE - CPWDE];$$

that is,

$$\begin{aligned} CF = {} & CPWR(1-f)(1-\tau) - CPWPC(1-\tau) \\ & - CPWEE - CPWDE + \tau CPWDR. \end{aligned} \qquad (11.8)$$

Equation (11.8) expresses the total cash flow to the corporation in terms of investment and revenue. The important point to remember is that each of the individual factors on the right-hand side of Eq. (11.8) is directly or indirectly dependent on production. Production in turn depends on the six parameters, q_{max}, t_D, t_p, t_{min}, t_{max}, and T, so that CF is a function of these

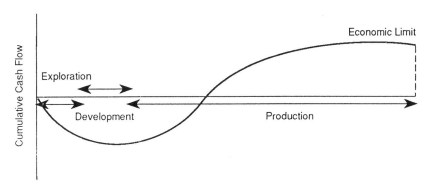

FIGURE 11.3 Cumulative cash flow with time for the simplified economic model of Figure 11.2.

parameters. Equation (11.8) provides a complex formula from which to maximize CF with respect to the six parameters for a given estimate of reserves and so obtain an assessment of the worth of the exploration project to the corporation. A sketch of expected cumulative cash flow with time is given in Figure 11.3.

The *total* estimated gains, G, to the corporation from the exploration opportunity are then given by

$$G = CPWR(1 - f)(1 - \tau), \qquad (11.9a)$$

while the total estimated costs to the corporation involved in the opportunity are

$$C = CPWPC(1 - \tau) + CPWEE + CPWDE - \tau CPWDR \qquad (11.9b)$$

so that the value, $V(\equiv G - C)$, is the same as the total cash flow, CF. Note that both the assumptions and parameters of the economic model, and of the scientific model, are now intimately intertwined in the estimates for gains, G, and total costs, C, of the opportunity. (Note that this total cost, C, is *not* the same as the cost estimate at risk in previous chapters.) The production model depends on both the reserve estimates and types of hydrocarbons estimated to be present, and the economic model depends, in turn, on the production model used, together with additional economic parameters.

Tracing the various parameter dependencies through any particular functional models of production and economics is not difficult, but is labor intensive, which is why different models are often computed numerically.

In particular, uncertainties in the values to use for the many parameters entering both the production and economic models make it quite difficult to see analytically which parameter values are causing the largest components of uncertainty in estimates of the value of the exploration opportunity.

IV. COMBINED PRODUCTION AND ECONOMIC MODELS

To keep the present exposition as simple as possible, consider that all factors of inflation, escalation, and discount are set to zero; that the production rate with time is completely dominated by the maximum value interval [so that $q_{max} \cong R/(t_{max} - t_{min})$] and $T \cong t_{max}$; and that the current dollar cost/STB of production $cb(t)$ is constant as are the fixed costs per unit time, $fp(t)$. Then:

$$CPWR = q_{max} \int_{t_{min}}^{t_{max}} S(t)\, dt, \tag{11.10a}$$

$$CPWPC = (q_{max}cb + fp)(t_{max} - t_{min}). \tag{11.10b}$$

The total constant-value exploration costs (from $t = 0$ to $t = t_{max}$) can be written

$$CPWEE = \langle \dot{E} \rangle t_{max}, \tag{11.10c}$$

where $\langle \dot{E} \rangle$ is the rate of exploration expenditure averaged to the end of the production lifetime; the total constant-value development expenditure can equally be written

$$CPWDE = \langle \dot{D} \rangle t_{max}. \tag{11.10d}$$

For a constant fractional depreciation relief, FD, one then has

$$CPWDR = FD(\langle \dot{E} \rangle + \langle \dot{D} \rangle)t_{max}. \tag{11.10e}$$

The total gains from the opportunity can then be estimated as

$$G = R(1-f)(1-\tau)\left[(t_{max} - t_{min})^{-1} \int_{t_{min}}^{t_{max}} S(t)\, dt\right], \tag{11.11a}$$

or, in terms of the time-averaged product selling price per bbl, $\langle S \rangle$, over the production lifetime, one can write

$$G = R\langle S \rangle(1-f)(1-\tau). \tag{11.11b}$$

Equally, the total costs for the opportunity can be estimated as

$$C = [R\,cb + fp(t_{max} - t_{min})](1-\tau) + (\langle \dot{E} \rangle t_{max} + \langle \dot{D} \rangle t_{max})(1 - FD\tau), \tag{11.11c}$$

or, in terms of the total fixed development cost, $FP[\equiv fp(t_{max} - t_{min})]$, the total exploration costs $E(\equiv \langle \dot{E} \rangle t_{max})$, and the total development costs $D(\equiv \langle \dot{D} \rangle t_{max})$, one can write

$$C = (R\,cb + FP)(1-\tau) + (E + D)(1 - \tau FD). \tag{11.11d}$$

The value, $V(\equiv G - C)$, can then be written

$$V = R\langle S\rangle(1 - f)(1 - \tau) - (1 - \tau)(R\,cb + FP) - (E + D)(1 - \tau FD), \quad (11.11e)$$

which includes producible reserve estimates, production, development, and exploration estimates, and time-averaged selling price estimate. The expected value of the opportunity, $E_1(V)$, is then given through

$$E_1(V) \equiv p_s G - C = p_s(1 - f)(1 - \tau)R\langle S\rangle - (1 - \tau)(R\,cb + FP) \\ - (E + D)(1 - \tau FD), \quad (11.12)$$

which includes the geological success probability, p_s, of finding hydrocarbons. Thus, even in this highly simplified model one has the geologic, production, and economic model factors intertwined.

Usually, in a fairly long-lived field, the total development costs over the field lifetime are proportional to the total sustained production so that, for illustrative purposes, one can write

$$D = \mu R, \quad (11.13)$$

where μ is the development cost per bbl of product. Then one has

$$G = R\langle S\rangle(1 - f)(1 - \tau), \quad (11.14a)$$
$$C = R[cb(1 - \tau) + \mu(1 - \tau FD)] + (1 - \tau)FP + E(1 - \tau FD). \quad (11.14b)$$

V. MINIMUM AVERAGE SELLING PRICE

For any specific choices of parameters in Eqs. (11.14) there is an absolute minimum value of $\langle S\rangle$ that is necessary, but *not* sufficient, to obtain a positive value of $V(\equiv G - C)$ because

$$G - C = R[\langle S\rangle(1 - f)(1 - \tau) - cb(1 - \tau) - \mu(1 - \tau FD)] \\ - (1 - \tau)FP - E(1 - \tau FD). \quad (11.15)$$

Thus the average selling price per bbl of product, $\langle S\rangle$, must certainly exceed

$$S_{\min} = [cb(1 - \tau) + \mu(1 - \tau FD)]/[(1 - f)(1 - \tau)] \quad (11.16)$$

if there is to be any hope of realizing a profit. To make $S_{\min}$ as low as possible, development costs per bbl of product, μ, and production costs per bbl of product, cb, must be kept as low as possible. However, even when $\langle S\rangle > S_{\min}$, one only has a necessary, but not sufficient, condition because to have $G \geq C$ requires in addition that

$$\langle S\rangle \geq S_{\min} + [(1 - \tau)FP + E(1 - \tau FD)] \times [R(1 - f)(1 - \tau)]^{-1} \equiv S_*. \quad (11.17)$$

Thus, either the anticipated reserves, R, have to be sufficiently large that total exploration costs, E, and total fixed development costs, FP, per bbl of product (E/R, FP/R) are negligible; or one must minimize FP and E to ensure the smallest value for S_*, so that the minimum average selling price, $\langle S \rangle$, per bbl of product can be kept as low as possible, yet still return a profit to the corporation *if* reserves were to be discovered. Alternatively, if the expected value $E_1(V)(\equiv p_s G - C)$ is required to be positive for the exploration opportunity, then the time-averaged selling price, $\langle S \rangle$, would be required to exceed the even larger minimum value of

$$\langle S \rangle \geq S_{\min} + p_s^{-1}(S_* - S_{\min}) \equiv S_1 \geq S_*. \quad (11.18)$$

Of course, because each of the parameters entering the combined production/economic model is uncertain, as are the parameters R and p_s from the geologic model, there will be a corresponding uncertainty in the minimum selling price required to make a profit. This aspect is considered next.

VI. PROBABLE PROFIT INCLUDING UNCERTAINTIES

The uncertainties in geological success probability, p_s, in reserve assessment, R, in selling price per unit of product, S, and also in each and every parameter entering geologic, production, and economic models influence the probability of a venture being profitable.

To illustrate succinctly the procedure for assessing likely profit when uncertainty prevails, consider again the simple model of the previous section. Let the average selling price be $\langle S \rangle$ and let the variance in this price be σ_s^2; let the average success probability be estimated as $\langle p_s \rangle$ with a variance σ_p^2; let the average estimated reserves be $\langle R \rangle$ with a variance σ_R^2. Assume, *for simplicity of illustration only,* that no other factors have any uncertainty at all. Then the average value, E_1, itself has an overall average of

$$\langle E_1 \rangle = \langle R \rangle \{\langle p_s \rangle [\langle S \rangle - S_{\min}] - [\langle S_* \rangle - S_{\min}]\}, \quad (11.19a)$$

where $\langle S_* \rangle$ is the average of S_* from Eq. (11.17), together with a variance, σ_E^2, given through

$$\sigma_E^2 = [\langle p_s \rangle (1 - \langle p_s \rangle) - \sigma_p^2][\langle R \rangle^2 + \sigma_R^2][\langle S \rangle^2 + \sigma_s^2]\zeta^2, \quad (11.19b)$$

where $\zeta = (1 - f)(1 - \tau)$.

The probability of obtaining a positive profit from the opportunity can then be written

$$P_+ = (2\pi\sigma_E^2)^{-1/2} \int_0^\infty \exp[-(V - \langle E_1 \rangle)^2/(2\sigma_E^2)]\, dV, \quad (11.20)$$

which can be rewritten in the form

$$P_+ = (2\pi)^{-1/2} \int_{-a}^{\infty} \exp(-x^2/2)\, dx, \tag{11.21}$$

where $a = \langle E_1 \rangle / \sigma_E$.

Note that if $a < 0$, corresponding to a negative expected average $\langle E_1 \rangle$, then the more negative a becomes, the lower is the probability, P_+, of a positive profit; that is, the smaller the average selling price, the less chance there is of making money. If the fluctuations, σ_s, in average selling price are extremely large, to the point that $|\langle E_1 \rangle|/\sigma_E \ll 1$, then $P_+ \cong 1/2$, so that the average selling price is not a big issue, but the fluctuations in selling price are then most certainly relevant because they control the chance of making a profit.

The point here is to illustrate that once one includes all fluctuations of all parameters in the geologic, production, and economic models, then one can calculate their influence on expected value and on the uncertainty of the expected value. Then the calculation of the probability of obtaining a profit can easily be given including the effects of each uncertainty.

More complex models change the technical details of the procedure but do not change the general logic path presented here with the simplified model.

Note that in the event that the average selling price is extremely large, so that one can approximate $\langle E_1 \rangle$ by

$$\langle E_1 \rangle \cong \langle R \rangle \langle p_s \rangle \langle S \rangle \tag{11.22a}$$

and σ_E^2 by

$$\sigma_E^2 = [\langle p_s \rangle(1 - \langle p_s \rangle) - \sigma_p^2][\langle R \rangle^2 + \sigma_R^2]\zeta^2 \langle S \rangle^2, \tag{11.22b}$$

then the ratio $\langle E_1 \rangle/\sigma_E$ becomes independent of the average selling price. The probability of a profitable outcome then depends only on the value of

$$a = \langle R \rangle \langle p_s \rangle \zeta^{-1}[\langle R \rangle^2 + \sigma_R^2]^{-1/2}[\langle p_s \rangle(1 - \langle p_s \rangle) - \sigma_p^2]^{-1/2} > 0, \tag{11.23}$$

which is dominated by the average success probability estimate and its uncertainties. Indeed, for a large average reserve estimate, with $\langle R \rangle \gg \sigma_R$, one has

$$a \cong \zeta^{-1} \langle p_s \rangle/[\langle p_s \rangle(1 - \langle p_s \rangle) - \sigma_p^2]^{1/2} > 0 \tag{11.24}$$

showing directly the strong dependence on $\langle p_s \rangle$.

Because $a > 0$, in this case there is a greater than 50% chance ($P_+ > 1/2$) that the venture will be profitable. But as average selling price becomes lower and more uncertain, and as uncertainty in reserves and success probability increases, there is less and less likelihood of a positive outcome ($P_+ < 1/2$).

In fact, even if there are no uncertainties in any of the parameters entering the geologic, production, or economic models there is still an uncertainty on probable profitability of the opportunity because $E_1(\equiv p_s G - C)$ also has the variance $\sigma^2 = p_s(1 - p_s)G^2$. Thus the probability of profit in the *absence* of uncertainties is

$$P_+ = (2\pi)^{-1/2} \int_{-a}^{\infty} \exp(-x^2/2)\, dx,$$

where, now,

$$a = (p_s - C/G)/[p_s(1 - p_s)]^{1/2}.$$

VII. NUMERICAL ILLUSTRATIONS

Consider a simple numerical example where base case values for geologic, production, and economic parameters are $1/bbl development cost ($\mu$); $2/bbl production cost ($cb$); tax fraction is 50% ($\tau = 0.5$); royalty fraction is 30% ($f = 0.3$); fractional depreciation relief is 50% ($FD = 0.5$); total exploration cost $E = \$50$ MM; total fixed development cost of \$70 MM ($\equiv FP$). Then, from Eqs. (11.14a) and (11.14b) one can write

$$G(\$MM) = 0.35RS,$$

$$C(\$MM) = 1.85R + 35 + 37.5 \equiv 1.85R + 72.5,$$

where the estimated reserves are in units of MMbbl. It then follows that

$$G - C = R(0.35S - 1.85) - 72.5.$$

Hence,

$$S_{min} = \$5.29/\text{bbl}, \tag{11.25a}$$

$$S_* = \$\left(5.29 + \frac{207}{R}\right)\bigg/\text{bbl}, \tag{11.25b}$$

and

$$S_1 = \$(5.29 + p_s^{-1} 207/R)/\text{bbl}. \tag{11.25c}$$

For an estimated reserve size of 100 MMbbl of recoverable oil, $S_* = \$7.36$/bbl; while at a typical success chance of $p_s \approx 0.25$, one has

$$S_1 \cong \$13.57/\text{bbl}. \tag{11.26a}$$

Thus, a minimum average selling price of about $S_1 \approx \$13-\$14$/bbl is required for break-even operations, provided there is no uncertainty on any parameter entering the estimate.

Suppose, then, that the success probability is uncertain by 20% ($\sigma_p = 0.2 p_s$); that the recoverable reserve assessment is uncertain by 30% ($\sigma_R = 0.3R$); and that there is some fractional uncertainty, h_s, on the actual average selling price, $\langle S \rangle (\sigma_s \equiv h_s \langle S \rangle)$.

Then from Eq. (11.19b) one has

$$\sigma_E^2 = \langle R \rangle^2 \langle S \rangle^2 [1 + h_s^2](0.7 \times 0.5)^2 \left(\frac{2.99}{4}\right),$$

while from Eq. (11.19a) one has

$$\langle E_1 \rangle = 25(\langle S \rangle - 5.29) - 207 \times 0.954 = 25 \langle S \rangle - 331.$$

For a positive value of $\langle E_1 \rangle$, one requires an average selling price in excess of

$$\langle S \rangle \geq \$(331/25)/\text{bbl} = \$13.25/\text{bbl}, \tag{11.26b}$$

which is slightly less than the estimate of \$13.57/bbl, which excluded uncertainty on recoverable reserves.

Then,

$$\langle E_1 \rangle = 25[\langle S \rangle - 13.25], \tag{11.27a}$$

$$\sigma_E = 30.3 \langle S \rangle (1 + h_s^2)^{1/2}. \tag{11.27b}$$

Thus, the ratio of $\langle E_1 \rangle$ to σ_E takes on the value

$$a = \langle E_1 \rangle / \sigma_E = 0.825[1 - 13.25/\langle S \rangle](1 + h_s^2)^{-1/2}. \tag{11.28a}$$

For an average selling price, $\langle S \rangle$, of around \$20/bbl, one has

$$a = 0.27(1 + h_s^2)^{-1/2}. \tag{11.28b}$$

If there were no fluctuation in the average selling price ($h_s = 0$), then the probability of a profitable outcome is

$$P_+(h_s = 0) = (2\pi)^{-1/2} \int_{-0.27}^{\infty} \exp(-x^2/2)\, dx$$

$$\cong 0.60.$$

Thus, there is a 60% chance of a profit given the uncertainties in success probability and reserve assessments. If the average selling price were to be uncertain by 50% ($h_s = 1/2$), then $a = 0.24$, and the corresponding value of profitable outcome is reduced to $P_+(h_s = 1/2) = 0.59$, an almost negligible reduction factor. The reason for this insensitivity is because the average selling price is already high (\$20/bbl), so that even comparatively large fluctuations in average selling price have only small effects. A factor of 2 uncertainty in selling price ($h_s = 2$) would yield $a = 0.01$, which reduces the probability of a profitable outcome to just over 50%. The point here is that once the average selling price is fairly large compared to the minimum, then there is always a fairly good chance of a profitable outcome. Criticality occurs when $\langle S \rangle$ is close to S_1, when all uncertainties in all parameters can have major impacts on the probable profitability.

12

BAYESIAN UPDATING OF AN OPPORTUNITY

I. INTRODUCTION

Once all geologic, production, and economic model assessments are completed, including estimated ranges of uncertainty, then one has available not only a sensitivity analysis of which factors and parameters are causing the largest uncertainties in a risk assessment of the opportunity, but also some idea of optimum working interest, the value of adding information, the optimum bid to make, and the worth of a particular opportunity in relation to other competing opportunities and in relation to available budget. Further, one then also has ranges of uncertainty of each of these quantities related to the uncertainties in model parameters.

At this stage in an exploration risk analysis, the corporate decision makers then commit to a strategy for spending money. The result is usually that several blocks are acquired and one goes through the evaluation of where to drill wells, how many to drill, etc.

The difficulty one now faces stems from the actual drilling undertaken: Either the company is successful in finding hydrocarbons in the borehole or it is not. In the case of a successful find, a post-drill review is made to compare the actual parameters (oil/gas type, reservoir petrophysical properties, economic worth, etc.) to those predicted; in the case of an unsuccessful well, a post-drill review is made to determine the causes of the failure. In either event, new information is available that is used to update the risk assessment of the opportunity in relation to decisions to (1) continue with the prior strategy of evaluation, (2) modify the strategy based on the information uncovered, or (3) abandon the opportunity. Thus the whole of the process of exploration risk analysis for the opportunity is reevaluated based on the well information. The updating changes prior assessments of probabilities based on the later information acquired, and so impacts all of the geologic, production, and economic assessments. This sort of updating of worth of an opportunity is an ongoing process as a field is explored, developed, and produced. For instance, in the late stages of decline of an oil field, there arises the question of whether it is economically worthwhile to sink another well versus, say, doing a water-flood injection with the existing wells. At all stages of investigation of an opportunity, the updating continues until the field is abandoned. This use of *a posteriori* information to update prior assessments of worth is usually referred to as Bayesian updating, after Bayes who formulated the basic concepts (see Jaynes, 1978, for an historical account of development of the method).

In this chapter the general method is presented within the framework of exploration risk analysis only. To include detailed aspects of Bayesian updating as they pertain to the diverse endeavors of a petroleum corporation (including refining, marketing, new nonpetroleum ventures, etc.) would make for a very long treatise indeed, and also one that is well beyond the authors' abilities.

II. GENERAL CONCEPTS OF BAYESIAN UPDATING

The sense of the main idea is best expressed through an example. Suppose that two radically different geological scenarios have been proposed for an exploration opportunity based on the available, incomplete data at hand; label the scenarios A and B. One might, for instance, consider that one scenario is a passive margin setting in quiet fluviodeltaic conditions, whereas the second scenario might consider that turbidite deposition is prevalent. Both geological scenarios cannot be correct simultaneously—but both could be incorrect simultaneously.

On the basis of each scenario, one performs a complete exploration risk analysis, thereby generating success probabilities, $p(S|A)$ and $p(S|B)$, which are conditional on each of scenarios A and B being assumed valid, respectively. One also assigns a probability of each scenario being correct, $p(A)$ and $p(B)$, respectively, with the sum being unity $[p(A) + p(B) = 1]$ when no other scenario is possible. Then the probability of success, $p(S)$, *irrespective of which scenario is correct,* is just

$$p(S) = p(S|A)p(A) + p(S|B)p(B). \qquad (12.1)$$

Now suppose that a successful outcome to drilling actually *does* occur. One wishes to update the probability of each scenario being correct based on that *a posteriori* information; that is, one wishes to calculate $p(A|S)$, the probability of scenario A being correct given that event S occurs and, correspondingly, $p(B|S)$. Bayes showed that

$$p(A|S)p(S) = p(S|A)p(A) \tag{12.2}$$

so that the updated probability of scenario A being correct given that S occurred is

$$p(A|S) = p(S|A)p(A)/[p(S|A)p(A) + p(S|B)p(B)]. \tag{12.3}$$

With a set of possible scenarios, A_i, $(i = 1, \ldots, N)$ the generalization of Eq. (12.3) is

$$p(A_i|S) = p(S|A_i)p(A_i) \bigg/ \left[\sum_{j=1}^{N} \{p(S|A_j)p(A_j)\} \right]. \tag{12.4}$$

The general argument does not have to be tailored only to success probability, but is also appropriate for any sort of event S. Thus, the argument could be based on the probability of cementation occurring in a formation, or on the probability of finding gas rather than oil, or on any other event that is then measured. In each case one starts with a set of different "states" A_i, $(i = 1, \ldots, N)$ and assigns the initial probability, $p(A_i)$, of the state A_i being the correct one; then one computes the exploration risk assessment of an event S occurring within the framework of state A_i, and then one uses later observations of the occurrence (or nonoccurrence) of event S to update the initial probability of state A_i being correct. The updated probabilities can then be used as initial probabilities and the process repeated, with new *a posteriori* information being sequentially added to update continuously the state probabilities. Thus, continuous updating of resource assessments and risk assessments of an opportunity can be made throughout the life of the opportunity—and the life is itself then determined by the economic worth of continuing to develop the opportunity based on all information used in a Bayesian update manner.

III. NUMERICAL ILLUSTRATIONS

The ability to use the Bayesian method to update probabilities of success based on new information is one of the most powerful tools available and operates in many settings, not just in the field of hydrocarbon exploration. Here, three illustrations of the procedure are given to provide some familiarity of the many possible uses; one example is from the disease-prevention area, the other two are from the hydrocarbon exploration theater.

A. TESTING FOR HIV

One of the major health concerns in the late twentieth century is the spread of AIDS. While no cure is available for AIDS, there is a test available to assess whether one has HIV, the precursor of AIDS.

Suppose, then, one takes the test with a negative result. Then either one does not have HIV, or the virus has not developed sufficiently to show up on the test, or the test is not 100% accurate. One can keep taking the test at prescribed time intervals. Suppose one does so, each time with a negative result. How many times should one take the test to ensure that the probability that one has HIV is less than, say, 1%? Bayesian methods permit one to use the negative results of each test to update the probability each time as follows.

Let the probability be $p(HIV)$ that one has HIV. Let $p(-\text{test}|HIV)$ be the probability that the test result is negative given that one *does* have HIV; and let $p(-\text{test}|NOHIV)$ be the probability that the test result is negative given that one *does not* have HIV. [If the test is perfect then $p(-\text{test}|NOHIV) = 1$; if the test is imperfect then $p(-\text{test}|NOHIV) \neq 1$.] Then the two "states" available are $A = HIV$, $B = NOHIV$. Use of Eq. (12.3) can be made, where $S =$ the event that the test result is negative. Then with $p(HIV|-\text{test})$ as the probability of one having HIV given that the test result is negative, the Bayes formula permits one to write

$$p(HIV|-\text{test}) = p(-\text{test}|HIV)p(HIV) \times [p(-\text{test}|HIV)p(HIV) \\ + p(-\text{test}|NOHIV)\{1 - p(HIV)\}]^{-1}. \qquad (12.5)$$

For brevity, denote $p(-\text{test}|HIV)$ by $1 - w$ and $p(-\text{test}|NOHIV)$ by $\in$ (with $\in = 1$ if the test is perfect); the probability that one has HIV given that the test is negative has been updated from $p(HIV)$ to $p(HIV|-\text{test})$.

If one kept taking the test, then after n negative result tests, the probability of one having HIV is

$$P_n \equiv p(HIV|-\text{test})_n = \frac{(1-w)^n p(HIV)}{(1-w)^n p(HIV) + \in [1 - p(HIV)]} \qquad (12.6)$$

As the number of negative results tests increases, the probability, P_n, that one has HIV becomes smaller and smaller. One can also write the result as an expression for the number, n, of tests as

$$n = \ln[\in P_n\{1 - p(HIV)\}/\{p(HIV)(1 - P_n)\}]/\ln(1 - w), \qquad (12.7)$$

which can be used to determine how many tests (with continuing negative results) to take in order to reach some preassigned level of confidence that one does *not* have HIV. For instance, suppose the test is 90% accurate so that $\in = 0.9$. Suppose the initial assessment that one has HIV is 50:50, so that $p(HIV) = 1/2$; and suppose the probability of a negative test result given that one *does* have HIV is only 5% (i.e., $w = 0.95$). The number of tests with negative results that one needs to take to ensure that there is only 1% chance ($P_n = 10^{-2}$) that one has HIV is then given by

$$n = \ln\{0.9 \times 10^{-2} (1 - 1/2)/[1/2(1 - 10^{-2})]\}/\ln(1 - 0.95), \qquad (12.8)$$

which works out to be $n = 1.6$; that is, after two tests with negative results, one is 99% sure that one does not have HIV.

If a test result ends up positive then one has, correspondingly.

$$p(HIV| + \text{test}) = \frac{wp(HIV)}{wp(HIV) + [1 - p(HIV)]\delta}, \quad (12.9)$$

where δ = probability of a positive test result given that one *does not* have HIV. If the test is 90% accurate then $\delta = 1 - \epsilon = 0.1$, for the numbers above. And then

$$p(HIV| + \text{test}) = \frac{0.95 \times 1/2}{(0.95 \times 1/2) + 0.1 \times (1 - 1/2)} = 0.905 \quad (12.10)$$

Thus, after one test, the probability of having HIV is

$$P_1 = p(HIV| + \text{test}) + p(HIV| - \text{test}),$$

which, for the numbers given, reduces to

$$P_1 = 0.905 + 0.053 = 0.958.$$

In the case of a negative test result, the initial probability of having HIV is *reduced* from 50% to 5.3%; while in the case of a positive test result, the initial probability of 50% is *increased* to 90.5%. Two negative test results will reduce the probability that one has HIV to less than 1%, whereas two positive test results will not raise the probability that one has HIV much above about 90%, because the limiting accuracy, δ, ($\cong 10\%$) of the test has already been reached.

B. TESTING FOR OIL FIELDS FROM BRIGHT SPOT OBSERVATIONS

The problems with any Bayesian update are effectively the same: One is interested in the probability of state A being correct given that either an event does occur or does not occur; and one is also interested in the probability of state A being correct regardless of whether an event occurs or does not occur.

In a general sense, the Bayesian updating can be put into a simple worksheet as shown in Figure 12.1. The point to note from Figure 12.1 is that the argument on Bayesian updating works both ways: In the language of the previous example one can work out the chance of having HIV given that a test is administered, or one can work out the probability of a positive or negative test result given the chance one has, or does not have, HIV.

A problem closely related in logic structure occurs in hydrocarbon exploration. Seismic evidence is often used to indicate the presence of an anomalously large reflection coefficient over a region ("bright spot"). It has been known for a long time that such bright spots can be caused by free-phase gas trapped in pores. And the gas does not have to be methane; it could be CO_2, H_2, S, N_2, etc. Thus, there is a finite probability, less than unity, that the presence of a bright spot is due to a hydrocarbon gas event. Now

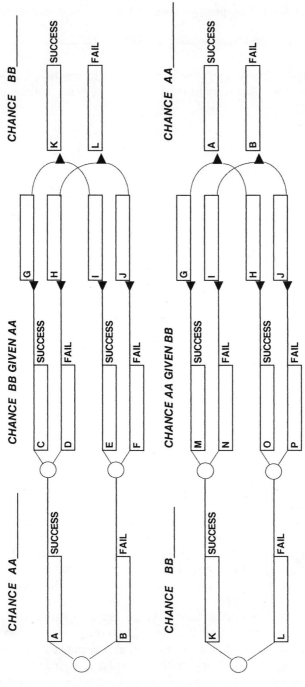

FIGURE 12.1 Bayesian chart flow diagram showing in general how one can compute chances with information updating.

it is also a fact that free-phase gas caps are often found overlying oil fields. Thus there is a probability of finding an oil field when a bright spot is measured on seismic. Figure 12.2 shows how the Bayesian update analysis is performed in such a case for both ways of looking at the problem: the chance of a bright spot given an oil discovery, and the chance of an oil discovery given a bright spot.

Note that, in both cases, one has to be concerned not only with the probability of, say, an oil discovery in the *presence* of a bright spot, but also in the *absence* of such an event. Equally, one can ask the converse question: Given that one has discovered an oil field, one can ask for the probability of there being an associated bright spot, which is also shown in Figure 12.2

C. DECISION TO ABANDON AN OPPORTUNITY

One of the difficult decisions in hydrocarbon exploration is to determine when to abandon an opportunity because the chance of finding an economic reservoir is too small. For instance, suppose that one has drilled three wells on a prospect, with the result that the first is a dry hole, the second shows a thin oil column, and the third is dry. Should one drill more wells to prove out the prospect and, if so, how many? Does one have sufficient information from the three wells already drilled to decide to abandon the opportunity? If one were to drill more wells, what fraction could be expected to be dry holes and what fraction could be expected to be oil bearing? Should one have drilled all of the existing wells or should one have abandoned the opportunity before all of the wells were drilled?

Bayesian updating procedures allow one to address these questions in a logical, rational manner, with quantitative estimates available of the probabilities of success or failure of each choice that one could make. Consider initially the situation *prior* to any wells being drilled. There must have been some exploration risk assessment made at that stage that helped the company to decide to go ahead with drilling. Let the initial probability of finding an economic reservoir be $p(ER)$, so that the probability of *not* finding an economic reservoir is $p(NER) \equiv 1 - p(ER)$.

Four conditional probabilities have to be considered:

1. The probability of drilling a dry hole given that an economic reservoir does exist, $p(DH|ER)$

2. The probability of drilling and encountering an oil column given that an economic reservoir does exist, $p(OC|ER)$

3. The probability of drilling a dry hole given that no economic reservoir exists, $p(DH|NER)$

4. The probability of drilling and encountering an oil column given that an economic reservoir does not exist, $p(OC|NER)$. [There can still be an

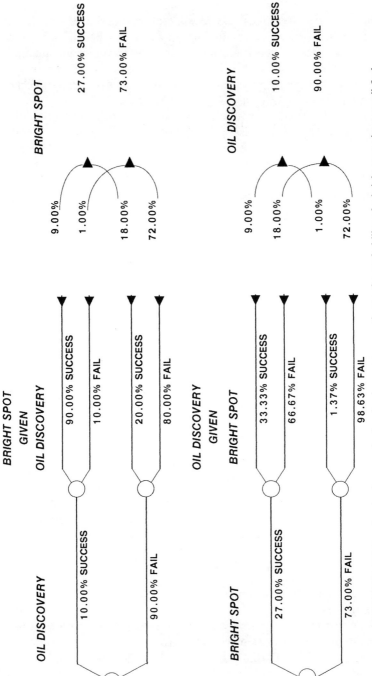

FIGURE 12.2 A seismic example showing how one can estimate the probability of a bright spot given an oil find and, equally, how one can compute the probability of an oil find given a bright spot.

oil-charged reservoir even if it is too small to be economically exploited so that $p(OC|NER)$ is not zero]

Initial estimates of these four conditional probabilities must also be provided from the original exploration risk assessment stage.

Suppose now that on the basis of available information, prior to any drilling, the exploration risk team makes the assessment of a 30% chance of drilling a dry hole and a 70% chance of encountering an oil column, if an economic reservoir exists [i.e., $p(DH|ER) = 0.3$; $p(OC|ER) = 0.7$]; while if there is no economic reservoir, the assessment calls for 80% chance of a dry hole and only 20% chance of finding an oil column [i.e., $p(DH|NER) = 0.8$; $p(OC|NER) = 0.2$]. Consider how matters are updated as each well is drilled and as the results of drilling are incorporated into the assessment of the prospect.

Assume also that the initial assessment of the prospect is also highly optimistic, so that the probability of encountering an economic reservoir is set at 60% [i.e., $p(ER) = 0.6$; $p(NER) = 0.4$].

1. After the First Well; a Dry Hole

Here one is interested in two updated probabilities: (1) $p(ER|1DH)$—the probability of an economic reservoir given that the first well is a dry hole and (2) $p(NER|1DH)$—the probability of no economic reservoir given that the first well is a dry hole.

From Bayes's general formula of Eq. (12.2) one has

$$p(ER|1DH) = p(DH|ER)p(ER) \times [p(DH|ER)p(ER) + p(DH|NER)p(NER)]^{-1}, \quad (12.11)$$

which, inserting the numerical values, yields

$$p(ER|1DH) = \frac{0.3 \times 0.6}{[0.3 \times 0.6 + 0.8 \times 0.4]} = 0.36 \equiv p_1(ER), \quad (12.12)$$

while the probability of no economic reservoir, given the first dry hole, is

$$p(NER|1DH) = p(DH|NER)p(NER) \times [p(DH|NER)p(NER) + p(DH|ER)p(ER)]^{-1}, \quad (12.13)$$

which, inserting the numerical values, yields

$$p(NER|1DH) = \frac{0.8 \times 0.4}{[0.8 \times 0.4 + 0.3 \times 0.6]} = 0.64 \equiv p_1(NER). \quad (12.14)$$

Thus, the initial assessment of a 60% chance of an economic reservoir has dropped to a 36% chance as a result of the first dry hole, while the chance of no economic reservoir has increased from 40% to 64%.

2. After the Second Well, an Oil Column Find

From Bayes's general formula of Eq. (12.2), one computes the updated probabilities of (1) an economic reservoir given that the second well yields an oil column find $p(ER|2OC)$, and (2) no economic reservoir given an oil column find, $p(NER|2OC)$. But one uses the updated probabilities $p_1(ER)$ and $p_1(NER)$, which already contain the changes in success chance brought about by incorporating the dry hole information from the first well. Thus,

$$p(ER|2OC) = p(OC|ER)p_1(ER) \times [p(OC|ER)p_1(ER) + p(OC|NER)p_1(NER)]^{-1} \quad (12.15)$$

which, inserting the numerical values, yields

$$p(ER|2OC) = \frac{0.7 \times 0.36}{(0.7 \times 0.36 + 0.2 \times 0.64)} = 0.66 = p_2(ER), \quad (12.16)$$

while the corresponding updated probability, $p(NER|OC)$, is given by

$$p(NER|2OC) = p(OC|NER)p_1(NER) \\ \times [p(OC|NER)p_1(NER) + p(OC|ER) \\ \times p_1(ER)]^{-1} = 0.34 = p_2(NER). \quad (12.17)$$

Thus, the finding of an oil column in the second well drilled has reversed the bleak picture that resulted from the first dry hole; now the probability of an economic reservoir has jumped from 36% to 66%, and the chance of not finding an economic reservoir has dropped from 64% to 34% as a result of the find.

3. After the Third Well, a Dry Hole

Following the same procedure as previously one can compute

$$p(ER|3DH) = p(3DH|ER)p_2(ER) \\ \times [p(3DH|ER)p_2(ER) \\ + p(3DH|NER)p_2(NER)]^{-1}, \quad (12.18)$$

which, inserting numerical values, yields

$$p(ER|3DH) = 0.42 = p_3(ER), \quad (12.19a)$$

while, correspondingly, one also obtains

$$p(NER|3DH) = 0.58 \equiv p_3(NER). \quad (12.19b)$$

Thus, the optimism, resulting from the previous well with its oil column, is ameliorated by the third well, a dry hole, with the chance of a successful economic reservoir now reduced from 66% to 42%, and the chance of not having an economic reservoir increased from 34% to 58%. These values are almost the same as the initial assessments of 60% and 40% but *reversed*. Thus, the original optimistic chance of 60% success is reduced to 42%, and the failure chance of 40% is increased to 58%.

4. Chances of Success

Based on the preceding updated probabilities, it is possible to assess the likelihood of encountering an economic reservoir on the opportunity.

Suppose that n wells could be drilled on the opportunity. What is the probability that k out of the n wells will result in finding an oil column? That probability is made up by considering two components: the probability of an oil column find given that an economic reservoir exists multiplied by the probability of an economic reservoir; and the probability of an oil column find given that there does not exist an economic reservoir multiplied by the probability that there does not exist an economic reservoir.

Thus, the probability of a well encountering an oil column, $p(OC)$ is

$$p(OC) = p(OC|ER)p(ER) + p(OC|NER)p(NER). \tag{12.20}$$

Prior to any wells being drilled one has

$$p(OC) = 0.7 \times 0.6 + 0.3 \times 0.4 = 0.54; \tag{12.21a}$$

after the first well (a dry hole) is drilled one has

$$\begin{aligned}p(OC) &= p(OC|ER)p_1(ER) + p(OC|NER)p_1(NER) \\ &= 0.7 \times 0.36 + 0.3 \times 0.64 = 0.45;\end{aligned} \tag{12.21b}$$

after the second well (an oil column find) is drilled one has

$$\begin{aligned}p(OC) &= p(OC|ER)p_2(ER) + p(OC|NER)p_2(NER) \\ &= 0.7 \times 0.66 + 0.3 \times 0.34 = 0.56;\end{aligned} \tag{12.21c}$$

while after the third well (a dry hole) is drilled one has

$$\begin{aligned}p(OC) &= p(OC|ER)p_3(ER) + p(OC|NER)p_3(NER) \\ &= 0.7 \times 0.42 + 0.3 \times 0.58 = 0.47,\end{aligned} \tag{12.21d}$$

Then, if one were interested solely in the probability of a well finding an oil column, one would write a success chance as $p(OC)$ and then note that the chance of k wells out of n yielding oil columns is

$$P_n(k) = \frac{n!}{k!(n-k)!} p(OC)^k [1 - p(OC)]^{n-k}. \tag{12.22}$$

However, one is interested in the chance that the k wells yielding oil columns will indicate the presence of an economic reservoir. Thus, one is interested in the probability of obtaining an economic reservoir given that k wells out of n did encounter oil columns multiplied by the probability that k wells out of n find an oil column; that is, one wants to compute

$$\begin{aligned}p(ER|k \text{ out of } nOC) &= p(k \text{ out of } nOC|ER)p(ER) \\ &\quad \times [p(k \text{ out of } nOC|ER)p(ER) \\ &\quad + p(k \text{ out of } nOC|NER)p(NER)]^{-1}.\end{aligned} \tag{12.23}$$

Now the probability of k wells out of n wells yielding an oil column, given that an economic reservoir exists, is just

$$P_n(k|ER) = \frac{n!}{k!(n-k)!} p(OC|ER)^k [1 - p(OC|ER)]^{n-k}, \qquad (12.24)$$

while the probability of k out of n wells yielding an oil column given that an economic reservoir does *not* exist is

$$P_n(k|NER) = \frac{n!}{k!(n-k)!} p(OC|NER)^k [1 - p(OC|NER)]^{n-k}. \qquad (12.25)$$

Thus, the probability of an economic reservoir given that k out of n wells encounter an oil column is

$$\begin{aligned}p(ER|k \text{ out of } nOC) = {}& p(OC|ER)^k [1 - p(OC/|ER)]^{n-k} p(ER) \\ & \times [p(OC|ER)^k [1 - p(OC|ER)]^{n-k} p(ER) \\ & + p(OC|NER)^k \times [1 - p(OC|NER)]^{n-k} \\ & \times p(NER)]^{-1}.\end{aligned} \qquad (12.26)$$

Because $p(OC|ER) = 0.7$, and $p(OC|NER) = 0.3$, one has

$$\begin{aligned}p(ER|k \text{ out of } nOC) = {}& 0.7^k \times 0.3^{n-k} p(ER) \\ & \times [0.7^k \times 0.3^{n-k} \times p(ER) + 0.3^k \\ & \times 0.7^{n-k} \times p(NER)]^{-1},\end{aligned} \qquad (12.27a)$$

which can be rewritten in the simpler form

$$p(ER|k \text{ out of } nOC) = [1 + (7/3)^{n-2k} \times p(NER)/p(ER)]^{-1}. \qquad (12.27b)$$

a. Minimum Acceptable Chance

Each corporation sets a minimum acceptable chance, *MAC*, of an opportunity being economic. Once the probability of an economic reservoir existing drops below the *MAC* then the decision is usually made to abandon the opportunity. Equating the *MAC* with the probability of finding an economic reservoir given that k wells out of n encounter an oil column, that is, $MAC = p(ER|k \text{ out of } nOC)$, one can write

$$(7/3)^{n-2k} p(NER)/p(ER) = (1 - MAC)/MAC, \qquad (12.28)$$

which enables a solution to be given for n as

$$n = 2k + \ln\{[(1 - MAC)/MAC]p(ER)/p(NER)\}/\ln(7/3). \qquad (12.29)$$

Thus, by using the Bayesian method on the results of the first three wells one can determine the number of wells to drill before abandonment should be considered based on both the minimum acceptable chance and on the updated information on $p(ER)$ and $p(NER)$ provided.

After the first well (a dry hole) one has no wells that encountered an oil column ($k = 0$) and $p_1(ER) = 0.36$, $p_1(NER) = 0.64$. If the minimum acceptable chance is 5%, that is, $MAC = 0.05$ then the number of wells that should be drilled is

$$n_1 = \ln\{(0.95/0.05)(0.36/0.64)\}/\ln(7/3) = 3.27. \qquad (12.30)$$

This value is to be compared with the initial assessment [i.e., before $p(ER)$ and $p(NER)$ were updated as a consequence of the dry hole results of the first well] of

$$n_0 = \ln(0.95/0.05)(0.6/0.4)\}/\ln(7/3) = 3.96. \qquad (12.31)$$

Thus, initially one could have anticipated drilling four wells before the MAC would have been achieved but, as a consequence of the negative results (dry hole) from the first well, one should drill only another three to test out the prospect.

After the second well (an oil column find) one has one well ($k = 1$) that has encountered an oil column, and $p_2(ER) = 0.66$, $p_2(NER) = 0.34$, so that

$$n_2 = 2 + \ln\{(0.95/0.05)(0.66/0.34)\}/\ln(7/3) = 6. \qquad (12.32)$$

Thus, a further four wells (one has already drilled two wells) are suggested to prove out the prospect. But after the third well (a dry hole) one still have only one well ($k = 1$) with an oil column find, and the updated probabilities are then $p_3(ER) = 0.42$, $p_3(NER) = 0.58$, so that

$$n_3 = 2 + \ln\{(0.95/0.05)(0.42/0.58)\}/\ln(7/3) = 5.59. \qquad (12.33)$$

Thus, only two more wells should now be authorized (one has already drilled three) and, after each, one should update the probabilities to determine if the prospect will prove out.

b. Answers to Questions Posed

The four questions asked at the beginning of Section III,C can now be addressed, based on the initial and updated probability assessments.

1. Should one drill more wells to prove out the prospect and, if so, how many? The answer is clearly to go ahead with the authorization to commit to two more wells, after the first three, with updating of the likelihood of finding an economic reservoir after the results of each well are known.

2. Does one have sufficient information from the three wells already drilled to decide to abandon the opportunity? The answer is "not yet," because there is still a 42% chance (down from the initial 60% estimate) of finding an economic reservoir.

3. If one were to drill more wells, what fraction could be expected to be dry holes and what fraction could be expected to be oil bearing? Note that this question does *not* equate "oil bearing" with economic reservoir status, so that one is interested in $p(OC)$, which, after the third well drilled, has dropped to 47% from the initial estimate of 60%. Thus, 53% of all wells

drilled could be expected to be dry, and 47% oil bearing *based on the Bayesian updating* from the three wells, with 29% oil bearing if an economic reservoir exists and 18% oil bearing if no economic reservoir is found.

4. Should one have drilled all of the existing wells or should one have abandoned the opportunity before all the wells were drilled? Based on a minimum acceptable chance (MAC) of 5%, the number of wells that should have been drilled varies from an initial estimate of $n_0 = 3.96$ before any drilling, to $n_1 = 3.27$ after the first well, to $n_2 = 6$ after the second well, and to $n_3 = 5.59$ after the third hole. Thus, there is no reason to have abandoned the prospect at all at any stage yet, and encouragement should be given to proceed with two more wells (after the first three) with probability updating using information from each of the new wells, just as was done for the first three wells.

IV. OIL PIPELINE SPILLS

Historically, the transport of crude oil has been by both ship and pipeline, depending on costs to transport. Oil spills from both ships and pipelines have occurred, with an average spill rate (for spills in excess of 1000 bbls per Bbbls transported) of about 0.5 spills/Bbbl transported for ship transport and 1.5 spills/Bbbl transported for pipeline transport. As pipelines age, they become subject to corrosion, both internally from the crude transported and externally from the weather fluctuations throughout the years. Thus, there is a greater chance that an older pipeline will have one or more spills than there is for a newer pipeline. Equally, crude oil transport by ship is subject to spills from storm weather conditions and also from human error or negligence.

Suppose, then, that a corporation has constructed an undersea pipeline for crude oil. The pumping of heavy crude through the pipeline causes some stress on the tie-downs, while the pumping of light crude, or of condensate, will produce a greater stress due to the lower density of the product. Engineering assessments indicate that the stress might reach critical levels to the point of snapping the tie-downs and causing sufficient buckling of the pipeline such that it ruptures, producing a spill. On the other hand, management is of the opinion that the engineering assessments are based on a too rapid rate of corrosion estimate for both the tie-downs and the pipeline, and so considers that, even if the tie-downs fail, the pipeline will have sufficient integrity that built-in safety factors will not be compromised and so no rupture will occur.

After some debate, the corporation decides that management has an 80% chance of being correct, and that the engineering staff were being unduly cautious, but gives them a 20% chance of being correct anyway, just in case.

However, not wanting to be the defendant in many lawsuits if the pipeline were to fail, the corporation also decides to monitor the tie-downs every year with an underwater TV camera dragged along the pipeline to inspect the tie-down failure rate. Management considers that there should be no more than one tie-down failure per year, while the engineers consider that three failures per year is more likely.

During the first year of full operation of the pipeline, with a mix of crudes continually being transported, the TV camera records two tie-down failures; in the second year no further failures are found; while in the third year five more failures occur; and in the fourth year a further two failures occur. The corporation then requests that an evaluation be done to assess whether the engineering position is likely correct, in which case only heavy crude would thereafter be transported in the pipeline. There would then be a loss of considerable income because the lighter crudes cost less to refine than the heavy crude, but with the likely lower possibility of a spill occurring—which would involve expensive cleanup, costly environmental lawsuits, as well as the expense of replacing the ruptured part of the pipeline.

To address this problem the corporation assumes that tie-down failure occurs randomly and independently, and can be addressed using a Poisson process relating the probability, $p(n)$, of n tie-down failures to a failure factor, λ, and elapsed time, t, through

$$p(n) = (\lambda t)^n \exp(-\lambda t)/n!. \qquad (12.34)$$

The average number of failures, $\langle n \rangle$, for the Poisson process is given by

$$\langle n \rangle = \lambda t, \qquad (12.35a)$$

while the mean square number of failures, $\langle n^2 \rangle$, is given by

$$\langle n^2 \rangle = \lambda t + (\lambda t)^2 \qquad (12.35b)$$

so that the variance, σ^2, in the mean is just

$$\sigma^2 = \langle n^2 \rangle - \langle n \rangle^2 = \lambda t. \qquad (12.35c)$$

The probability of no failures in a time t, $p(0)$, is just

$$p(0) = \exp(-\lambda t), \qquad (12.36)$$

which decreases as elapsed time increases, while the probability of $(n + 1)$ or more failures, $P(\geq n + 1)$, is given by

$$P(\geq n + 1) = 1 - \exp(-\lambda t) \sum_{m=0}^{n} \frac{(\lambda t)^m}{m!} \qquad (12.37a)$$

so the probability of $(n + 1)$ or fewer tie-down failures, $P(\leq n + 1)$, is just

$$P(\leq n + 1) = 1 - P(\geq n + 1). \qquad (12.37b)$$

These results will be used later in this section. For the management position of 1 tie-down failure/yr it follows that management assigns $\lambda_M = 1$ yr^{-1}; while the engineering group correspondingly assigns $\lambda_E = 3$ yr^{-1}.

Consider then the Bayesian updating process applied to the 4 years of observations of tie-down failure. The initial probability assignment that management is correct is $P_0(M) = 0.8$, and that the engineers are correct is $P_0(E) = 0.2$.

A. FIRST YEAR UPDATING

The observations indicate two tie-down failures. For the Poisson process, the probability, π_2, of having just two failures and no more or less in the period to $t = 1$ is just the probability of having two failures multiplied by the probability of not having three or more, multiplied by the probability of not having one or less, that is,

$$\pi_2 = p(2)[1 - P(\geq 3)][1 - P(\leq 1)], \tag{12.38a}$$

which can be written (for $t = 1$) as

$$\pi_2 = \frac{1}{2}\lambda^2(1 + \lambda + \lambda^2/2)(1 - e^{-\lambda})\,e^{-2\lambda}. \tag{12.38b}$$

If management is correct, then $\lambda = 1$. The conditional probability $\pi(2/M)$ of just two failures in the first year given that management is correct is obtained by setting $\lambda = 1$ in Eq. (12.38b) to yield

$$\pi(2/M) = 2.5\,e^{-2}(1 - e^{-1}) \equiv 0.107/2. \tag{12.39a}$$

Correspondingly, if the engineers are correct then $\lambda = 3$, so that

$$\pi(2/E) = 0.157/2. \tag{12.39b}$$

Then, from the Bayes formula, Eq. (12.2), the probability that management is correct in light of the two failure events in the first year is updated to

$$p_1(M) = \frac{\pi(2/M)p_0(M)}{[\pi(2/M)p_0(M) + \pi[2/E)p_0(E)]}, \tag{12.40a}$$

which is

$$p_1(M) = 0.837\ (83.7\%), \tag{12.40b}$$

while the probability the engineering staff are correct is, correspondingly, updated to

$$p_1(E) = 0.163\ (16.3\%). \tag{12.40c}$$

B. SECOND YEAR UPDATING

In the second year of operation no further tie-down failures occur. So the probability of having no failures in the second 1-year period is now given by

$$\pi_0 = \exp(-\lambda). \tag{12.41}$$

Again, with $\lambda = 1$ for management, and $\lambda = 3$ for engineering, the conditional probabilities of just two failures after 2 years are

$$\pi_2(2/M) = 0.359, \tag{12.42a}$$

$$\pi_2(2/E) = 0.046, \tag{12.42b}$$

so that the probability that management is correct is now updated to

$$p_2(M) = \frac{\pi_2(2/M)p_1(M)}{[\pi_2(2/M)p_1(M) + \pi_2(2/E)p_1(E)]} \tag{12.43a}$$

$$= 0.977 \ (97.7\%), \tag{12.43b}$$

while the engineering probability of being correct is

$$p_2(E) = 0.023 \ (2.3\%). \tag{12.43c}$$

C. THIRD YEAR UPDATING

In this year a further five tie-down failures occur, for a total of seven in 3 years. For the Poisson process the probability of obtaining just five failures in the third 1-year period is

$$\pi_5 = p(5)[1 - P(\geq 6)][1 - P(\leq 4)] \tag{12.44a}$$

$$= \frac{(\lambda t)^5}{5!} \exp(-2\lambda t) \left[1 + \lambda t + \frac{(\lambda t)^2}{2!} + \frac{(\lambda t)^3}{3!} + \frac{(\lambda t)^4}{4!} + \frac{(\lambda t)^5}{5!} \right]$$

$$\times \left[1 - \exp(-\lambda t) \left\{ 1 + \lambda t + \frac{(\lambda t)^2}{2!} + \frac{(\lambda t)^3}{3!} \right\} \right] \tag{12.44b}$$

so that

$$\pi_3(5/M) = 1.29 \times 10^{-4} \tag{12.45a}$$

and

$$\pi_3(5/E) = 0.042. \tag{12.45b}$$

The probability that management is correct is then updated to

$$p_3(M) = \frac{\pi_3(5/M)p_2(M)}{[\pi_3(5/M)p_2(M) + \pi_3(5/E)p_2(E)]} = 0.115 \ (11.5\%), \tag{12.46a}$$

while the probability the engineering staff is correct is

$$p_3(E) = 0.885 \ (88.5\%). \tag{12.46b}$$

D. FOURTH YEAR UPDATING

In this year a further two failures occur so that

$$\pi_4(2/M) = 0.0535 \text{ and } \pi_4(2/E) = 0.0785, \tag{12.47}$$

and the updated probabilities are

$$p_4(M) = 0.081 \ (8.1\%), \tag{12.48a}$$

$$p_4(E) = 0.929 \ (92.9\%). \tag{12.48b}$$

What is happening is that the few tie-down failures (two) in the first 2 years when the pipeline is new favor management's position, but as the pipeline ages the failures become more frequent and the total of seven failures in years 3 and 4 swings the probabilities strongly in favor of the engineering position that the tie-down rate of failure is indicative of high stress, as they had predicted.

The recommendation to the corporation at the end of year 4 should be that only heavy crude should be pumped through the pipeline for a while until one can see if the tie-down failure rate reduces any, as should be the case if the engineering estimates of lower stress with heavy crude are correct. Of course, ongoing TV monitoring of the failures should be maintained so that one can continually update the probability of further failures.

E. PROBABILITY OF A SPILL

The physical connection between pipeline rupture and prior tie-down failure is not simple. From a purely empirical approach a more rapid connection can be made as follows. Use the fact that a pipeline transporting crude has a statistical probability of rupture given by

$$P_n(B) = \frac{(\Lambda B)^n}{n!} \exp(-\Lambda B), \tag{12.49}$$

where Λ is the spill rate in units of spills per Bbbls transported, B is the cumulative number of barrels (in units of Bbbl) transported, and n is the number of spills, that is, $P_n(B)$ is the probability of n spills after B units are transported. For a steadily flowing pipeline, the flow rate, R, in Bbbls/year is related to B by

$$B = Rt \quad \text{or } t = B/R. \tag{12.50}$$

Consider the tie-down failure. The probability of m tie-downs failing in time t is

$$p_m(t) = \frac{(\lambda t)^m}{m!} \exp(-\lambda t), \tag{12.51}$$

where λ is the average rate of tie-down failure per year. Replacing time, t, in Eq. (12.51) with B/R, enables one to write

$$p_m(B) = \frac{(\mu B)^m}{m!} \exp(-\mu B), \tag{12.52}$$

where $\mu = \lambda/R$.

In this way the tie-down probability of failure is related to the cumulative barrels transported, as is the spill rate probability. Thus, one can eliminate the variable B between the two equations to provide an expression directly connecting the rupture probability to the tie-down failure probability.

One simple way to illustrate this point is as follows. From Eq. (12.51) one can write the cumulative probability, P_{TD}, of one or more tie-down failures as

$$P_{TD} = 1 - \exp(-\mu B), \tag{12.53a}$$

which can be rewritten as

$$B = -\mu^{-1} \ln[1 - P_{TD}]. \tag{12.53b}$$

Then, the probability of n ruptures is

$$P_n = \frac{1}{n!} \Lambda^n \mu^{-n} [-\ln(1 - P_{TD})]^n (1 - P_{TD})^{\Lambda/\mu}, \tag{12.54}$$

which relates the probability of n ruptures directly to the probability of one or more tie-down failures. If P_{TD} is small compared to unity, then P_n is small because the logarithmic factor is then close to log1, which is zero. Because the probability of one or more tie-down failures increases with time, so, too, P_n increases as P_{TD} increases.

The peak of P_n occurs when

$$P_{TD} = 1 - \exp(-n\mu/\Lambda) \tag{12.55}$$

and at this value of P_{TD}, P_n takes on the value $(n^n/n!) \exp(-n)$.

As P_{TD} systematically increases, first the probability of one rupture peaks, then two, then three, etc., in sequence. The probability of *no* rupture occurring, P_0, is given by

$$P_0 = \exp(-\Lambda B) \equiv (1 - P_{TD})^{\Lambda/\mu} \tag{12.56}$$

so that P_0 tends to zero as P_{TD} increases; there is then less and less chance of avoiding a rupture.

The number of barrels that can be transported before the probability of no rupture crosses a particular threshold is

$$B_{crit} = -\Lambda^{-1} \ln P_0 \qquad (12.57)$$

and the corresponding critical probability for one or more tie-down failures is then

$$P_{TDcrit} = 1 - P_0^{\mu/\Lambda}. \qquad (12.58)$$

For instance, suppose one is transporting 0.01 Bbbls/yr through a pipeline ($R = 0.01$). Use the engineering estimate of an average of three tie-down failures/yr so that $\lambda = 3$. Also use the historical estimate of 1.5 spills/Bbbl transported so that $\Lambda = 1.5$. Then $\mu = \lambda/R = 300$, and $\mu/\Lambda = 200$.

If one wants to have a 90% guarantee that the pipeline will *not* rupture during its lifetime then $P_0 = 0.9$. From Eq. (12.57) it then follows that $B_{crit} = \frac{2}{3} \times 0.1 = 0.067$. The pipeline can then operate for 6.7 yr before crossing this threshold. At the same time, from Eq. (12.58), the probability of one or more tie-down failures is

$$P_{TD} = 1 - 0.9^{200} = 1 - 7.06 \times 10^{-10};$$

that is, it is almost certain that one or more tie-down failures will occur. The average number of failures in 6.7 years is given by

$$\langle n \rangle = \lambda t = 20.1$$

and the uncertainty on this average value is

$$\sigma = [\langle n^2 \rangle - \langle n \rangle^2]^{1/2} = (\lambda t)^{1/2} = 4.5$$

so that one anticipates 20 ± 4.5 tie-down failures over the period of 6.7 years.

Thus, one can use both the Bayesian updating procedure and the failure estimates to evaluate not only how long a pipeline should be kept operating at a given flow rate, but also how many tie-down failures should occur before the likelihood of rupture is large enough to warrant major repairs to the pipeline itself and/or to the tie-downs.

13

OPTIONS IN EXPLORATION RISK ANALYSIS

I. INTRODUCTION

In the stock market the option to buy a stock at a later time at some fixed price requires money. For instance, if a stock is currently trading at $50/share and one anticipates the stock price will rise to $80/share, then the cost of the option to buy shares at some future time for $50 each should be no more than $30/share ($80 − 50) because, if the option is exercised, the total costs per share are $50 *plus* the option cost of $30. If the stock price actually drops below $50/share then the option contract to buy at $50 is worth zero, and one has lost $30/share. The stock market question is this: How much should one actually pay for the option? Guessing correctly leads to spectacular financial gain; guessing incorrectly leads to financial ruin (e.g., the recent examples of the bankruptcies of Orange County, California, and Barings Bank, London). Notice that the critical ingredient here is to provide some means of assessing the probability that the share price will rise to $80 during the time period for which one has purchased the option.

In the hydrocarbon exploration field of endeavor, a somewhat similar option problem arises as follows. Suppose an exploration opportunity is available. Corporate assessment of the opportunity is made with the result that the probability of success (i.e., hydrocarbons being found) is estimated at p_s (with a corresponding probability of failure of $p_f = 1 - p_s$). Total gains to the corporation, if the project is successful, are estimated at G, while costs to finance the project are C. Conventionally, the decision to become involved in the project is based on the expected value, EV, which,

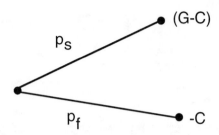

FIGURE 13.1   Decision-tree diagram for an exploration opportunity excluding optioning.

for the decision-tree representation of Figure 13.1, is $EV = p_s G - C$. A positive EV will, on average, make the project worthwhile, that is, $G \geq C/p_s$ is necessary for $EV \geq 0$. Note that the success probability branch of the decision tree of Figure 13.1 has a high channel value, HCV, of $HCV = p_s(G - C)$, which is positive under the weaker constraint $G \geq C$.

The reason why a corporation might want to take an option on an opportunity is to wait until more information becomes available before committing to the decision to drill. For instance, in a situation where an opportunity is one of the first prospects in a new play trend, there is often very little known (as opposed to surmised) about the potential gain, and the chances of success are also often poorly determined. Thus, until the estimates of costs, gains, and success chances are firmed up in the future by drilling information from other operators, the corporation would prefer to hold off on the decision to drill a particular prospect.

On the other hand, the corporation must make a decision *now,* either on the basis of currently available information, or on the basis of some methods for estimating anticipated future values of success probability, gains and projects costs, as to the maximum amount it will pay out now for the exclusive option to drill later. Such option costs then add to the total project costs if the corporation eventually decides to drill the prospect, thereby reducing estimated gains, unless future information provides a massive shift in the estimates of chances of success or gains.

However, if the current estimated chance of success is already very close to 100%, then one would not want to pay very much for an option because it is highly unlikely that future information will drive the successs probability down very far; one would in that instance likely prefer to drill the prospect.

Equally, if the expected value, EV, of the project is small, then paying a large option price will drive down the new expected value (i.e., including both the option costs and project costs) and, if the option price paid exceeds EV, the new expectation is for a negative EV, so that the project is unlikely to ever prove profitable on average.

Thus, the maximum option worth, *based on available information,* is limited at both the extreme of a highly successful chance of success and the extreme of a low expected value. The problem is to evaluate how the maximum option worth that one must pay out now depends on currently available information or on anticipated future information, so that a future decision to drill or not drill provides the best return (or least loss) to the corporation when one does not yet have available the future information. Recently, Dixit and Pindyck (1994) have persuasively argued qualitatively that an options approach should be taken to capital investment in any business field because "Opportunities are options—rights but not obligations to take some action in the future" and "as soon as you begin thinking of investment opportunities as options, the premise" (that investment decisions can be reversed if conditions change or, if they cannot be reversed, that they are now-or-never propositions) changes. Irreversibility, uncertainty, and the choice of timing alter the investment decision in critical ways." From an oil industry perspective Dixit and Pindyck eloquently provide the rationale for considering an option position rather than dealing only with net present value considerations.

The ability to calculate the maximum worth of an option to a hydrocarbon exploration opportunity is what this chapter is about.

II. AN EXAMPLE OF CURRENT OPTION VALUE AND FUTURE DECISIONS

A prospect generator has come to a corporation with a prospect to drill. He wants his usual override and would like the well spudded in 3 months (before other key acreage is acquired and drilled by others). Because this prospect is one of the first in a new play trend, very little is known about the potential gain but it is estimated to range from a 10% chance of more than $200,000 to a 90% chance of less than $2,000,000 after post-drilling costs and the override. The mean estimated gain is $1,000,000. The cost to drill is $100,000 and the chance of success is estimated at 35%. The expected value is $250,000, based on the mean estimated gain (if successful) of $1,000,000 ($0.35 \times 10^6 - 0.1×10^6).

Rather than take the prospect as proposed and commit to early drilling, the corporation offers $100,000 for the option to drill later during a 5-year lease term. The option is accepted. The corporation then waits 3 years and, in the meantime, several wells are drilled on trend by other companies that indicate the range of the gain is now $500,000 to $2,500,000, with a mean of $1,500,000. The drilling costs are higher at $150,000 and the chance of success is higher at 50%. The corporation could drill the well based on the new *prospect EV* of $600,000, or an *EV* (including the option cost) of

$500,000. However, a successful operator in the area (with even more information about chance, cost, and size) offers the corporation $500,000 for the option. The question is this: What is the best decision to make, based on both the initial information and the 3-year later information, in order to maximize gains to the corporation in relation to total costs?

Four possibilities need to be considered. With the initial information supplied by the prospect generator the corporation could choose one of these options:

1. To drill immediately and not option
2. To option, but to then drill immediately
3. To option, but to wait 3 years and then drill
4. To sell the option at the end of the 3 years.

Shown in Table 13.1 are the ratios of return to costs for the four different scenarios. It is clear from Table 13.1 that there is little point in optioning and then drilling immediately; effectively one has just doubled the cost of the project by so doing. Thus, possibility 2 is to be avoided.

Between possibilities 1 and 3 it is apparent from Table 13.1 that the information that becomes available at the 3-year stage is not sufficient to offset the price of $0.1 MM paid initially for the option. The ratio of earnings/total costs drops under possibility 3 relative to possibility 1, implying that one paid too much for the option initially relative to the improve-

TABLE 13.1

State	Initial information		Three-year later information	
Drill, no option	$EV = \$0.25$ MM $G = \$1.00$ MM $p_s = 0.35$ Option $= \$0.0$ MM $C = \$0.1$ MM	$EV/C = 2.5$ $p_s G/C = 3.5$	—	—
Option, then drill immediately	$EV = \$0.15$ MM $G = \$1.00$ MM $p_s = 0.35$ Option $= \$0.1$ MM $C = \$0.1$ MM	$EV/(C + \text{option})$ $= 0.75$ $p_s G/(C + \text{option})$ $= 1.75$	—	—
Option, then drill 3 years later	—	—	$p_s = 0.5$ $G = \$1.5$ MM $C = \$0.15$ MM Option $= \$0.1$ MM $EV = \$0.5$ MM	$EV/(C + \text{option})$ $= 2.0$ $p_s\, G/(C + \text{option})$ $= 3.0$
Option, then sell option 3 years later	—	—	$p_s = 1.0$ $G = \$0.4$ MM Option $= \$0.1$ MM Max. option worth $= \$0.15$ MM $C = \$0.0$ MM	$EV/\text{option} = 4.0$ $p_s G/\text{option} = 4.0$

ment in knowledge gained 3 years later, which, of course, one did not know at the time of paying for the option.

Possibility 4 has the highest ratio of earnings/total costs, implying that one makes the right decision in selling the option to the more successful operator. If one had chosen *not* to sell the option, then one would have been better off drilling the prospect from the start without paying an option cost of $0.1 MM.

A different point of view is to note that successful wells on the trend over the 3-year period have typically had a 50% success rate and have had drilling costs of around $0.15 MM. Given that one *did* commit $0.1 MM to the option position 3 years ago, what sort of potential gains are needed now in order to make the earnings/cost ratio larger than it would have been if one had just drilled the prospect initially? Because $EV = \frac{1}{2}G -$ 0.25 MM, while project costs plus option cost are at $0.25 MM, it follows that the ratio of EV/costs, if one had drilled directly, would be 2.5 (see Table 13.1) so that if $(\frac{1}{2}G - 0.25)/0.25$ is to exceed 2.5, potential gains in excess of $1.75 MM are now required to be estimated based on the later information; and such gains are available because, while the mean gains are estimated at $1.5 MM, there is a 90% chance of gains of $2.5 MM or less. Thus the problem here is that using only mean values completely overlooks the ranges of variation, which can have a significant influence on profitable decision making.

An alternative viewpoint is to ask a slightly different question: If, at the initial option stage, one anticipates that the success probability *will* rise to 0.5, the gains will increase to $1.5 MM, and the costs will increase to $0.15 MM in the 3 years after paying for the option, what should the option price be in order for the earnings/total cost for possibility 3 to be larger than that for possibility 1? The new expected value, including an option cost amount O_w (in $MM), with $p_s = 0.5$, $G = \$1.5$ MM, and project costs of $0.15 MM, is $EV_1 = \$0.6$ MM $- O_w$; the total costs of the project (including the option cost) are now $C_1 = \$0.15$ MM $+ O_w$. Thus, the ratio of earnings/cost is $EV_1/C_1 = (0.6 - O_w)/(0.15 + O_w)$. If EV_1/C_1 is to exceed the value 2.5 [the original value of the ratio of EV (=$0.25 MM) to costs ($0.1 MM)], it then follows that $O_w \leq \$64,300$. Thus, at an investment of %0.1 MM one overpaid for the option relative to *anticipated* improvements in knowledge.

Clearly, the aim of the option decision at the initial stage is not only to figure out the maximum option worth one *could* pay based on then available information, but also to estimate an actual option price (or range) that one *should* pay if anticipated future information is to lead to an improved ratio of earnings/costs.

The difficulty, as usual, is that one does not have available future information at the time the option decision has to be made. Thus one has to evaluate the option amount either on the basis of available information at the time the option decision must be made, or on "what if" potential futureward scenarios, in order to estimate the likelihood of an option being worthwhile.

For instance, in the "what if" category one can include the set of models that posit behaviors in deterministic or stochastic manners for the evolution with time of gains, costs, or success probability. Effectively, such models are predicting the future. For example, the two-commodity exchange option model put forward by Kensinger (1987) assumes that the ratio of future values to present-day values has a variance independent of time, and that the ratio is log-normally distributed for all time. This assumption permits one to calculate the variance from present-day information, and to then predict *exactly* the average expected value at any future time. This model is a variant of the Black-Scholes (1973) economic prediction model, which makes the key assumptions that continuously varying price fluctuations are both uncorrelated and are log-normally distributed.

Thus, again one has a future predictability based on present-day assessments of the variance. Effectively, one is limiting what the future is permitted to do. As pointed out by Bouchaud and Sornette (1994, 1995, 1996; Riley, 1996) such imposed model limitations do not compel the future to follow the model, and they argue for a more appropriately chosen type of behavior that also permits massive "swings" in price assessments and also permits stock market "crashes." In any event, such model behaviors also involve some future predictability of prices in either deterministic or stochastic manners.

III. OPTION VALUE BASED ON AVAILABLE INFORMATION ONLY

The problem faced by an exploration group for a potentially lucrative opportunity is whether to option or not, whether to use only currently available information in making the option decision, or whether to use predictive future models of anticipated behavior in assessing the option worth.

Two different viewpoints can be taken. One can argue that present-day values of success probability, gains, and costs already provide an estimate of the uncertainty of the expected value $EV\ (\equiv p_s G - C)$ of the opportunity through the variance, $p_s(1 - p_s)G^2$. And one can also note that the expected value only measures the statistical outcome from an infinite number of trials of well drilling, and that the expected value is never realized for any single well drilled—either a well is successful or it is not. Thus the uncertainty in the expected value then can be viewed as taking on the role of measuring departures due to actually achievable outcomes from the statistical average that the expected value considers. Thus, because the expected value, EV, is *not* one of the possible outcomes, we are required to have a variance, σ^2, such that $EV + \sigma$ represents approximately the success value $p_s(G - C)$ and such that $EV - \sigma$ is similarly close to the failure value $-p_fC$.

Under this viewpoint there are *two* option worths: one that measures the maximum option amount one should pay relative to the expected value if one is interested in the likelihood of maximizing potential gains; and a second that measures the maximum option amount one should pay if one is interested in the likelihood of minimizing potential losses.

The second viewpoint notes that an option paid gives one the right to delay the decision to drill, or not to drill until a future time when conditions may have changed to the point where one can better decide whether it is more profitable to drill or to abandon the opportunity. Of course, the problem here is that one does not know what the conditions will be in the future, so some models of evolution of success probability, and of evolution of gains and costs, are needed in order to estimate ranges of uncertainty to attach to an option amount; and the amount to pay now has also to be determined.

Because there is already uncertainty on estimates of the current success probability, gains and costs due to the geologic, production, and economic models used (see earlier chapters), the imposition of future deterministic and/or stochastic prediction models for evolution of p, G, and C adds yet more degrees of uncertainty to the problem of determining an option amount to pay.

In this section of the chapter, concentration is given to assessment of an option amount to pay based only on currently available information. One could argue that there is little point to providing such a calculation because options are inherently supposed to be of value only as future conditions change, but such an argument is incorrect. The point is that current conditions are themselves uncertain [namely, the variance $\sigma^2\{\equiv p_s(1 - p_s)G^2\}$ around the expected value $EV(=p_sG - C)$]. Future information may do nothing to change the expected value, but may narrow down the current variance. In such situations, the exercise of a previously acquired option makes for a sharper evaluation of the worth of either going ahead with development of the opportunity or of abandonment. Equally, one would not consider that the current estimates of gains, costs, and success probability for an exploration opportunity are themselves precisely known either, leading to extra uncertainties in the expected value and in the variance around the average expected value.

While the decision to proceed with an investment, or to defer action until a more plausible alternative is identified, is heavily influenced by the expected value and the risk-adjusted value, no competent decision maker would proceed without having available the 10% and 90% (p_{10} and p_{90}) confidence values (or proxies for them such as $EV - \sigma$ and $EV + \sigma$, respectively). The assessment of the $p_{90} - p_{10}$ range of uncertainty augments the expected value information and is based on uncertainty in estimates of the expected value and on uncertainty in estimates of the underlying variables, which constitute both the expected value and the variance, through the various combinations used to calculate EV and RAV (see earlier chapters). Thus, some method should be available for assessing maximum option worth based on anticipated departures from the current expected value.

As remarked earlier, two option amounts can be calculated: O_p, which is based on a maximal gains scenario; and O_n which is based on a minimal losses scenario; and they are not the same ($O_p \neq O_n$).

A. MAXIMAL GAINS OPTION, O_p

1. High-Gain Scenario

The optioning position that a corporation can take with respect to the opportunity can be set up as follows. Let the corporation be prepared to pay up to a maximum option amount, O_p, for the privilege of investing in the opportunity at some future time with a buy-in working interest of 100% and total project costs of C. Then the total corporate costs, if the option is exercised, are $C + O_p$, while the gains remain fixed at G. Thus, the decision-tree diagram of Figure 13.1 is then modified as shown in Figure 13.2, and the expected value to the corporation is reduced to $EV_1 = (p_s G - C) - O_p$.

The expected high gains to the corporation are given by the success probability branch of the decision tree of Figure 13.2, namely, $HCV \equiv p_s[(G - C) - O_p]$. The maximum option amount, O_p, is then just the difference between HCV and EV_1, provided both $HCV \geq 0$ and $EV_1 \geq 0$. Then,

$$O_p = p_s[(G - C) - O_p] - [(p_s G - C) - O_p], \qquad (13.1)$$

which gives the maximum option worth explicitly as

$$O_p = C(1 - p_s)/p_s \qquad (13.2)$$

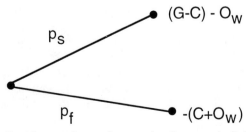

FIGURE 13.2 Decision-tree diagram for an exploration opportunity including optioning.

provided

$$G \geq C/p_s^2, \tag{13.3}$$

which is necessary to keep $EV_1 \geq 0$. For high-risk projects ($p_s \ll 1$), the requirement $G > C/p_s^2$ is a much higher expectation of G than $G > C/p_s$, which is all that is needed to keep EV positive in the absence of paying the maximum option amount. Note that the gains cancel out of expression (13.2) for O_p, directly reflecting the fact that it is the increased costs that lower HCV and EV_1. When the maximum option worth is given by Eq. (13.2) it follows by direct substitution of Eq. (13.2) into the formula $EV_1 = (p_sG - C) - O_p$ that the expected value of the project is lowered to

$$EV_1 = p_s(G - C/p_s^2), \tag{13.4a}$$

which is positive as long as $G > C/p_s^2$, and the success probability branch worth is then given by

$$HCV = p_s(G - C/p_s) = p_sG - C, \tag{13.4b}$$

the old expected value. If the option is exercised at the maximum option worth, the total project costs (TPC) to the corporation are then given by

$$TPC = C + O_p = C/p_s. \tag{13.5}$$

Note also that O_p exceeds C whenever $p_s \leq 0.5$, and O_p is very large indeed for small values of p_s.

2. Partial Option Amounts

The estimates of option worth so far have been based on the assumption that the corporation is prepared to pay the maximum option worth, O_p, and that G is sufficiently large to still support that level of cost. However, a corporation may balk at the total option price and be prepared to pay only a fraction f of the maximum O_p for the same working interest of 100% in an opportunity of project costs C. In such a situation the expected value is $E_2 = (p_sG - C) - fCp_f/p_s$, while the success probability branch yields $HCV_2 = p_s(G - C) - fC(1 - p_s)$. Then $E_2 \geq 0$ and $HCV_2 \geq 0$ as long as $f \leq 1 - p_f^{-1}[1 - p_s^2G/C]$ provided $p_s^2G \leq C$, and $f \leq 1$ in $p_s^2G \geq C$, which requires only that $p_sG > C$ to yield $E_2 \geq 0$. The extra requirement that f be a fraction less than unity implies

$$f \leq \min\{1, 1 - p_f^{-1}(1 - p_s^2G/C)\}. \tag{13.6a}$$

For instance, consider the situation in which $p_s < 0.5$, $G = \$40$ MM, and $C = \$10$ MM. Then,

$$f \leq \min\{1, 1 - p_f^{-1}(1 - 4p_s^2)\}. \tag{13.6b}$$

Note that $1 - p_f^{-1}(1 - 4p_s^2)$ takes on the value 1 when $p_s = 1/2$ and, for values of p_s less than $1/2$, $f \leq 1 - p_f^{-1}(1 - 4p_s^2)$. As p_s decreases, the fraction of the maximum option worth that the corporation can take drops to zero at $p_s = 1/4$, which is where $p_s G = C$, and zero expected value EV then occurs.

Thus, the corporate maximal option position is then restricted to fO_p, which limitation is an option commitment of less than or equal to $[1 - p_f^{-1}(1 - p_s^2 G/C)]O_p$ in $p_s^2 G \le C$. Effectively the corporation can option up to an amount $p_s G - C$ (i.e., the expected value) in $G < C/p_s^2$, and up to $C(1 - p_s)/p_s$ in $G > C/p_s^2$, while still relating both $EV \ge 0$ and $HCV \ge 0$. Of course, if the corporation is prepared to gamble aggressively on an anticipated high gain opportunity (i.e., one with $G \gg C/p_s^2$), then it can choose to option at a much higher amount, but in that case it cannot keep $EV \ge 0$.

3. Returns to Investment Ratio

One of the criteria used to assess the profitability of an opportunity is the ratio of expected value to cost (known casually but universally in America as the "bang for the buck," BFB). In the case of an option amount fO_p this BFB is given by

$$BFB = (p_s G - C - fO_p)/(C + fO_p), \qquad (13.7)$$

which is, of course, smaller than the BFB in the absence of optioning. A measure of maximum option amount is to ask whether the high side potential ($p_s G - C - fO_p + \sigma$), including option amount, provides a BFB that exceeds the BFB in the absence of optioning, that is, the maximum O_p is given by

$$(p_s G - C - O_p + \sigma)/(C + O_p) \ge (p_s G - C)/C, \qquad (13.8)$$

where $\sigma = G[p_s(1 - p_s)]^{1/2}$.

Inequality (13.8) can be rewritten as

$$O_p \le C[(1 - p_s)/p_s]^{1/2}, \qquad (13.9)$$

which differs slightly from the value in Eq. (13.2) by the presence of the square root factor because σ is different than the difference between high channel value and expected value, which was used to derive the absolute maximum option amount to pay of Eq. (13.2). Thus, the BFB provides an option measure of profitability to within a standard error, σ, around the expected value. The corresponding constraint on gains is that $p_s G - C - O_p + \sigma \ge 0$, which is just $G \ge C/p_s$, a less stringent constraint than the previous $G \ge C/p_s^2$, which was based solely on the difference between HCV and EV.

B. MINIMAL LOSS OPTION, O_n

A precisely similar logic path exists for assessing the option amount, O_n, to hedge against a loss situation. The total low channel value costs, LCV, of Figure 13.2 are $LCV = -(1 - p_s)(C + O_n)$, while the expected value, EV, is, again $(p_sG - C - O_n)$. Thus, the minimal loss option amount, O_n, is limited to $EV - LCV$, which can be written

$$O_n \leq (1 - p_s)(C + O_n) + (p_sG - C - O_n), \tag{13.10}$$

yielding

$$O_n \leq p_s(G - C)/(1 + p_s); \tag{13.11}$$

and, at the maximum O_n value, one has

$$EV_1 = p_sG - C - O_n = (1 + p_s)^{-1}(p_s^2G - C) \tag{13.12}$$

so that, once again, one can only option at the maximum amount if $G > C/p_s^2$ in order to keep $EV_1 \geq 0$; else only a fraction f_n of the maximum O_n can be taken in $C/p_s < G < C/p_s^2$, with

$$f_n = \min\{1, (1 + p_s)p_f^{-1}(p_sG - C)/(G - C)\}. \tag{13.13}$$

Equally, the *lowering* of the *BFB* by including the minimal losses option should be limited, so that one would write

$$\frac{(p_sG - C) - \sigma}{C} \leq \frac{(p_sG - C - O_n)}{(C + O_n)}, \tag{13.14}$$

which limits O_n to the value

$$O_n \leq C(p_f/p_s)^{1/2}[1 - (p_f/p_s)^{1/2}]^{-1}. \tag{13.15}$$

This *BFB* corresponds to keeping values within σ of the expected value.

C. NUMERICAL ILLUSTRATIONS FOR MAXIMAL GAINS OPTION

1. Current Value Option Worths

Consider an exploration opportunity in which the potential gains, G, are estimated at \$100 MM, project costs, C, are estimated at \$10 MM, the success probability is 50% ($p_s = 0.5$), and in which the corporation is prepared to take a 100% working interest. The question is what is the maximum option worth, O_w? First note that $Gp_s^2/C = 2.5$, which exceeds unity, so that an optioning position up to the maximum can indeed be taken.

Then the maximum option worth is $O_w = \$10 \times (1 - 0.5)/0.5$ MM = \$10 MM; the total project costs to the corporation if the option is exercised are $TCP = \$(10 + 10)$ MM = \$20 MM, where the first \$10 MM represents 100% of the option value and the second \$10 MM represents the original project costs. The expected value is $EV_1 = \$30$ MM, and $HCV = \$40$ MM. If no option position had been taken (i.e., if $O_w \equiv 0$), then $EV = \$40$ MM, so that the option exercise has reduced EV from \$40 MM to \$30 MM, but HCV (with $O_w = 0$) would then have been \$45 MM. Thus, if one did *not* include the maximum option worth in the assessment of HCV or EV, then one would have had \$(45 − 40) MM ≡ \$5 MM as an estimate of the maximum option worth. In fact, however, the option is maximally worth $10 MM, because both *HCV* and *EV* are reduced by inclusion of option costs, but *EV* is reduced faster than *HCV*, thereby increasing the option worth. If the corporate assessment of the opportunity reflects the likely success probability, gains, and costs accurately, then the corporation should be prepared to pay up to a maximum of $10 MM for the option to buy into the opportunity at 100% working interest at a future time. The option worth of $10 MM is only 25% of the estimated *HCV* of $40 MM and only $33\frac{1}{3}$% of EV_1, so that the option costs are a small fraction of potential high side gains above *EV*, as measured by the difference between *HCV* and EV_1.

Because option worth is sensitive to costs, success probability, and gains (through the requirement $G \geq C/p_s^2$ to option at the maximum), and because p_s, G, and C are uncertain at the assessment stage of an exploration opportunity, it makes corporate sense to evaluate the sensitivity of maximum option worth to variations in gains, costs, and the chance of success.

Figure 13.3 plots fO_W (in $MM) as the gains, G (also in $MM), vary for fixed values of $p_s = 0.5$ and costs of $C = \$10$ MM. Note that in $G \leq \$40$ MM the maximum option worth cannot be taken because then one would have $EV_1 \leq 0$. Accordingly the option worth is restricted to less than or equal to $p_s G - C$ in $G < C/p_s^2$. Once G is greater than C/p_s^2 then, because

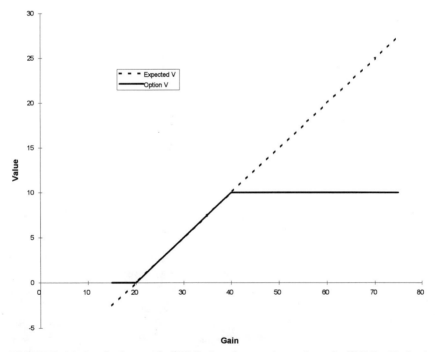

FIGURE 13.3 Option worth ($MM) plotted versus increasing gain ($MM) with fixed values of $p_s = 0.5$ and $C = \$10$ MM.

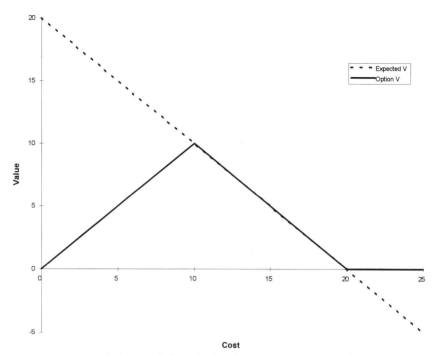

FIGURE 13.4 Option worth ($MM) plotted versus increasing costs ($MM) with fixed values of $p_s = 0.5$ and $G = \$40$ MM.

O_W is independent of G, it follows that the maximum option worth stays fixed at $10 MM if one wants to keep both HCV and EV positive including option costs.

Figure 13.4 plots the option worth fO_W (in $MM) as the costs of the project vary for a fixed value of $p_s = 0.5$ and for a fixed gain of $G = \$40$ MM. Note that the requirement $G \geq C/p_s^2$ is then only satisfied for $C \leq \$10$ MM, so that the maximum option worth increases linearly with increasing costs at low costs (reflecting directly the *relative* increase of HCV compared to EV_1), but drops to $p_s G - C$ at $C \geq \$10$ MM.

Figure 13.5 plots the option worth fO_W (in $MM) as the success probability systematically increases for fixed values of gains, $G = \$40$ MM, and costs, $C = \$10$ MM. In this case note that the worth of the option is $p_s G - C$ for $p_s \leq (C/G)^{1/2} = 0.5$ and, at higher values of p_s, the maximum option worth declines with increasing p_s from a value of $10 MM at $p_s = 0.5$ because, as $p_s \to 1$, there is less and less possibility of a failure, so that the difference between HCV and EV_1 systematically lessens, making the maximum option worth that much less.

The point of these illustrations has been to show how to figure the maximum worth of an option to an exploration opportunity based on currently available information, and also to examine the sensitivity of the

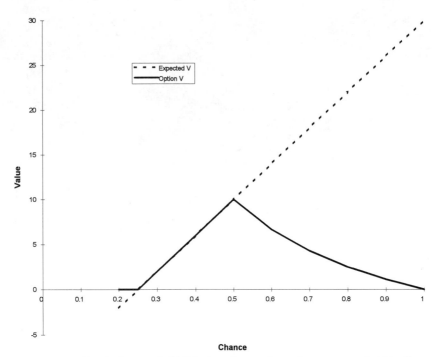

FIGURE 13.5 Option worth ($MM) plotted versus increasing success probability (fraction) with fixed values of $G = \$40$ MM and $C = \$10$ MM.

maximum option worth to changes in parameter values of the opportunity influencing the option worth. The exhibited rapid variations of the maximum option worth as gains, costs, and success probability all vary, and the shaping of the option worth curve with each variable, provide indications of the sensitivity, and so of the chances for guessing correctly (high financial gain) or incorrectly (financial loss of at least the option payment) the likely worth of an exploration opportunity to a corporation.

D. RISK TOLERANCE FACTORS

Inclusion of a corporate risk tolerance, RT, in evaluating an option position, allowing for the uncertainties in both the expected value, EV, and in the risk-adjusted value, RAV, can also be carried through in a similar manner. Basically the risk tolerance factor forces replacement of expected value, EV, by the risk-adjusted value, RAV, calculated according to one of the models (exponential, hyperbolic, parabolic, etc.) connecting risk tolerance to parameters of the opportunity. The novelty here is that the working interest, W, the corporation could take in the opportunity is also changed by the option cost compared to the situation of not optioning the opportunity. The case of a parabolic RAV formula is worked through here in detail to illustrate the procedure; other model RAV formulas can be handled in like manner.

Recall that the parabolic RAV formula is given by

$$RAV(WV, WC) = WE\left(1 - \frac{1}{2}Ev^2W/RT\right) \quad (13.16)$$

with

$$E = (p_sV - p_fC), \quad v^2 = \{(p_sV^2 + p_fC^2)/E^2 - 1\}, \quad \text{and} \quad V = G - C.$$

Most exploration economic investigations carried out assume that one *will* indeed invest in an opportunity at the amount WC. However, it may happen that one is not convinced that the success probability, p_s, is as high as some initial estimate indicates, or that costs are really at the value C, or that the intrinsic gain is really at the estimated value G. (Or it may be that one does not have enough cash on hand at a given time to commit the amount CW to the project and one wishes to have time to save that money.) One would prefer to wait before choosing to commit costs of WC to see if the opportunity really does develop along the lines of the anticipated estimates, or whether significant changes occur in costs C, gain estimates G, or success probability p_s as time progresses. Thus one would like the option of deciding to buy into the opportunity at some future time when the evaluation of the project is clearer.

However, the exploration of the opportunity cannot proceed without total project costs C being available. Thus, other involved parties must bear the total costs unless the extra funding WC is made available. Accordingly, for the option to invest in the opportunity at a cost of WC at some future date, a *total* amount O_p is charged. The total cost to the corporation, should the corporation decide to be involved in the project, is then increased to $O_p + WC$. Should the project prove profitable, the gain to the corporation is reduced from $GW - WC$ to $g = GW - O_p - WC$.

Clearly, the success probability branch of the relevant decision tree yields a worth of the project to the corporation as

$$p_s(GW - WC - O_p). \quad (13.17)$$

The question to be addressed is this: What is the maximum value of O_p that the corporation should pay now for the option of being able to buy in later at WC?[1]

The problem addressed is similar to that of buying and selling a stock option. Zisook (1994) has noted "Suppose I want an option to buy a stock

---

[1] An alternative problem can be handled in like manner: If a corporation already has the right to drill now, how much should it pay now for the option to delay drilling until further information becomes available to clarify the worth of the opportunity better?

that's presently priced at $50 a share for the same price a year from now. I want to know how much I should pay for that privilege—what the value of that option is. For example, if I think the price will soar to $75, the option is worth $25 a share; if I think the price will sink to $25, the option is worth nothing. What I don't do is try to predict through market research—which can range from intuition to sophisticated mathematical analysis—what the stock price will be in a year." In the case of an exploration project one has the *RAV* corresponding to the expected worth of the project at a working interest W, and one has $(GW - O_p - WC)$ as the high value of the project with p_s the probability that the high value will occur. Thus, the maximum value of the option, O_p, is then the difference between Eq. (13.17) and the *RAV* (including the *total* costs to the corporation), that is,

$$O_p = p_s g - RAV[g; W(C + O_p)] \qquad (13.18)$$

provided $g \geq 0$ and $RAV \geq 0$.

In the case of the parabolic *RAV* formula it follows that

$$O_p = \left[W p_f \left[C + \frac{1}{2} p_s W G^2 / RT \right] \right] p_s^{-1}, \qquad (13.19)$$

provided $O_p < WV$ and that $WE - \frac{1}{2} p_s p_f W^2 G^2 / RT > O_p$, both of which criteria can be combined to require that

$$p_s^2 V - p_f C(1 + p_s) > (1/2) W G^2 p_f p_s (1 + p_s) / RT \qquad \text{or else } W = 0 \qquad (13.20)$$

in order to have positive values for both *RAV* and g.

Inequality (13.20) requires

(1) $\qquad V \geq p_f (1 + p_s) p_s^{-2} C \equiv V_* \qquad (13.21\text{a})$

and then

(2) $\qquad 0 \leq W \leq \min\{1, 2RTG^{-2}[p_f p_s (1 + p_s)]^{-1}[p_s^2 V - p_f (1 + p_s) C]\} = W_{\max}. \qquad (13.21\text{b})$

When both $V \geq V_*$ and $0 \leq W \leq W_{\max}$ then

$$O_p = W p_f p_s^{-1} \left[C + \frac{1}{2} p_s G^2 W / RT \right]; \qquad (13.22)$$

otherwise, the maximum option should not be taken, because as V approaches V_* from above, O_p approaches zero from above. The requirement $V \geq V_*$ can also be written in the familiar form

$$G \geq C / p_s^2. \qquad (13.23)$$

Thus, if the anticipated gain G of the project is not high enough ($\geq C/p_s^2$) relative to anticipated costs C and success probability p_s, then the maximum option value O_p cannot be taken. In that case only a fraction, f, of the maximum option value can be taken, with the fraction limited by the value of f which returns a value of zero for the RAV. When the maximum option value is given by Eq. (13.22) it follows that the RAV is then given by

$$RAV = p_s^{-1} W \left\{ p_s^2 G - C - \frac{1}{2} p_s W (1 - p_s^2) G^2 / RT \right\}, \qquad (13.24)$$

which has a maximum value at

$$W = W_* \equiv (p_s^2 G - C) RT G^{-2} (1 - p_s^2)^{-1} p_s^{-1} \qquad (13.25a)$$

with

$$RAV_{max} = \frac{1}{2} (p_s^2 G - C)^2 RT / [G^2 (1 - p_s^2) p_s^2]. \qquad (13.25b)$$

For $W = W_*$, one has

$$O_p = \frac{(p_s^2 G - C) RT G^{-2} (1 - p_s)[p_s^2 G + C(1 - 2p_s^2)]}{2 p_s^2 (1 - p_s^2)^2}, \qquad (13.26)$$

provided $W_* \leq W_{max}$, with W_{max} given through Eq. (13.21b).

For comparison, note that if one did *not* choose to option the opportunity, then the optimum working interest in

$$W_{opt} = (p_s G - C) RT G^{-2} p_s^{-1} (1 - p_s)^{-1} \qquad (13.27a)$$

and the corresponding optimum RAV is

$$RAV_{opt} = \frac{1}{2} (p_s G - C)^2 RT / [G^2 p_s (1 - p_s)]. \qquad (13.27b)$$

Then,

$$W_*/W_{opt} = (p_s^2 G - C)/[(1 + p_s)(p_s G - C)] \qquad (13.28a)$$

and $W_*/W_{opt} \leq 1$ for $G \geq C$, so that a smaller optimum working interest would be taken if one exercised the option to proceed. Equally, the RAV ratio is

$$RAV_{max}/RAV_{opt} = (p_s^2 G - C)^2 (p_s G - C)^{-2} p_s^{-1} (1 + p_s)^{-1}, \qquad (13.28b)$$

but this ratio is not always less than unity because the shape of the RAV versus working interest curve is adjusted by inclusion of the option amount itself.

E. A NUMERICAL ILLUSTRATION

To exhibit the behavior of maximum option values under practical considerations, this section shows how the option can be useful in assessing costs and working interest involvement, and the parabolic RAV formula is used for illustrative purposes only.

Consider that an exploration project has a 50% chance of succeeding ($p_s = 0.5$) with a cost of $C = \$10$ MM, a potential gain estimated at $G = \$100$ MM, and a corporate risk tolerance of $RT = \$100$ MM. Then, at a fixed working interest W the parabolic RAV formula yields

$$RAV = 2.5W(12 - 15W) \text{ \$MM} \qquad (13.29\text{a})$$

together with

$$O_p = 5W(2 + 5W) \text{ \$MM,} \qquad (13.29\text{b})$$

provided $0 \le W \le 0.8$ in order to keep both $RAV > 0$ and $WV > O_p$. Note that the high value of the project *without* the option cost is $p_s G - C \equiv \$40$ MM, while the maximum RAV occurs at a working interest of $W_* = 0.4$ when $RAV_{\max} = \$6$ MM and $O_p(W = W_*) = \$8$ MM.

The high value worth of the project at 40% working interest is then $p_s G - C - O_p = \$32$ MM. Thus, the maximum option value to have a chance of realizing the potential gains of the project is only 25% (\$8 MM/\$32 MM) of the high value worth. The total cost involvement, if the project is invested in at the optimum RAV working interest of 40%, is $WC + O_p \equiv \$12$ MM, so that, while the maximum option value increases the cost dramatically (from \$4 MM for WC alone to \$12 MM including the option); nevertheless the potential gains are still sufficiently large (\$32 MM) that the 25% of the expected gains that the maximum option value represents is a worthwhile venture.

The point about this illustration is that the RAV represents only a *mean* risk-adjusted estimate of potential for the exploration project. Nowhere in the RAV formula is allowance made for the variation of either the potential gains (or potential losses) of the project; that is, the variance around the RAV measures the riskiness. The value of the maximum option spells out the procedure for including the risk of optioning the high-end profitability of the exploration project in order to increase returns to the corporation.

F. EFFECTS OF VARIABLE GAINS, COSTS, WORKING INTEREST, AND SUCCESS PROBABILITY

So far gains, G, costs, C, working interest, W, risk tolerance, RT, and success probability, p_s, have all been treated as being *precisely* known parameter values in the exploration opportunity. Yet, in any exploration assessment it is likely that any and all of these parameters are uncertain. To illustrate the influence of variations in G, C, p_s, W, and RT on the maximum option worth O_p, a suite of Monte Carlo simulations was run to yield cumulative probability plots of O_p together with plots of the relative importance of each of the parameters in contributing to the uncertainty in O_p.

This uncertainty of information available at the onset of an exploration project parallels a similar problem in stock market options. For example, Zisook (1994) has noted "I begin with an estimated range for stock price— something substantially easier than predicting where the market will go. Let's say I think the stock price will be between $25 and $75 a share. I don't know whether the price will go up or down, but there is a probability distribution for the stock price that gives me the average value of the option. Suppose the average value turns out to be $8 a share. If a customer wants to buy the option, we'll offer to sell it for $9, or if a customer wants to sell an option, we'll offer to buy it for $7. We don't really care if the customer buys or sells; we just want to make $1 on the transaction."

Notice that Zisook (1994) is using the average value of an option as the yardstick by which to measure the decision to buy or sell the option. In his case knowledge of the probability distribution of the stock price plays an important role because of the influence not only on the average value of the option but also on fluctuations in the probable option value, and it is the fluctuations that can lead to vast financial gain or financial ruin.

In the case of an exploration opportunity the factors leading to uncertainty in the value of the maximum option worth are the variations in p_s, C, G, RT, and W. To illuminate the influence of the various factors in controlling the value of the option, in this section we take the parameters p_s, G, C, RT, and W as culled from *uniform* distributions centered at the mean values $p_s = 0.5$, $G = \$100$ MM, $C = \$10$ MM, $RT = \$100$ MM, and $W = 0.5$, but allowing for (1) $\pm 10\%$ variation and (2) $\pm 20\%$ variation of each around the mean values. In addition, both the Cozzolino (exponential) *RAV* formula and the parabolic *RAV* formula were used to assess the cumulative probability distribution of the maximum option value in order to evaluate the dependence of the option value probability on the choice of method for adjusting risk. Parameter values were chosen by random Monte Carlo techniques and a total of 500 random runs done for each variable.

1. Results for ±10% Variation in Parameters

The cumulative probability for the maximum option value is presented in Figure 13.6a for the Cozzolino *RAV* formula and in Figure 13.6b for the parabolic *RAV* formula. The first point to note is that, even by eye, the two curves are very similar, indicating that there is a rugged stability of the maximum option value against different choices in *RAV* formula. Such stability is of importance when the decision to option an exploration project is being considered because then less concern has to be spent on the *method* of arriving at an option value and more concern can be given to the mean value, and range of values, of the likely option worth.

One method of assessing the behavior of the maximum optional value under fluctuations in parameter values is to use the volatility, v, defined as

$$v = (P_{90} - P_{10})/P_{mean},$$

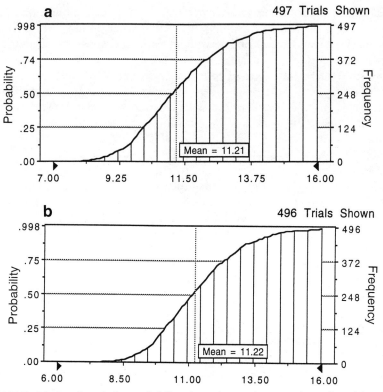

FIGURE 13.6 Cumulative probability charts for maximum option value with ±10% parameter ranges for p_s, G, C, RT, and W, with mean values chosen as described in text. The abscissa is option value, O_w, in \$MM. (a) Using the Cozzolino RAV formula. (b) Using the parabolic RAV formula.

where P_{mean} is the mean value of the option worth (occurring at the cumulative probability of about 68%), and P_{90} and P_{10} measure the 90% and 10% cumulative probability values, respectively. A low value ($\nu \ll 1$) of volatility implies a highly stable estimate of mean option value, while a high volatility ($\nu \gg 1$) implies a strong dependence on fluctuations in parameter values. Thus, from Figure 13.6a it is 90% (10%) certain that a maximum option value of *less than* (greater than) \$13.4 MM is appropriate, and it is 90% (10%) certain that a maximum option value of *greater than* (less than) \$9.4 MM is appropriate, with a mean option value of \$11.21 MM. In the case of the Cozzolino *RAV* formula it then follows that the volatility of the option worth is

$$\nu = (13.4 - 9.4)/11.21 = 0.36, \qquad (13.30a)$$

implying a fairly insensitive degree of uncertainty of the mean option worth to the ±10% fluctuations in the parameter values, G, C, RT, p_s, and W

controlling the option. Equally, reading off from Figure 13.6b for the parabolic RAV formula, one has a mean of $11.22 MM for the maximum option value, while the P_{90} and P_{10} values occur at $13.5 MM and $9.36 MM, respectively, for a volatility estimate of

$$\nu = (13.5 - 9.36)/11.22 = 0.37 \qquad (13.30b)$$

which is virtually identical to that obtained from the optional value probabilities using the Cozzolino RAV formula, as is the mean value.

Thus, in either event, the suggestion is that, at the 90% confidence level, the maximum option value is in about the range $11.22^{+2.28}_{-1.86}$ MM, reflecting directly the range of uncertainty.

While each of the parameters p_s, G, C, RT, and W influencing the maximum option value were chosen to vary equally by $\pm 10\%$ around their mean values, and to be drawn from uniform probability distributions, there is not the same influence of each parameter on the cumulative probability distribution of the option value because of the nonlinear manner in which the option value depends on the parameters. Thus, the variance in the option value probability distribution is controlled to greater and lesser extents by the different parameters.

The relative importance (%) of the variations in each parameter in contributing to the variance in the maximum option value is given in Figure 13.7a (13.7b) using the Cozzolino (parabolic) RAV formula to assess the option. To be noted from both parts of Figure 13.7 is that uncertainties in working interest, W, and success probability, p_s, are roughly equally dominant in contributing to the variance in the maximum option value and, together, account for almost 70% of the uncertainty. The potential gain, G, comes in at about 23% contribution, while uncertainties in risk tolerance, RT, and costs, C, together contribute only about 6–8% of the total.

Thus, if effort were to be expended in attempting to narrow the range of uncertainty around the mean option value, then that effort is best spent on doing a better job to narrow the uncertainty on the chances of the project succeeding and on defining more tightly the working interest. If it is felt that neither the mean values of these parameters nor their uncertainties can be better constrained than current information allows, then doing a better job on assessing the range of uncertainty of the potential gains, G, is the next best choice.

The relative importance does not determine *if* an improvement is needed in the mean values of individual parameters and their ranges influencing the option worth, but it does determine which parameters should be improved if the decision is made to attempt to narrow the uncertainty on the cumulative probability distribution of the maximum option value.

2. Results for ±20% Variation in Parameters

As the uncertainty on each of the parameters p_s, G, C, RT, and W increases there is a greater chance overall that more individual realizations

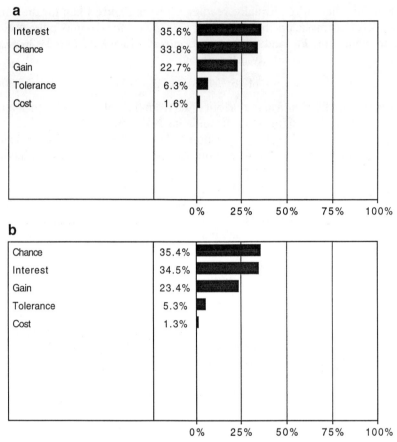

FIGURE 13.7 Relative importance of contributions of the ±10% parameter ranges for p_s, G, C, RT, and W, with mean values as described in text, to the variance in the maximum option value. Note the dominance of chance (p_s) and interest (W) in controlling the option value variance. (a) Using the Cozzolino RAV formula. (b) Using the parabolic RAV formula.

will lead to a lower value for the maximum option worth because of the twin requirements that $g > 0$ and $RAV > 0$. This point is illustrated in Figure 13.8a using the Cozzolino RAV formula and in Figure 13.8b using the parabolic RAV formula. The mean for the cumulative probability of the option value has dropped to $9.01 MM (Cozzolino RAV) and to $8.72 MM (parabolic RAV).

In addition, the P_{90} and P_{10} values are now at $14.10 MM and $0.04 MM, respectively, using the Cozzolino RAV, and at $14.06 MM and $0.03 MM, respectively, for the parabolic RAV, yielding volatility estimates of

$$\nu(\text{Cozzolino}) = 1.56, \tag{13.31a}$$

$$\nu(\text{parabolic}) = 1.61, \tag{13.31b}$$

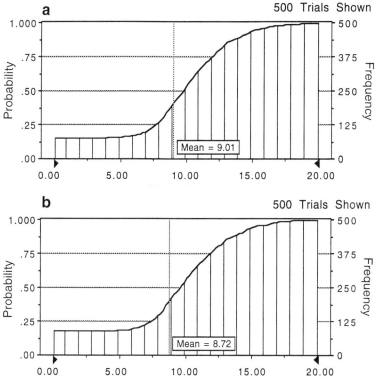

FIGURE 13.8 As for Figure 13.6 but now with parameter ranges for p_s, G, C, RT, and W increased to $\pm 20\%$ around their respective mean values. (a) Using the Cozzolino RAV formula. (b) Using the parabolic RAV formula.

both of which represent very large ranges of uncertainty on the mean option value, MOV, so that, at the 90% probability level, one has

$$MOV(\text{Cozzolino}) = \$9.01^{+5.09}_{-8.97} \text{ MM}, \quad (13.32a)$$

$$MOV(\text{parabolic}) = \$8.72^{+5.34}_{-8.69} \text{ MM}. \quad (13.32b)$$

Thus, a rapid shift in the maximum option value is caused by a doubling of the uncertainties on the parameters p_s, G, C, W, and RT entering the basic formulas for option value, which is why the option value in the stock market is so highly volatile.

In the present case, the increase of uncertainty in the basic parameters to $\pm 20\%$ has caused a change in volatility by a factor of 4, and has also decreased the mean of the option value by about 20% in both the Cozzolino and parabolic RAV cases.

This high degree of sensitivity (to changes in parameter ranges) of the maximum option value probability for both the Cozzolino and parabolic RAV formulas, and the rugged sameness of results from both the Cozzolino and parabolic formulas, indicate that it is not the differences in the use of various *RAV* formulas that is causing the sensitivity, but rather the high degree of nonlinear dependence of the maximum option value on the parameters entering the calculation.

In this case, too, the main contributions to the uncertainty in the maximum option value have been investigated, as measured by the contributions to the variance in the cumulative probability distribution for option value.

Figure 13.9a (Cozzolino *RAV* formula) and Figure 13.9b (parabolic *RAV* formula) now show that there is a change in the dominant contributions to the variance as compared to the previous case of 10% variations in parameters. The gain, *G*, and working interest, *W*, now control about 85–95% of the relative importance to the variance, with uncertainty in gain about twice as dominant as uncertainty in working interest. Thus the relative

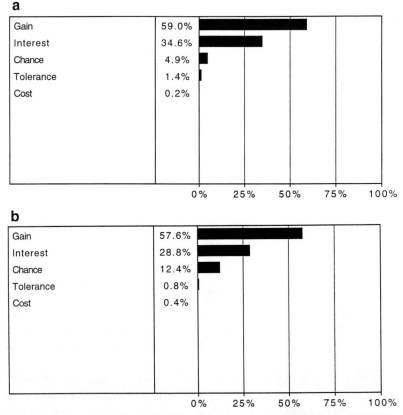

FIGURE 13.9 As for Figure 13.7 but now with parameter ranges for p_s, *G*, *C*, *RT*, and *W* increased to ±20% around their respective mean values. Note that the relative importance of gain (*G*) and interest (*W*) now account for about 85–95% of the total variance in option value. (a) Using the Cozzolino *RAV* formula. (b) Using the parabolic *RAV* formula.

importance argues that *if* effort is to be expended to narrow the uncertainty in the mean option value, then that effort is best expended on trying to improve the ±20% uncertainty on anticipated gain, with a lesser effort on determining better the uncertainty in working interest; all other factors are of less significance at this stage of uncertainty.

The points to be made with these two examples are that the uncertainties caused by use of different formulas (Cozzolino versus parabolic) are negligible in comparison to the sensitivity of the option value variance and of the mean option value to uncertainties in the basic parameters; in addition, the changes in both mean and variance of the option values are not equally dependent on equal fractional variances of each of p_s, G, C, RT, and W due to the nonlinearity of the formulas determining O_W.

The relative importance chart indicates which of the various parameters is causing the greatest degrees of uncertainty in the variance and, therefore, where greatest effort should be expended to narrow the range of uncertainty around the central mean value of each parameter.

The high degree of sensitivity of the maximum option value to changes in parameter ranges, *centered on fixed mean values of the parameters*, provides a measure of the straddle in maximum option values that should be taken at the 90% probability interval. In this way a better job can be done on deciding whether an option is an appropriate course of action to maximize corporate gain.

G. EFFECTS OF MASSIVE CHANGES IN GAINS, COSTS, AND SUCCESS PROBABILITY

So far analysis of the maximum option value under both deterministic and probabilistic considerations has made the assumptions that gain, G, costs, C, success probability, p_s, and risk tolerance, RT, either are known or sit squarely in some prescribed range of values. But it can happen that as time proceeds there is a significant reevaluation of the project (due possibly to the acquisition of new information) or of the risk tolerance of the corporation (due possibly to new management or to a change in corporate capital asset value). But, for whatever reasons, G, C, p_s, RT, and/or W change to values outside of the original prescribed range, say, to G_1, C_1, p_{s1}, RT_1, and W_1. In that case the maximum option worth is no longer as calculated previously, but is changed because of the changes in G, C, p_s, RT, and W.

This problem has a close parallel in the stock market where, as Zisook (1994) has noted, "Suppose I have just bought an option. If the stock price were to change, the probability distribution would readjust itself, and the value of my option would shift. If I calculate that the option value will move 50 cents for every $1 change in the stock price, I will sell half as many stock shares as the option was for. Then every time the stock price does go up, the loss incurred by selling stock before the shift exactly cancels the gain in the value of the option I bought. The gains and losses are reversed if the stock price decreases, but the result is the same: Everything comes out even. This is called statistical arbitrage."

In the oil exploration arena it is most often the case that one is given the opportunity to buy in at a *fixed* working interest W rather than at a fixed cost, WC. In such a case the maximum option worth changes as G, C, p_s, and RT shift outside their initially prescribed ranges. At a *fixed* working interest, W, the change in the worth of the option is

$$\Delta O_p = O_p(1) - O_p(0), \tag{13.33}$$

where $O_p(1)$ is the worth of the option calculated with G_1, C_1, p_{s1}, RT_1 replacing G, C, p_s, RT, respectively, and $O_p(0)$ is the option worth calculated with G, C, p_s, and RT. If $\Delta O_p < 0$, then the worth of the option is reduced, while $\Delta O_p > 0$ increases the worth of the option. Thus, the ability to decide to invest in the opportunity as a result of changes in the parameters G, C, p_s, and RT controlling the value of the project, and the ability to decide to walk away from the opportunity at the cost of losing the option money, can then easily be controlled.

An alternative is to adjust the working interest that one takes in a project in order to keep total costs $TC \equiv (WC + O_p)$ fixed. Thus, consider the use of the parabolic RAV, for which the maximum option value is given by Eq. (13.22), so that total costs are

$$TC = Wp_s^{-1}\left(C + \frac{1}{2}p_f p_s G^2 W/RT\right). \tag{13.34}$$

Suppose that original anticipated project costs C increase by the amount ΔC, which is considerably larger than the range of uncertainty originally estimated for C. Then, all other factors being equal, the working interest one can take in the project must decrease to allow for the increase in C. Let W decrease to $W - \Delta W$ in order to hold TC fixed. Then,

$$\Delta W = \frac{1}{2\alpha}\{C + 2\alpha W + \Delta C - [(C + 2\alpha W + \Delta C)^2 - 4\alpha W \Delta C]^{1/2}\}, \tag{13.35a}$$

where

$$\alpha = \frac{1}{2}p_f p_s G^2/RT. \tag{13.35b}$$

For example, consider again the situation with $W = 0.5$, $p_s = 0.5$, $G = \$100$ MM, $RT = \$100$ MM, and $C = \$10$ MM. Then,

$$\Delta W = 4 \times 10^{-2}\{22.5 + \Delta C - [(22.5 + \Delta C)^2 - 25\Delta C]^{1/2}\}. \tag{13.36}$$

Suppose the initial cost estimate of $10 MM is suddenly doubled to $20 MM so that $\Delta C = C = \$10$ MM; then $\Delta W = 0.16$, so that the working interest that can be taken decreases from 50% to 34% in order to hold total costs at a fixed amount.

Note that, at a fixed working interest, the maximum option value *increases* by the amount $Wp_f p_s^{-1} \Delta C$ as the costs increase by ΔC. The reason for this behavior is that while the high value worth $p_s[W(G - C) - O_p]$ decreases as C increases, the RAV also decreases with increasing C and at a faster rate than the high value worth. Because the maximum option value, O_p, is the difference between high value worth and RAV, it follows that the option worth then rises.

Note also that as the success probability, p_s, increases then because

$$O_p = \frac{1}{2} W^2 (1 - p_s) G^2 / RT + W(1/p_s - 1)C, \qquad (13.37)$$

it follows that the maximum option worth decreases with increasing p_s. The reason, of course, is that the higher the success probability the closer RAV approaches the high value worth, making an option less appealing. Note that the option worth varies quadratically with anticipated gain, G, so that a doubling of G would make an option that much more appealing.

In short, because of the high degree of nonlinearity of maximum option value on the parameters p_s, G, C, RT, and W, there can be considerable swings in the option value worth as the parameters change. It is this high degree of volatility to option values that makes them such powerful devices for massive corporate gain (or massive corporate ruin) unless they are used with caution.

H. OPTIONS AND PORTFOLIO BALANCING

With a portfolio of hydrocarbon exploration opportunities, each with its own chances of success, estimated gains, and estimated costs, the problem arises of maximizing the *total RAV* for *all* opportunities when there is a fixed total corporate budget of B. Effectively one has to determine the best working interest, W_i, to take in the ith opportunity so that the total RAV is at its maximum value and the budget, B, is then expended on the total costs, $TC = \sum_{i=1}^{N} (W_i C_i + O_{wi})$ for the N opportunities. This problem was solved analytically in closed form in the *absence* of any option commitments earlier in this volume.

However, when an option position can be taken in each and every project, then that cost must be paid, in addition to the amount $W_i C_i$ for the project costs if one chooses to invest in the opportunity at some future time. Thus, a greater fraction of the budget is then expended on each opportunity, so that a smaller working interest can be taken in each than in the case of not optioning any of the opportunities provided, however, that all options are exercised—otherwise, the cost saved goes to other projects.

The question then arises: What is the working interest that should be taken in each opportunity subject to a fixed budget, B, so that the total

RAV for the sum of all projects is maximized when allowance is made for an option cost, O_{pi}, to be taken in each project? This problem is addressed here using the parabolic RAV formula, for which the maximum option worth is given by

$$O_p = W p_f p_s^{-1} \left[C + \frac{1}{2} p_s G^2 W / RT \right], \quad (13.38a)$$

where $p_f = 1 - p_s$.

The maximum option cost, O_{pi}, for the ith opportunity also has to be limited by

$$0 \leq O_{pi} \leq W_i V_i \quad (13.38b)$$

in order for the success probability branch of the corresponding decision tree to yield a positive return, and it must also be arranged that *each opportunity in the portfolio has an $RAV_i \geq 0$*, including the option cost, as shown earlier. With these constraints understood one can proceed as follows.

Consider N opportunities, with parameter values p_{si}, G_i, C_i, and W_i for the ith opportunity. Let B be the total budget, and let the ith opportunity be optioned at the maximum amount of O_{pi}, given by Eq. (13.38a) for parabolic weighting. Detailed computations are carried through here for the parabolic RAV formula, whereas Appendix B provides the corresponding results for the Cozzolino RAV formula. In addition, for conservative option amounts, less than the maximum for each opportunity, the corresponding results are presented in Appendix C.

The aim here is to provide expressions for W_i for each opportunity given the constraints of a fixed budget and of fixed values for p_{si}, C_i, and G_i for each opportunity. The corporate risk tolerance is taken at the fixed value of RT. The total costs for all N projects, including option costs, are required to balance the budget so that

$$\sum_{i=1}^{N} \{W_i C_i + O_{pi}\} \leq B. \quad (13.39)$$

Using Eq. (13.38b) for the option cost, Eq. (13.39) can then be written

$$\sum_{i=1}^{N} \left\{ W_i C_i [1 + p_{fi} p_{si}^{-1}] + \frac{1}{2} W_i^2 p_{fi} G_i^2 / RT \right\} \leq B. \quad (13.40)$$

For the parabolic RAV formula for each opportunity, the total RAV of all N projects, including the option costs of O_{pi} per project, is given by

$$RAV = \sum_{j=1}^{N} W_j p_{sj}^{-1} \left[p_{sj}^2 G_j - C_j - \frac{1}{2} p_{sj} p_{fj} G_j^2 W_j / RT \right]. \quad (13.41)$$

The aim is to maximize RAV, as given by Eq. (13.41), with respect to W_i for all i, subject to the constraint of a fixed budget, B, as given through Eq. (13.40).

The procedure for achieving the maximum total RAV is as follows. Regard RAV as a function of the variable W_k for the kth opportunity. Then RAV has an extremum when the partial derivative

$$\frac{\partial RAV}{\partial W_k} = 0 \quad \text{for all } k. \tag{13.42}$$

Now, from Eq. (13.40) one can regard the budget constraint as providing an expression for W_i in terms of W_k as far as the partial derivative in Eq. (13.42) is concerned.

Then the optimum working interest, W_k, for the kth opportunity is determined from

$$0 = \frac{\partial RAV}{\partial W_k} = p_{sk}^{-1}[p_{sk}^2 G_k - C_k - p_{sk}p_{fk}G_k^2 W_k/RT]$$
$$+ p_{si}^{-1}[p_{si}^2 G_i - C_i - p_{si}p_{fi}G_i^2 W_i)/RT]\frac{\partial W_i}{\partial W_k}. \tag{13.43}$$

But, using the constraint of Eq. (13.40),

$$\frac{\partial W_i}{\partial W_k}\{C_i p_{si}^{-1} + W_i p_{fi}G_i^2/RT\} = -\{C_k p_{sk}^{-1} + W_k p_{fk}G_k^2/RT\}, \tag{13.44}$$

which, when used in Eq. (13.43), yields

$$\begin{array}{l} p_{sk}^{-1}[p_{sk}^2 G_k - C_k - p_{sk}p_{fk}G_k^2 W_k/RT] \times \\ [C_k p_{sk}^{-1} + W_k p_{fk}G_k^2/RT]^{-1} = (k \rightarrow i) \quad \text{for all } i, k \end{array} \tag{13.45}$$

where the term $(k \rightarrow i)$ on the right-hand side of Eq. (13.45) means that the left-hand side is repeated with subscript k replaced by subscript i. However, i and k were arbitrary choices. The only way Eq. (13.45) can be valid for *all* i and k is if each side is some absolute dimensionless constant, H, say, independent of i and k. Then,

$$\begin{array}{l} p_{sk}^2 G_k - C_k - p_{sk}p_{fk}G_k^2 W_k/RT = H[C_k + W_k p_{sk}p_{fk}G_k^2/RT] \\ \text{for all } k, \end{array} \tag{13.46}$$

which provides a relation connecting W_k for the kth opportunity in terms of a constant H, yet to be determined. Equation (13.46) can be rewritten in the form

$$W_k = \min\{RT(p_{sk}p_{fk}G_k^2)^{-1}(1+H)^{-1}[p_{sk}^2 G_k - C_k(1+H)], 1\}, \tag{13.47}$$

which yields W_k in terms of H.

Direct substitution of Eq. (13.47) into the budget condition, Eq. (13.40), yields

$$1 + H \geq \left(\sum_{i=1}^{N} (p_{si}^2/p_{fi})\right)^{1/2} \left[2B/RT + \sum_{i=1}^{N} (C_i^2/G_i^2)(p_{si}^2 p_{fi})^{-1}\right]^{-1/2} > 0, \qquad (13.48)$$

which expresses H directly in terms of the budget, B, provided $0 \leq W_k \leq 1$ for all k.

Consider Eq. (13.47) in a little more detail. So that $W_k \leq 1$, Eq. (13.47) requires

$$H \geq p_{sk}^2 G_k / [C_k + p_{sk} p_{fk} G_k^2 / RT] - 1. \qquad (13.49a)$$

If inequality (13.49a) is *not* satisfied then $W_k = 1$.

So that Eq. (13.47) provides $W_k \geq 0$, it is required that

$$H \leq p_{sk}^2 G_k / C_k - 1. \qquad (13.49a)$$

Equation (13.49b) has the overriding requirement

$$p_{sk}^2 G_k > C_k \qquad (13.50)$$

or else it is impossible to have $1 + H$ greater than unity at the maximum option cost O_{pk}. In that case only a fraction λ_k of the maximum option cost can be taken in the kth project. This situation is considered in more detail in Appendix A. When Eq. (13.47) yields $0 \leq W_k \leq 1$ for all projects, the total RAV for all N projects is then given by

$$RAV = \frac{1}{2} RT(1 + H)^{-2} \sum_{i=1}^{N} G_i^{-2} (p_{si}^2 p_{fi})^{-1} [p_{si}^2 G_i - C_i(1 + H)]$$
$$\times [p_{si}^2 G_i(1 + 2H) - C_i(1 + H)], \qquad (13.51)$$

which yields RAV_i for the ith project as

$$RAV_i = \frac{1}{2} RT(1 + H)^{-2} G_i^{-2} (p_{si}^2 p_{fi})^{-1} [p_{si}^2 G_i - C_i(1 + H)]$$
$$\times [p_{si}^2 G_i(1 + 2H) - C_i(1 + H)], \qquad (13.52)$$

and $RAV_i > 0$ (for $0 \leq W_i < 1$) when also

$$H \leq (p_{si}^2 G_i - C_i)/(2p_{si}^2 G_i - C_i). \qquad (13.53)$$

In addition, the maximum option cost of the ith opportunity is then

$$O_{pi} = \frac{1}{2} W_i p_{si}^{-1} (1 + H)[p_{si}^2 G_i - C_i(1 - 2p_{fi})(1 + H)]. \qquad (13.54)$$

In order for the kth project to yield $0 \leq W_k \leq 1$, Eqs. (13.49a) and (13.49b) require

$$1 + H_{k,\min} \equiv p_{sk}^2 G_k / [C_k + p_{sk} p_{fk} G_k^2 / RT] \leq 1 + H \leq p_{sk}^2 G_k / C_k$$
$$\equiv 1 + H_{k,\max}. \qquad (13.55)$$

For $H \leq H_{k,\min}$, then $W_k = 1$; for $H \geq H_{k,\max}$, then $W_k = 0$. When $0 < W_k < 1$, the requirement for the kth project to yield a positive value of RAV_k is that

$$H \leq H_{*,k} \equiv (p_{sk}^2 G_k - C_k)/(2p_{sk}^2 G_k - C_k) \leq 1. \tag{13.56}$$

Now $H_{*,k} < H_{k,\max}$ in $p_{sk}^2 G_k \geq C_k$, so that the joint requirement of a positive RAV for the kth opportunity and of having $0 \leq W_k \leq 1$ is

$$1 + H_{k,\min}, \leq 1 + H \leq 1 + H_{*,k}. \tag{13.57}$$

Any opportunity with $p_{sk}^2 G_k < C_k$ is not to be considered further at the *maximum* option cost because $W_k \leq 0$, and any project with $H_{*,k} < H$ is also rejected because RAV_k would then be negative.

Of the remaining opportunities, first check to determine if

$$1 + H_{*,k} \geq 1 + H_{k,\min} \quad \text{for all remaining } k. \tag{13.58}$$

If this inequality is *not* satisfied, then those opportunities are discarded because, for them, a positive RAV_k cannot be achieved with $0 \leq W_k \leq 1$.

Of the remaining opportunities, N, say, all satisfy

1. $\quad p_{sk}^2 G_k \geq C_k$ \hfill (13.59a)
2. $\quad H_{*,\min} \geq H_{k,\min}$ \hfill (13.59b)

and are viable at this stage. The question then is this: Is H constrained by the range (13.57) in order that the kth project be considered (i.e., is $W_k \leq 1$?).

To answer this question, organize the remaining N opportunities in order of decreasing values of $H_{k,\min}$ with

$$H_{N,\min} > H_{N-1,\min} > H_{N-2,\min} > \cdots > H_{1,\min}. \tag{13.60}$$

Suppose a value of H is chosen with

$$H_{M,\min} > H > H_{M-1,\min}. \tag{13.61}$$

Then the working interests W_k of the opportunities $k = M, M+1, \ldots, N$ are all to be set to unity, while the remaining working interests for $k = 1, \ldots, M-1$ are given by Eq. (13.47).

Then consider the budget constraint

$$B \geq \sum_{i=1}^{N} (W_i C_i + O_{wi})$$

$$= \frac{RT}{2(1+H)^2} \left\{ \sum_{i=1}^{M-1} \mu_i (p_{si}^2/p_{fi}) - (1+H)^2 \mu_i \sum_{i=1}^{M-1} \frac{C_i^2 G_i^{-2}}{(p_i^2 p_{fi})} \right\} \tag{13.62}$$

$$+ \sum_{i=1}^{N} [C_i p_{si}^{-1} + \frac{1}{2} p_{fi} G_i^2/RT] \mu_i,$$

where μ_i is unity as $RAV_i > 0$, or $\mu_i = 0$ as $RAV_i \leq 0$. Equation (13.62) can be rewritten in the form

$$1 + H \geq \left(\sum_{i=1}^{M-1} \mu_i p_{si}^2/p_{fi} \right)^{1/2} \left\{ \frac{2}{RT} \left[B - \sum_{i=M}^{N} \mu_i \left(C_i p_{si}^{-1} + \frac{1}{2} p_{fi} G_i^2/RT \right) \right] \right. \\ \left. + \sum_{i=1}^{M-1} C_i^2 G_i^{-2} (p_{si}^2 p_{fi})^{-1} \mu_i \right\}^{-1/2}. \tag{13.63}$$

As H decreases, M decreases toward unity. However, the requirement of a positive RAV for each of the opportunities has not yet been enforced, that is, μ_i has not been determined.

Now the kth opportunity has

$RAV_k =$

$$\begin{cases} W_k p_{sk}^{-1} [p_{sk}^2 G_k - C_k - \frac{1}{2} p_{sk} p_{fk} G_k^2 W_k/RT], & \text{for } 0 \leq W_k < 1 \quad (13.64a) \\ p_{sk}^{-1} \left[p_{sk}^{-1} G_k - C_k - \frac{1}{2} p_{sk} p_{fk} G_k^2/RT \right], & \text{for } W_k = 1. \quad (13.64b) \end{cases}$$

Hence, for the k opportunities with $N \geq k \geq M$, check to see if Eq. (13.64b) yields a positive RAV_k. If so, then set $\mu_k = 1$; if not, then set $\mu_k = 0$. For the k opportunities with $1 < k \leq M - 1$, assume initially that all the $\mu_k = 1$. Then calculate H from Eq. (13.63). Use the calculated value of H to determine W_k ($0 \leq W_k \leq 1$) from Eq. (13.47) and then check the sign of RAV_k from Eq. (13.64b). If the sign is positive, return $\mu_k = 1$; if negative, set those $\mu_k = 0$. Then recalculate H from Eq. (13.63) and repeat the loop.

Finally, calculate the maximum option cost, O_{pi}, of the ith opportunity from

$$O_{pi} = p_{fi} p_{si}^{-1} \left[C_i + \frac{1}{2} p_{si} G_i^2/RT \right] \quad \text{for } W_i = 1 \quad (13.65a)$$

and from

$$O_{pi} = p_{fi} p_{si}^{-1} W_i \left[C_i + \frac{1}{2} p_{si} W_i G_i^2/RT \right] \quad \text{for } 0 \leq W_i \leq 1. \quad (13.65b)$$

In this way the portfolio of available opportunities is balanced with respect to both a fixed budget and with respect to a maximum option cost for each opportunity. The RAV for the total portfolio is optimized under those constraints, and the individual RAVs for each opportunity (including project costs and maximum option costs) are kept positive. In addition, the best working interest to take in each opportunity is determined, as are the individual option costs.

In performing these calculations two caveats have to be borne in mind: The first is that the maximum positive option cost, O_{pi}, must be less than or equal to $W_i V_i$; and the second is that $RAV_i \geq 0$.

One simple way to handle all of the above problems at once is to (1) choose a value for H; (2) calculate O_{pi} from Eq. (13.54) (subject to $0 \leq O_{pi} \leq W_i V_i$); (3) calculate RAV_i (≥ 0) from Eq. (13.52); (4) calculate W_i from either Eq. (13.47) if $0 \leq W_i \leq 1$, or set $W_i = 0$ if Eq. (13.47) yields $W_i \leq 0$, or set $W_i = 1$ if Eq. (13.47) yields $W_i \geq 1$; (5) use the calculated values for all opportunities in Eq. (13.62) to write down the budget value, B, in terms of H; and (6) vary H and so construct a look-up table of B versus H. In this way it is far easier to perform the necessary technical manipulations with H varying and to then finally read off the corresponding budget value, B. This procedure is also computationally quick.

The best working interest to take in each hydrocarbon exploration opportunity available in a portfolio when dealing with a fixed corporate budget, B, and a corporate-mandated risk tolerance, RT, is limited by the direct exploration costs associated with each opportunity. In addition, there is also a corporate cost associated with each opportunity irrespective of whether the decision is made to drill or not. That cost is the amount one must pay for the option to participate in an opportunity. This option cost can occur in many guises: It can be the amount one bids at a lease sale for a block; it can be the amount one corporation charges another for the right to decline to participate or not at a future time in an ongoing venture; it can be the amount one pays in earnest money for the right of first refusal on exploring an opportunity; it can be a farm-in or farm-out premium, and so on. But the point is that if one goes ahead with development of an opportunity, then the option cost adds to the total project costs, thereby shifting risk-adjusted value and optimum working interest evaluations.

For a portfolio of opportunities the tasks of interest are to determine the maximum amount one should pay for an option for each opportunity, the best working interest to then take in each opportunity so that the total portfolio risk-adjusted value is maximized, and to also determine a quantitative procedure that will permit the evaluations to be made under a fixed total budget constraint and a given risk tolerance.

All of these concerns have been addressed simultaneously in this subsection. In addition, should a corporation choose to option at less than the maximum cost, or use a risk-adjusted formula different than the parabolic dependence on working interest used in the body of the text, then Appendixes B and C spell out the details of these technical modifications. The point of these computations is to provide quantitative, reproducible measures of use to corporate decision makers who must, most often, operate under the constraints of fixed budgets and fixed corporate risk tolerance controls.

I. PORTFOLIO BALANCING WITH COMMON WORKING INTEREST

As a variation on a theme, one particular class of problems in portfolio balancing is to determine the optimal working interest to be taken in a variety of opportunities but with the following bureaucratically imposed constraints:

1. The *same* working interest, W, shall be taken in each and every opportunity.
2. Optioning the portfolio of opportunities is permitted, but one must then option each opportunity and, furthermore, if one wishes to develop the exploration opportunities at a later date then one will *still* take a common working interest in each and every opportunity.

The effect of these rules is, apparently, to ensure that poor prospects are drilled rather than just good prospects.

The question is this: What is the working interest that one should take in the set of all opportunities both in the absence and in the presence of an optioning choice?

1. Working Interest in the Absence of an Option Choice

For the N projects, $(i = 1, \ldots, N)$, each with its own estimated success probability, p_{si} (failure probability $p_{fi} = 1 - p_{si}$), gain G_i, and costs C_i, the parabolic risk aversion value for each project, for a corporate risk tolerance of RT, is

$$RAV_i = WE_i \left[1 - \frac{1}{2} E_i v_i^2 \, W/RT \right], \tag{13.66}$$

where W is *common* to all projects.

The total RAV for all N projects is

$$RAV = W \left(\sum_{i=1}^{N} E_i \right) - \frac{1}{2} W^2 / RT \sum_{i=1}^{N} (E_i^2 \, v_i^2). \tag{13.67}$$

Equation (13.67) has a maximum ($\partial RAV/\partial W = 0$) when

$$W = RT \sum_{i=1}^{N} (p_{si} G_i - C_i) \Big/ \sum_{i=1}^{N} (p_{si} p_{fi} G_i^2). \tag{13.68}$$

Provided Eq. (13.68) yields $0 \leq W \leq 1$ then the total RAV is

$$RAV = \frac{\frac{1}{2} RT \left(\sum_{i=1}^{N} [p_{si} G_i - C_i] \right)^2}{\left(\sum_{i=1}^{N} (p_{si} p_{fi} G_i^2) \right)} \quad (13.69)$$

$$= \frac{1}{2} W \sum_{i=1}^{N} E_i,$$

which is intrinsically positive so that it is worthwhile, on average, to take an interest in the total portfolio of projects even though some of the opportunities will fail.

The requirements of a positive value for W in $0 \leq W \leq 1$ are, from Eq. (13.68), that first

$$\sum_{i=1}^{N} (p_{si} G_i - C_i) > 0; \quad (13.70a)$$

that is, the average project proves to be of positive worth and, when inequality (13.70a) is in force, then also one must have

$$RT < \frac{\sum_{i=1}^{N} (p_{si} p_{fi} G_i^2)}{\sum_{i=1}^{N} (p_{si} G_i - C_i)} \equiv RT_*. \quad (13.70b)$$

Thus, the risk tolerance must be kept low enough. If $RT > RT_*$ then W should be set to unity, with a total RAV then of

$$RAV(W = 1) = \sum_{i=1}^{N} [p_{si} G_i - C_i - p_s p_{fi} G_i^2/RT] \quad \text{for } RT \geq RT_*, \quad (13.71)$$

which is again intrinsically positive when inequality (13.70a) is in force. The total budget required is then

$$B = W \sum_{i=1}^{N} C_i. \quad (13.72)$$

If inequality (13.70a) is reversed (i.e., the average project is not economically viable), then there is no common working interest that will lead to a positive total RAV in the absence of an option.

2. Working Interest in the Presence of an Option Choice

The maximum option cost (at a common working interest, W) for the ith opportunity under the parabolic RAV formula is

$$O_{pi} = W p_{fi} p_{si}^{-1} \left(C_i + \frac{1}{2} p_{si} G_i^2 W/RT \right). \quad (13.73)$$

The total budget costs are then

$$B = W\left(\sum_{i=1}^{N} C_i p_{si}^{-1}\right) + \frac{1}{2}(W^2/RT)\left(\sum_{i=1}^{N} G_i^2 p_{fi}\right), \quad (13.74)$$

while the total RAV is reduced to

$$RAV = W\left[\sum_{i=1}^{N} p_{si}^{-1}(p_{si}^2 G_i - C_i)\right] - \frac{1}{2}(W^2/RT)\sum_{i=1}^{N} p_{fi} G_i^2. \quad (13.75)$$

The total RAV now has a maximum with respect to W on $W = W_1$ where

$$W_1 = RT\sum_{i=1}^{N} p_{si}^{-1}(p_{si}^2 G_i - C_i)/\left(\sum_{i=1}^{N} p_{fi} G_i^2\right) \quad (13.76)$$

provided $0 \leq W_1 \leq 1$, and at the value $W = W_1$, one has the maximum

$$RAV_1 = \frac{\frac{1}{2}RT\left(\sum_{i=1}^{N} p_{si}^{-1}(p_{si}^2 G_i - C_i)\right)^2}{\left(\sum_{i=1}^{N} p_{fi} G_i^2\right)} \geq 0, \quad (13.77)$$

together with the budget requirement

$$B_1 = \frac{RT}{\left(\sum_{i=1}^{N} p_{fi} G_i^2\right)}\left[\sum_{i=1}^{N} p_{si} G_i\right]\sum_{i=1}^{N}[p_{si}^{-1}(p_{si}^2 G_i - C_i)]. \quad (13.78)$$

To ensure $0 \leq W_1 < 1$, Eq. (13.76) requires

$$\sum_{i=1}^{N} p_{si} G_i > \sum_{i=1}^{N} C_i/p_{si} \quad (13.79)$$

and, when inequality (13.79) is in force, the risk tolerance must be kept sufficiently low ($RT < RT_1$), where

$$RT_1 = \frac{\sum_{i=1}^{N} p_{fi} G_i^2}{\sum_{i=1}^{N} p_{si}^{-1}(p_{si}^2 G_i - C_i)}. \quad (13.80)$$

If $RT > RT_1$ and also inequality (13.79) is in force, then one should set $W_1 = 1$ together with

$$RAV_1(W_1 = 1) = \sum_{i=1}^{N} p_{si}^{-1}(p_{si} G_i - C_i) - \frac{1}{2RT}\sum_{i=1}^{N} p_{fi} G_i^2, \quad (13.81)$$

which is positive, and the budget requirement is then

$$B_1(W_1 = 1) = \frac{1}{2RT}\sum_{i=1}^{N} G_i^2 p_{fi} + \sum_{i=1}^{N} C_i p_{si}^{-1}. \tag{13.82}$$

If inequality (13.79) is violated, then $W_1 < 0$ and no common working interest or option position should be taken in the portfolio of available opportunities at the maximum option amount.

J. BIDS AND OPTIONS

Option methods can also be used in assessing leasing situations, as the following illustration exhibits. Suppose that an opportunity has been evaluated yielding estimated gains, G, if successful, estimated costs, C, to drill, and an estimated success probability, p_s, for an expected value of $EV = p_s G - C$. One could go ahead and estimate a sealed bid amount to make *now*, prior to the closure data for bid receipt, based on the methods given in earlier chapters. However, a seismic survey is under way over the tract of interest, and may be completed and interpreted before the lease sale closure date for bids.

If one waits, let the survey results resolve a fraction, f, of the success probability, p_s. If the resolution is positive, then the success probability increases to $p_s(1 + f)$; if the resolution is negative, then the success probability decreases to $p_s(1 - f)$.

The new expected value if the resolution is positive is

$$EV_1 = p_s(1 + f)G - C, \tag{13.83a}$$

and in the negative resolution situation is

$$EV_1 = p_s(1 - f)G - C. \tag{13.83b}$$

Now, because one does not know how the resolution will come out in the future, let the probability of a positive resolution be p_r, and so $(1 - p_r)$ is the probability of a negative resolution.

The new average expected value is then given by

$$EV(\text{new}) = p_r \times EV_1 + (1 - p_r) \times EV_2 = p_s G - C + p_s G f(2p_r - 1) \tag{13.83c}$$

so that the difference between $EV(\text{new})$ and the original EV is

$$EV(\text{new}) - EV = 2p_s f G(p_r - 1/2).$$

The difference can be positive or negative depending on whether the positive resolution probability, p_r, is greater or less than 0.5—as might have been anticipated.

If the survey is indeed finished and interpreted in time before bid closure date, then one has *absolute certainty* on both p_r and f. However, if it becomes clear that the survey results will be delayed beyond the bid closure date, then one has to estimate somehow whether the resolution probability will be greater or less than 0.5; or one has to make a bid based on the original *EV*. The basic option problem is reduced to estimating the resolution probability, p_r, on the basis of available information. Once such an estimate is somehow made, then one can determine the bid to make based on the estimate of *EV*(new). The option that one has here is to bid now, based on the original *EV*, or to hold off on the bid in the hope that the new information will be available in time: A higher bid could be made if $p_r < 0.5$. In the event that the future information is *not* available in time, then one has to estimate the value of p_r based on other considerations (e.g., historical precedent or a model for the future), which would, hopefully, be updated as future information does become available, so that Bayesian procedures also operate in this arena of exploration economic assessment.

A second way of bidding that can be influenced by the optioning approach is as follows. Suppose that a corporation decides to put in a bid for a block. If the bid is successful then the corporation has acquired the block. If the corporation acquires and then drills immediately, then one is treating the bid as an investment cost that is at risk. (This situation only works as stated in situations where drilling is an option. In many countries a well, or more than one, are obligations. Such obligation costs for drilling can be factored in as part of the bid. Then the question devolves into one of after the obligatory wells are completed, what should the corporation do?) On the other hand, the corporation could choose to wait after acquiring the block until additional information on success probability, which will come from adjacent drilling activity, is available, when a fraction f of the success probability is resolved. In this case the corporation is treating the bid as an option because, until the additional information is available, the corporation has not spent the drilling costs. If the resolution *increases* the success probability then, at that time, the corporation can choose to go ahead with the drilling or to farm out or abandon the opportunity; if the resolution *decreases* the success probability the corporation has the same three choices. The point is that the corporation treats drilling on the acquired block as an option and not an obligation to drill. The question in this case is how much to bid on an option approach versus on an obligation approach to the block.

At the stage *prior* to resolution being achieved on part of the success probability, there is no extra cost associated with the option value because there has not yet been a decision made to drill or not drill—the option value is "free." The difference between the option approach to bidding and the conventional expected value approach previously discussed is to calculate the *additional* value of the free option.[2]

Figure 13.10 provides a compendium of the parameters of both the option approach and the expected value approach prior to resolution of the fraction, f, of the success probability by additional information.

The expected value, $EV(\text{new})$, of the resolution is

$$EV(\text{new}) = p_R EV_1 + (1 - p_R) EV_2 = EV + p_s fG(2p_R - 1), \quad (13.84)$$

while the option value, $OV(\text{new})$, of the resolution is

$$OV(\text{new}) = p_R OV_1 + (1 - p_R) OV_2 = OV - p_s fC(2p_R - 1). \quad (13.85)$$

The variance of the resolved opportunity, $\sigma(\text{new})^2$, is

$$\sigma(\text{new})^2 = \sigma^2[1 + f(2p_R - 1)] [1 - p_s (1 - p_s)^{-1} f(2p_R - 1)], \quad (13.86)$$

where

$$\sigma^2 = p_s(1 - p_s)G^2. \quad (13.87)$$

The change in expected value is

$$\Delta EV = EV(\text{new}) - EV = p_s fG(2p_R - 1), \quad (13.88)$$

while the change in option value is

$$\Delta OV = OV(\text{new}) - OV = -p_s fC(2p_R - 1). \quad (13.89)$$
$$\text{Hence, } \Delta(OV)/\Delta EV = -C/G, \quad (13.90)$$

which is independent of f, p_s, or p_R!

As EV increases (decreases) so OV decreases (increases). Now, because $p_s G > C$ in order to have the original $EV(\equiv p_s G - C)$ positive, it follows that

$$\frac{\Delta OV}{\Delta EV} = - [C/(p_s G)]p_s, \quad (13.91)$$

which is less than unity in magnitude, so that absolute changes in OV, $|\Delta OV|$, are always smaller than absolute changes, $|\Delta EV|$, in expected value. Thus, it should always be better to bid using an option approach than an expected value approach because there is less volatility and, presumably, less *total* cost involved prior to the resolution point being reached. To check this contention, note that the magnitude of $|\Delta OV|$ cannot exceed $p_s fC$, no matter how p_R varies.

---

[2] An alternative viewpoint is to note that if one were to contemplate drilling at some time after resolution, then one should factor that probability into the bid estimate based on an option approach, because the bid amount would then be at risk together with drill costs. This viewpoint is considered in Appendix A to this chapter.

No Resolution of p_s

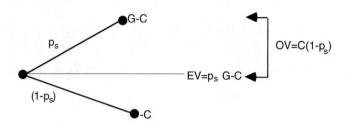

Variance on EV: $\sigma^2 = p_s(1-p_s)G^2$

Resolution of the fraction f of p_s
(a) Positive (Resolution Probability, p_R)

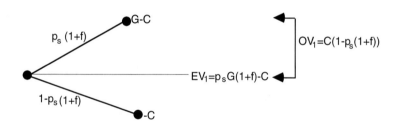

(b) Negative (Resolution Probability, $1-p_R$)

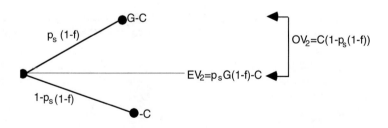

Variance on EV_2: $\sigma_2^2 = G^2 p_s(1-f)(1-p_s(1-f))$.

FIGURE 13.10 Parameters for the option approach and expected value approach to bidding.

For a fractional bid, b, of either EV (an expected value approach) or of $EV + |\Delta OV|$ (an option approach) [with $b = 0.35$ as a common practical rule (Capen et al., 1971)], then:

1. Total costs, TC_{EV} in the expected value approach are

$$TC_{EV} = C + bEV \quad (13.92)$$

because one will bid and then drill immediately.

2. Total costs, TC_{OA}, in the option approach are

$$TC_{OA} = b(EV + |\Delta OV|) \quad (13.93)$$

because one does *not* expend any drilling cost money at all until the resolution point is reached, and then only if the resolution result is favorable enough in terms of "bang for the buck."

The difference between *committed* costs is

$$\Delta TC = TC_{EV} - TC_{OA} \quad (13.94a)$$

that is,

$$\Delta TC = C(1 - bp_s f) > 0 \quad \text{i.e., } TC_{EV} > TC_{OA}. \quad (13.94b)$$

One saves the amount ΔTC by using an option strategy. Thus, regardless of how the resolution comes out, one can afford to bid more with an option approach than with an expected value approach.

The bid difference is given by

$$\Delta B = B_{OA} - B_{EV} = Cp_s f b > 0, \quad (13.95)$$

where

$$B_{EV} = bEV \quad (13.96a)$$

$$B_{OA} = b(EV + |\Delta OV|). \quad (13.96b)$$

After resolution is achieved from additional adjacent drilling information, one knows f, and also whether the resolution is positive ($p_R = 1$) or negative ($p_R = 0$). At that point one has to evaluate the worth of the opportunity again, but now with the updated data. If one chooses to drill, then the bid cost B_{OA} must be added to the total drilling cost, C, for a total investment at risk of $C + B_{OA}$. The expected value of the opportunity is now

$$EV_1 = p_s(1 + f)G - C - B_{OA} \quad (13.97)$$

if the resolution is positive; and

$$EV_2 = p_s(1 - f)G - C - B_{OA} \quad (13.98)$$

if the resolution is negative.

The expected return on total investment (the "bang for the buck") is then

$$R_1 = EV_1/(C + B_{OA}) \quad \text{or} \quad R_2 = EV_2/(C + B_{OA}) \quad (13.99)$$

depending on whether the resolution is positive or negative, while in the expected value approach the expected return on total investment is

$$R_0 = EV/(C + B_{EV}). \tag{13.100}$$

So one can then choose to drill if R_1 (or R_2, respectively) exceeds unity, and one can see if the increase in resolution of the success probability is sufficient to justify having taken an option approach by determining if R_1/R_0 (or R_2/R_0, respectively) is larger than unity.

A slightly more robust measure of worth is to calculate the probability, P_+, of a positive worth, using the expected value and the variance of the expected value, as done previously; or to calculate the probability of a worth in excess of total costs. In either case one can evaluate worth using not only the expected value, but also the uncertainty on the expected value. For example, with total costs of TC, expected value, E, and variance, σ^2, the probability of obtaining a worth to the opportunity in excess of the total sunk costs of TC is given by

$$P_+ = (2\pi\sigma^2)^{-1/2} \int_{TC}^{\infty} \exp[-(x-E)^2/(2\sigma^2)]\,dx, \tag{13.101a}$$

which can be written

$$P_+ = (2\pi)^{-1/2} \int_{a}^{\infty} \exp(-u^2/2)\,du \tag{13.101b}$$

with $a = (TC/\sigma)(1 - R)$, where the rate of return ratio is $R = E/TC$.

Thus, one can calculate P_+ under both the expected value approach and the option approach to determine if the worth probability is increased or decreased after resolution.

Because one already has values for all parameters except the fraction, f, of the success probability that is to be resolved, one can, as a variant on a theme, also invert the argument to ask: What fraction, f, must be resolved such that P_+ is greater with the option approach than with the expected value approach?

IV. OPTION VALUES BASED ON FUTURE MODELS

The investigations presented so far have been predicated on the fact that only currently available information is used to provide estimated option values. However, there exists a considerable body of literature, mainly in the business area, which tries to provide present-day option values based on future predictions of behavior. The future models range from deterministic to stochastic, with different underlying assumptions connecting the proposed evolution to present-day information. Consider, as an example, the two-commodity exchange model put forward by Kensinger (1987).

A. TWO-COMMODITY EXCHANGE MODEL

The essence of the model is to have an option in some time period to exchange an initial commodity (I) for a second commodity (S). When the option matures, the payoff will be the maximum of $I - S$ or zero.

Kensinger (1987) uses a variant of the Black-Scholes option pricing model (which assumes no temporal correlation in price fluctuations, and that the fluctuations are log-normally distributed) to write the value of the option at present, V_0, as

$$V_0 = \alpha(t) S_0 - \gamma(t) I_0, \tag{13.102}$$

where S_0 and I_0 represent the present prices for the two commodities, where t represents the time to maturity from the present for the option, and where $\alpha(t)$ and $\gamma(t)$ are coefficients describing the effects of the volatility of each commodity and the correlation between them. Kensinger (1987) writes

$$d_1 = \left[\ln(S/I) + \frac{1}{2}\sigma^2 t\right]/(\sigma t^{1/2}), \tag{13.103a}$$

$$d_2 = d_1 - \sigma t^{1/2} \equiv \left[\ln(S/I) - \frac{1}{2}\sigma^2 t\right]/(\sigma t^{1/2}), \tag{13.103b}$$

with

$$\alpha(t) = N(d_1); \quad \gamma(t) = N(d_2), \tag{13.104}$$

where $N(x)$ is the normal probability density function, and where

$$\sigma^2 = \sigma_I^2 + \sigma_S^2 - 2\sigma_I\sigma_S\rho_{IS} \tag{13.105}$$

is the instantaneous variance of the ratio S/I, with ρ_{IS} the correlation coefficient for the price changes. The extra assumption is clearly made here that σ^2 is independent of time, else the replacement of $\sigma^2 t$ by $\int_0^t \sigma^2(\tau)\,d\tau$ would be needed.

Basically the normal Gaussian distribution assumption causes a spreading of the distribution function as time proceeds, so that there is some uncertainty on the worth of the option. The normal distribution coefficients, α and γ, being offset by $\pm\frac{1}{2}\sigma t^{1/2}$ from the value $\ln(S/I)$, attempt to bracket this swing of uncertainty probabilities for the *average* option worth. Missing from the evaluation is the intrinsic uncertainty on the present-day commodity prices, S_0 and I_0. For N option points in an opportunity, each maturing at times $t_1, t_2, \ldots, t_N$, the mean option value is then given by the generalization

$$V_N = S_0 \sum_{i=1}^N \alpha(t_i) - I_0 \sum_{i=1}^N \gamma(t_i), \tag{13.106}$$

while the expected value, EV, is just

$$EV = V_N - C, \tag{13.107}$$

where C is the initial opportunity costs.

B. A NUMERICAL ILLUSTRATION

Consider a field development opportunity with a lease costing $2000. The lease expires 2 years from the purchase date if no platform is set. The platform production per year for the 2-year interval is *estimated* to be 1000 bbls/year.

The current price of undeveloped oil in the ground is $6/bbl, and the current price of crude oil in a storage tank is $7/bbl. Let the rate of increase of value of undeveloped oil be *estimated* at 10%/yr; and let the same *estimate* of 10%/yr be taken for storage tank crude.

Let the standard deviation for both undeveloped oil and for storage tank crude be *estimated* at 30%/yr with an *estimated* correlation coefficient of 80% between the rates of change. And, for simplicity of illustration, assume that there are no operating costs in excess of those incorporated in the value calculation.

The question to be addressed is this: Does one set the platform and produce now, or does one choose to wait a year and then set the platform and produce, or does one wait until almost the end of the leasing period before one sets the platform?

For the option to delay setting the platform for 1 year, the relevant values are $d_1 = 0.907$ and $d_2 = 0.716$, leading to an option value/bbl of $1.44/bbl. For the option to delay 2 years, the relevant values are $d_1 = 0.709$, $d_2 = 0.44$, yielding an option value/bbl of $1.304/bbl. The option worth is maximal when the delay is over the 2-year period, giving a value (including lease costs) of $448 ($1000 \times 1.304 + 1000 \times 1.144 - 2000$), relative to going ahead directly with setting the platform now. The storage tank value is then $7.70/bbl at the end of the first year, and $8.47/bbl at the end of the second year under the 10% price inflation assumption; while the corresponding undeveloped, in-ground, oil values are $6.60 and $7.26, respectively.

The problems that really need to be addressed with such futuristic models for options are several-fold: First, within the framework of a particular model, to what extent are both the uncertainties in mean values being incorporated and, in addition, the uncertainties in underlying parameters of the model—which control the worth of the calculation.

Second, to what extent are the intrinsic assumptions of the model constraining the possible outcomes; there is little point in using a model if its assumptions are so at odds with the system one is trying to model that the model is a far cry from reality.

Third, the model response is just that—a model. What is needed is a comparison of different model responses, so that one can determine to what extent there is a robust character to the outcomes from different models, and to what extent the outcomes are peculiarly or sensitively dependent on the models chosen.

In essence, one is interested in the uniqueness, resolution, sensitivity, and precision of outputs from the models; and these four quantities are beholden to greater or lesser extents to the model assumptions, the model parameters and their uncertainties, and to the data quality, quantity, and sampled distribution frequency used to constrain a model.

Each of the factors much be evaluated so that one can assess the relative importance of each in influencing any output. Without such an analysis one can, all too often, be involved in making costly decisions without a solid base of support, with the attendant consequence of spectacular failure, when a little more homework could have helped mitigate that likelihood.

V. APPENDIX A: INCLUDING OPTION VALUE IN COSTS

When option value is included in costs, then the pattern of development is identical to that in the text, with only technical differences due to the changes of cost. See Figure 13.11 where

$$F(x) = \begin{cases} 1 & \text{for } G > Cp_s^{-2}(1+x)^{-2} \\ \dfrac{[p_s G(1+x) - C]p_s(1+x)}{(C[1 - p_s(1+x)])} & \text{for } C \geq G p_s^2(1+x)^2 \geq Cp_s(1+x). \\ 0 & \text{for } Gp_s(1+x) \leq C \end{cases}$$

Then

$$EV(\text{new}) = p_R EV_1 + (1 - p_R) EV_2;$$

that is,

$$\begin{aligned} EV(\text{new}) = EV &+ fGp_s(2p_R - 1) \\ &- C[(1 - p_R)F(-f) - p_R F(f) - (1 - p_s)p_s^{-1} F(0) \\ &+ [p_s(1 - f^2)]^{-1} [p_R(1 - f)F(f) + (1 - p_R)(1 + f)F(-f)], \end{aligned}$$

where $EV = p_s G - C[1 + F(0)(1 - p_s)/p_s]$, and $OV(\text{new}) = p_R OV_1 + (1 - p_R) OV_2$; that is,

$$\begin{aligned} OV(\text{new}) = C[p_s(1 - f^2)]^{-1} &[p_R(1 - f)F(f) + (1 - p_R)(1 + f)F(-f) \\ &- (1 - f^2)p_s\{p_R F(f) + (1 - p_R)F(-f)\}]. \end{aligned}$$

One can also compute $\sigma^2(\text{new})$ if needed. One can again go through the same logic as done in the body of the text to write the bid as

$$B_{OA} = b(EV + |OV(\text{new}) - OV|),$$

$$B_{EV} = bEV,$$

No Resolution

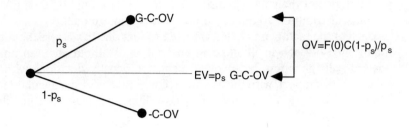

Resolution

(a) Positive (Resolution Probability p_R)

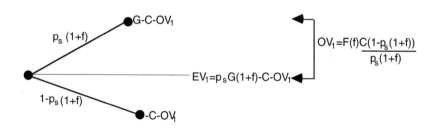

(b) Negative (Resolution Probability $1-p_R$)

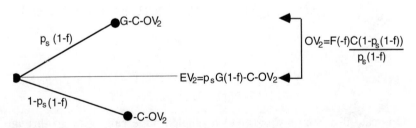

FIGURE 13.11 Parameters for the option approach and expected value approach to bidding.

in the two cases of an option approach (OA) and an expected value approach (EV) yielding

$$\Delta B = b\,|OV(\text{new}) - OV|,$$

and total costs in both cases as

$$TC_{EV} = C + OV + B_{EV}$$
$$TC_{OA} = b(EV + |OV(\text{new}) - OV|)$$

to obtain

$$\Delta TC = TC_{EV} - TC_{OA} = C + OV - b|V(\text{new}) - OV|.$$

Clearly, for $\Delta TC > 0$ the option approach, including costs, is to be preferred. The cumulative probability of success can then be computed per the method given in the chapter.

VI. APPENDIX B: PORTFOLIO BALANCING USING THE COZZOLINO FORMULA

In the case of the Cozzolino (1977a, 1978) RAV formula, the maximum option worth is such that the total budget, B, must now balance as

$$B = \sum_{i=1}^{N} \{W_i\, C_i + \lambda_i\, p_{si}^{-1}\, RT \ln[p_{si} \exp\{-W_i(G_i - C_i)/RT\} \quad (3.108)$$
$$+ p_{fi} \exp(W_i C_i / RT)] + \lambda_i W_i (G_i - C_i)\},$$

while the total RAV for all N opportunities is now

$$RAV = -\sum_{i=1}^{N} \lambda_i W_i (G_i - C_i) - RT \sum_{i=1}^{N} (1 + \lambda_i p_{si}^{-1}) \quad (13.109)$$
$$\times \ln\{p_{si} \exp[-W_i(G_i - C_i)/RT] + p_{fi} \exp(W_i C_i / RT)\}.$$

Using the constraint equation (13.108) to express W_i in terms of W_k implies that RAV has a maximum with respect to W_k when

$$[1 + \lambda_k p_{sk}^{-1}(1 + H)]\,[p_{sk} \exp(-W_k G_k / RT) + p_{fk}]^{-1}$$
$$\times [p_{sk} \exp(-W_k G_k / RT)(G_k - C_k) - p_{fk} C_k] \quad (13.110)$$
$$= \lambda_k G_k (1 + H) + C_k [H(1 - \lambda_k) - \lambda_k] \quad \text{for all } k,$$

where H is an arbitrary constant independent of k.

Equation (13.110) can be solved explicitly to yield

$$W_k = (RT/G_k) \ln\{p_{sk} p_{fk}^{-1}(1 + H)^{-1}\,[G_k\{p_{sk} + \lambda_k p_{fk}(1 + H)\}$$
$$- C_k(1 + H)(p_{sk} + \lambda_k p_{fk})] \quad (13.111)$$
$$\times [\lambda_k G_k p_{sk} + C_k(p_{sk} + \lambda_k p_{fk})]^{-1}\},$$

provided $0 \leq W_k \leq 1$.

Substitution of expression (13.111) into expression (13.108) for the budget, B, then provides the relation for H in terms of B. Again, the limits of H that permit equation (13.111) to yield $0 \leq W_k \leq 1$ have to be evaluated and, outside of those limits, W_k has to be set to zero or unity, respectively.

VII. APPENDIX C: PORTFOLIO BALANCING WITH CONSERVATIVE OPTIONING

The body of the text worked through the balancing of a portfolio of opportunities when the full option cost, O_{pi}, was paid for each opportunity. However, some corporations are more conservative and will choose to take only a fraction, λ_i, $(0 \leq \lambda_i \leq 1)$ of the maximum option cost. Occasionally such choices of λ_i are based on prior corporate historical patterns of successes and failures under optioning. But, in any event, the question arises as to the optimal working interest, W_i, that should be taken in each project if the option cost is only $\lambda_i O_{pi}$, with λ_i given, and when the budget is fixed.

The total costs of all projects, including option amounts, balance against a budget B when

$$\sum_{i=1}^{N} (W_i C_i + \lambda_i O_{pi}) = B, \tag{13.112}$$

which, under the parabolic RAV formula, yields

$$\sum_{i=1}^{N} W_i \{C_i p_{si}^{-1} (p_{si} + \lambda_i p_{fi}) + \frac{1}{2} W_i \lambda_i p_{si} p_{fi} G_i^2 / RT\} = B. \tag{13.113}$$

For the parabolic RAV formula for each opportunity, the total RAV of all N projects, including the option cost of $\lambda_i O_{pi}$ per project, is

$$RAV = \sum_{i=1}^{N} W_i p_{si}^{-1} [p_{si}^2 G_i - C_i (p_{si} + \lambda_i p_{fi}) \tag{13.114}$$
$$- \frac{1}{2} (G_i^2 W_i / RT) p_{fi} p_{si} (p_{si} + \lambda_i p_{fi})].$$

For fixed values of λ_i, and using the budget constraint equation (13.113) to relate W_i to W_k, RAV has a maximum with respect to W_k when

$$p_{sk}^{-1} \{p_{sk}^2 G_k - C_k (p_{sk} + \lambda_k p_{fk}) - (G_k^2 W_k / RT) p_{fk} p_{sk} (p_{sk} + \lambda_k p_{fk})\}$$
$$= H[C_k p_{sk}^{-1} (p_{sk} + \lambda_k p_{fk}) + W_k \lambda_k p_{sk} p_{fk} G_k^2 / RT] \quad \text{for all } k, \tag{13.115}$$

where H is an absolute constant, independent of k. Then, from Eq. (13.115) one has

$$W_k = RT p_{sk}^{-1} \{p_{sk}^2 G_k - C_k (p_{sk} + \lambda_k p_{fk})(1 + H)\}$$
$$\times \{G_k^2 p_{fk} [p_{sk} + \lambda_k p_{fk} + \lambda_k H p_{sk}]\}^{-1}, \tag{13.116}$$

provided $0 \leq W_k \leq 1$. Then, by substituting Eq. (13.116) for W_k into Eq. (13.113), one relates the constant H to the budget B, just as in the body of the text.

Again one has to determine the limits of H that will yield $0 \leq W_k \leq 1$, $\lambda_k O_{pk} \leq W_k V_k$, and $RAV_k \geq 0$, in Eq. (13.116), and to then set W_k to zero or unity, respectively, outside of those limits on H. The procedure is similar to that recorded and does not need to be repeated here.

Epilogue

Because the exploration for hydrocarbons is not only expensive but also a risky venture scientifically and economically, corporations involved in that area are always keen to assess their risk of loss and potential gains, and to be guided in their decision-making process by the estimates made. And yet, because of the uncertainty, the tools and procedures used to assess the risk are themselves not perfect: There are always inherent assumptions in any measures of scientific worth or economic value. Part of the problem of providing an exploration risk analysis is to examine the dependence of the analysis on the intrinsic assumptions made. An analysis that records a rapid change in assessment as the intrinsic assumptions are but slightly changed is so beholden to having available sufficiently detailed and sufficiently precise information that it is doubtful that reality will ever be modeled correctly. On the other hand, an analysis that records few to no changes in assessments even when parameters of the model change massively is equally suspect.

Somewhere between these two extremes, models have to be constructed that account for known situations but also predict within uncertainty bands the probable outcomes of risk assessment analysis based on the incomplete and inaccurate information available. At the same time the models must be capable of being updated as new information becomes available.

Thus one is always concerned with the uniqueness, precision, resolution, and sensitivity of model outputs in relation to intrinsic assumptions and their variations, to parameters in the model and their variations, and with respect to the quality, quantity, and sampled distribution frequency of data used as control information on the model.

At the same time one must keep firmly in mind the fact that one knows hydrocarbon exploration has a less than perfect success percentage and, even when hydrocarbons are found, it is not always the case that an *economic* find has been made. One must also keep in mind that there is often very little time to perform whatever calculations are requested by the corporation to aid the decision makers, so the ability to do such calculations quickly and simply is of considerable value.

The main factors that cause difficulties in assessing the worth of an opportunity, either in isolation from all competing opportunities or under controlled budget conditions, are relatively easy to identify. But when multiple opportunities have to be evaluated simultaneously, or when a budget has to be doled out over a range of time with opportunities arriving and being evaluated for funding at different times in the budget cycle, then it becomes more complex to evaluate (or even find!) the optimal path to follow to maximize potential profits for the corporation—or to minimize potential losses—both of which are, quite often, different paths.

Further, the number of possible choices that one could make is legion, and each has some associated reward or penalty. The task is, as always, to determine the rewards or penalties of each choice in relation to all other choices and in relation to corporate controls and corporate risk.

The main goal of this book has been to illustrate how these sorts of problems can be brought to some semblance of order and objectivity. In this way one is not as beholden to personal prejudice in making decisions, but has a rational way of producing assessments in a reproducible manner and with prescribable uncertainties. One also can use the methods to determine which factors are providing the largest contributions to the uncertainty of risk assessments. Thus, one can determine where to put further effort to reduce uncertainty of assessments and whether the effort is of worth relative to other costs, gains, and probability considerations.

We hope the general procedures laid out here will be adapted and applied by others, more able than ourselves, to address quantitatively problems of concern throughout the hydrocarbon exploration arena and, more speculatively, in other disciplines as well. If such considerations had been included then this volume would have been very long indeed.

There is always the problem of having to make risk assessments knowing that the information available is flawed or limited, and knowing that there are very likely going to be expensive failures. But without perfect information and perfect models of the future, such uncertainties are always going to be present.

The problems of political, social, and civil risk are areas that have not been addressed; attention focused almost exclusively on financial risk. This focus should not be construed as a statement that those other risks are unimportant. Quite the contrary can be true—they are often the most important factors in dealing with an exploration opportunity, and also in the determination of repatriation of corporate profit. But dealing with such risks is well beyond the scope of the present volume and they are best considered by others.

Perhaps the main point to make, as corporations grapple with the daily problems of hydrocarbon exploration, is that there are available quantitative ways to examine uncertainty, and to use the results in an objective, reproducible manner to improve the chances of corporate profit in the uncertain business of exploration—and that has been the goal throughout this volume.

APPENDIX

NUMERICAL METHODS AND SPREADSHEETS

I. INTRODUCTION

The methods presented so far for addressing concerns of scientific and economic risk in hydrocarbon exploration—and the numerical examples given to illustrate salient points—have been performed either analytically or with simple numerical insertions into algebraic formulas and then hand calculation of results done. While such procedures are fine for pedagogical exposition of procedures, they become laborious if one wishes to carry out a complete exploration risk assessment many times for many different areas under many conditions.

In addition, the methods and procedures developed and applied have an underlying precept which is to avoid, to the extent possible, either the need for large computers or the need for massive numbers of Monte Carlo simulations. The whole raison d'être has been geared to reducing principles and procedures to the level that they can be addressed, to a very large extent, with simple spreadsheet analysis on a PC-level machine. This appendix provides many of the numerical algorithms, in spread-sheet form where possible, so that interested readers can generate quickly versions of the procedures for use in their own applications.

Many of the spreadsheets require only a few cells of code such as Chapter 8 (Nind's formula). Others require hundreds of thousands of cells of code. It is our intent to provide the necessary information to allow the reader, with moderate familiarity with Excel, to duplicate the spreadsheets. While careful, and in some cases extensive, typing of the formulas into the appropriate cells will accomplish this objective, the reader might prefer to simply contact one of the authors via the Internet to get a copy of the spreadsheets of interest. (Jim MacKay is the "programmer" for most of the spreadsheets and he can be contacted at the following Internet addresses: *mackaja@texaco.com* or *mackay@prodigy.net*.)

II. OVERALL FORMAT

The individual spreadsheets are in order of how they were discussed in each chapter. Where possible, the values used in the "numerical example" for each chapter are used as variables in the spreadsheets. The spreadsheet is displayed using gridlines and cell references to assist in identifying each cell. Following the spreadsheet is a list of the "contents" of each cell, starting with the first column. Under each column heading are the row numbers followed by the contents of the cell. At the end of the list of columns and rows is a list of cell names if used in that particular sheet. These names must be assigned for the formulas to work.

In some cases, references to colors indicate input cells (usually blue) and output cells (usually red). These colors do not appear on the copies of the spreadsheets that follow. In some cases critical formulas are displayed on the spreadsheet as text. These are used as references and are not necessary for the program to run.

The spreadsheets are designed to "stand alone" where possible. The graphs on the sheets assist in the display and are not described in the accompanying text.

III. INDIVIDUAL CODES

CHAPTER 2: VARIABLE WORKING INTEREST

The spreadsheet of Figure A.1 is used to calculate the optimum working interest and the apparent risk tolerance using both the exponential and hypertangent formulas described in Chapter 2. The graph compares these two calculations at a range of working interests, varying from 0 to 100% in 10% increments.

Four input variables are required to calculate the optimum working interest: chance, NPV, cost, and risk tolerance. An additional variable, selected working interest, is required to calculate the apparent risk tolerance. The minimum critical cells necessary for the calculations are:

1. The input cells: B1, B2, B3, B5, and B18.
2. The output cells: B21 and B23

The rest of the cells are used for the graph or for clarification. Note that the input cells are "named references" in the spreadsheet.

INDIVIDUAL CODES

	A	B	C	D	E
1	CHANCE	20%			
2	NPV (millions)	$90.000			
3	COST (millions)	$10.000			
4	EXPECTED VAL (mm) @100%	$10.000			
5	RISK TOLERANCE (mm)	$20.000			
6	WORKING INTEREST	rav-coz	rav-tanh		
7	100%	($5.571)	($3.136)		
8	90%	($4.593)	($2.579)		
9	80%	($3.628)	($1.983)		
10	70%	($2.688)	($1.350)		
11	60%	($1.785)	($0.685)		
12	50%	($0.943)	($0.007)		
13	40%	($0.203)	$0.639		
14	30%	$0.377	$1.146		
15	20%	$0.703	$1.313		
16	10%	$0.639	$0.909		
17	0%	$0.000	$0.000		
18	SELECTED WI	50.00%			
19	EV AT SELECTED WI	$5.000			
20	RAV AT SELECTED WI	($0.943)			
21	APPARENT RISK TOLERANCE	$61.658			
22					
23	OPTIMUM WORKING INTEREST	16.22%			
24	EV AT optimum WI	$1.622			
25	RAV AT optimum WI	$0.734			

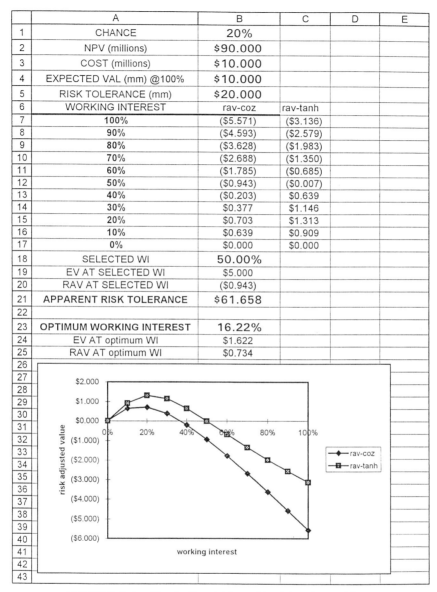

FIGURE A.1 Chapter 2 variable working interest spreadsheet.

Column A:
Rows 1 – 25:
1. CHANCE
2. NPV (millions)
3. COST (millions)
4. EXPECTED VAL (mm) @100%
5. RISK TOLERANCE (mm)
6. WORKING INTEREST
7. 1
8. 0.9
9. 0.8
10. 0.7
11. 0.6
12. 0.5
13. 0.4
14. 0.3
15. 0.2
16. 0.1
17. 0
18. SELECTED WI
19. EV AT SELECTED WI
20. RAV AT SELECTED WI
21. APPARENT RISK TOLERANCE
22.
23. OPTIMUM WORKING INTEREST
24. EV AT optimum WI
25. RAV AT optimum WI

Column B:
Rows 1 – 25:
1. 0.2
2. 90
3. 10
4. =NPV*CHANCE-COST*(1-CHANCE)
5. 20
6. rav-coz
7. =-RT*LN(1-(((CHANCE*(RT*(1-EXP(-NPV*$A7/RT)))+(1-CHANCE)*(RT*(1-EXP(COST*$A7/RT))))))/RT)
8. =-RT*LN(1-(((CHANCE*(RT*(1-EXP(-NPV*$A8/RT)))+(1-CHANCE)*(RT*(1-EXP(COST*$A8/RT))))))/RT)
9. =-RT*LN(1-(((CHANCE*(RT*(1-EXP(-NPV*$A9/RT)))+(1-CHANCE)*(RT*(1-EXP(COST*$A9/RT))))))/RT)
10. =-RT*LN(1-(((CHANCE*(RT*(1-EXP(-NPV*$A10/RT)))+(1-CHANCE)*(RT*(1-EXP(COST*$A10/RT))))))/RT)
11. =-RT*LN(1-(((CHANCE*(RT*(1-EXP(-NPV*$A11/RT)))+(1-CHANCE)*(RT*(1-EXP(COST*$A11/RT))))))/RT)
12. =-RT*LN(1-(((CHANCE*(RT*(1-EXP(-NPV*$A12/RT)))+(1-CHANCE)*(RT*(1-EXP(COST*$A12/RT))))))/RT)
13. =-RT*LN(1-(((CHANCE*(RT*(1-EXP(-NPV*$A13/RT)))+(1-CHANCE)*(RT*(1-EXP(COST*$A13/RT))))))/RT)
14. =-RT*LN(1-(((CHANCE*(RT*(1-EXP(-NPV*$A14/RT)))+(1-CHANCE)*(RT*(1-EXP(COST*$A14/RT))))))/RT)
15. =-RT*LN(1-(((CHANCE*(RT*(1-EXP(-NPV*$A15/RT)))+(1-CHANCE)*(RT*(1-EXP(COST*$A15/RT))))))/RT)
16. =-RT*LN(1-(((CHANCE*(RT*(1-EXP(-NPV*$A16/RT)))+(1-CHANCE)*(RT*(1-EXP(COST*$A16/RT))))))/RT)

17. =-RT*LN(1-(((CHANCE*(RT*(1-EXP(-NPV*$A17/RT)))+(1-CHANCE)*(RT*(1-EXP(COST*$A17/RT))))))/RT)
18. 0.5
19. =B18*B4
20. =-RT*LN(1-(((CHANCE*(RT*(1-EXP(-NPV*WI/RT)))+(1-CHANCE)*(RT*(1-EXP(COST*WI/RT))))))/RT)
21. =(COST+NPV)*WI/LN((CHANCE*NPV)/((1-CHANCE)*COST))
22.
23. =RT/(COST+NPV)*LN((CHANCE*NPV)/((1-CHANCE)*COST))
24. =EV*B23
25. =-RT*LN(1-(((CHANCE*(RT*(1-EXP(-NPV*B23/RT)))+(1-CHANCE)*(RT*(1-EXP(COST*B23/RT))))))/RT)

Column C:
Rows 1 – 25:
1.
2.
3.
4.
5.
6. rav-tanh
7. =-RT*LN(1+(1-CHANCE)*TANH(A7*COST/RT)-CHANCE*TANH(A7*NPV/RT))
8. =-RT*LN(1+(1-CHANCE)*TANH(A8*COST/RT)-CHANCE*TANH(A8*NPV/RT))
9. =-RT*LN(1+(1-CHANCE)*TANH(A9*COST/RT)-CHANCE*TANH(A9*NPV/RT))
10. =-RT*LN(1+(1-CHANCE)*TANH(A10*COST/RT)-CHANCE*TANH(A10*NPV/RT))
11. =-RT*LN(1+(1-CHANCE)*TANH(A11*COST/RT)-CHANCE*TANH(A11*NPV/RT))
12. =-RT*LN(1+(1-CHANCE)*TANH(A12*COST/RT)-CHANCE*TANH(A12*NPV/RT))
13. =-RT*LN(1+(1-CHANCE)*TANH(A13*COST/RT)-CHANCE*TANH(A13*NPV/RT))
14. =-RT*LN(1+(1-CHANCE)*TANH(A14*COST/RT)-CHANCE*TANH(A14*NPV/RT))
15. =-RT*LN(1+(1-CHANCE)*TANH(A15*COST/RT)-CHANCE*TANH(A15*NPV/RT))
16. =-RT*LN(1+(1-CHANCE)*TANH(A16*COST/RT)-CHANCE*TANH(A16*NPV/RT))
17. =-RT*LN(1+(1-CHANCE)*TANH(A17*COST/RT)-CHANCE*TANH(A17*NPV/RT))
18.
19.
20.
21.
22.
23.
24.
25.

Named Cells:
B1 = CHANCE
B2 = NPV
B3 = COST
B5 = RT

CHAPTER 2: VARIABLE RISK TOLERANCE

The spreadsheet of Figure A.2 is used to display graphically the change in risk-adjusted value as the risk tolerance is changed for both the exponential and hypertangent formulas. The risk tolerance values assigned range from $10 to $110 in $10 increments. These can be modified to fit your preferences. All cells are required to complete the graph.

	A	B	C	D	E	F	G
1	CHANCE	20%					
2	NPV (millions)	$90.000					
3	COST (millions)	$10.000					
4	EXPECTED VAL (mm) @100%	$10.000					
5							
6	RISK TOLERANCE (mm)	rav-coz	rav-tanh				
7	$10	($7.769)	($3.431)				
8	$20	($5.571)	($3.136)				
9	$30	($3.572)	($1.697)				
10	$40	($1.887)	($0.013)				
11	$50	($0.507)	$1.598				
12	$60	$0.620	$3.009				
13	$70	$1.547	$4.189				
14	$80	$2.317	$5.153				
15	$90	$2.965	$5.933				
16	$100	$3.516	$6.563				
17	$110	$3.989	$7.074				
18							
19	**rav-coz formula**						
20	=-A17*LN(1-(((CHANCE*(A17*(1-EXP(-NPV*1/A17)))+(1-CHANCE)*(A17*(1-EXP(COST*1/A17))))))/A17)						
21	**rav-htan formula**						
22	=-A17*LN(1+(1-CHANCE)*TANH(1*COST/A17)-CHANCE*TANH(1*NPV/A17))						
23							
24							
25							
26							
27							
28							
29							
30							
31							
32							
33							
34							
35							
36							
37							
38							
39							
40							
41							
42							
43							

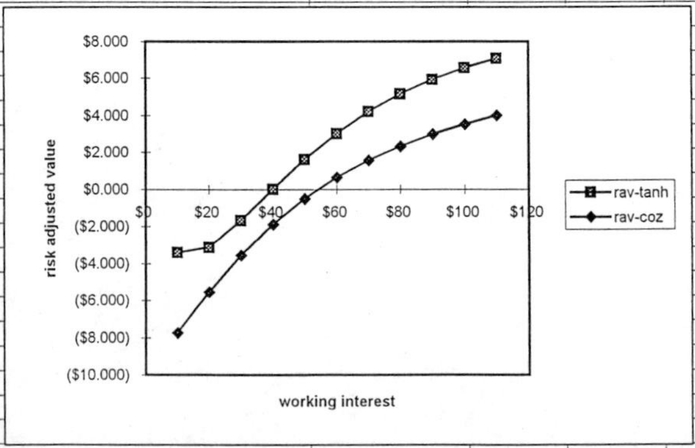

FIGURE A.2 Chapter 2 variable risk tolerance spreadsheet.

VARIABLE RT SPREADSHEET

Column A:
Rows 1 – 17:

1. CHANCE
2. NPV (millions)
3. COST (millions)
4. EXPECTED VAL (mm) @100%
5.
6. RISK TOLERANCE (mm)
7. $10
8. $20
9. $30
10. $40
11. $50
12. $60
13. $70
14. $80
15. $90
16. $100
17. $110

Column B:
Rows 1 – 17:

1. 0.2
2. 90
3. 10
4. =NPV*CHANCE-COST*(1-CHANCE)
5.
6. rav-coz
7. =-A7*LN(1-(((CHANCE*(A7*(1-EXP(-NPV*1/A7)))+(1-CHANCE)*(A7*(1-EXP(COST*1/A7))))))/A7)
8. =-A8*LN(1-(((CHANCE*(A8*(1-EXP(-NPV*1/A8)))+(1-CHANCE)*(A8*(1-EXP(COST*1/A8))))))/A8)
9. =-A9*LN(1-(((CHANCE*(A9*(1-EXP(-NPV*1/A9)))+(1-CHANCE)*(A9*(1-EXP(COST*1/A9))))))/A9)
10. =-A10*LN(1-(((CHANCE*(A10*(1-EXP(-NPV*1/A10)))+(1-CHANCE)*(A10*(1-EXP(COST*1/A10))))))/A10)
11. =-A11*LN(1-(((CHANCE*(A11*(1-EXP(-NPV*1/A11)))+(1-CHANCE)*(A11*(1-EXP(COST*1/A11))))))/A11)
12. =-A12*LN(1-(((CHANCE*(A12*(1-EXP(-NPV*1/A12)))+(1-CHANCE)*(A12*(1-EXP(COST*1/A12))))))/A12)
13. =-A13*LN(1-(((CHANCE*(A13*(1-EXP(-NPV*1/A13)))+(1-CHANCE)*(A13*(1-EXP(COST*1/A13))))))/A13)
14. =-A14*LN(1-(((CHANCE*(A14*(1-EXP(-NPV*1/A14)))+(1-CHANCE)*(A14*(1-EXP(COST*1/A14))))))/A14)
15. =-A15*LN(1-(((CHANCE*(A15*(1-EXP(-NPV*1/A15)))+(1-CHANCE)*(A15*(1-EXP(COST*1/A15))))))/A15)
16. =-A16*LN(1-(((CHANCE*(A16*(1-EXP(-NPV*1/A16)))+(1-CHANCE)*(A16*(1-EXP(COST*1/A16))))))/A16)
17. =-A17*LN(1-(((CHANCE*(A17*(1-EXP(-NPV*1/A17)))+(1-CHANCE)*(A17*(1-EXP(COST*1/A17))))))/A17)

Column C:
Rows 1 – 17:

1.
2.
3.
4.
5.
6.   rav-tanh
7.   =-A7*LN(1+(1-CHANCE)*TANH(1*COST/A7)-CHANCE*TANH(1*NPV/A7))
8.   =-A8*LN(1+(1-CHANCE)*TANH(1*COST/A8)-CHANCE*TANH(1*NPV/A8))
9.   =-A9*LN(1+(1-CHANCE)*TANH(1*COST/A9)-CHANCE*TANH(1*NPV/A9))
10. =-A10*LN(1+(1-CHANCE)*TANH(1*COST/A10)-CHANCE*TANH(1*NPV/A10))
11. =-A11*LN(1+(1-CHANCE)*TANH(1*COST/A11)-CHANCE*TANH(1*NPV/A11))
12. =-A12*LN(1+(1-CHANCE)*TANH(1*COST/A12)-CHANCE*TANH(1*NPV/A12))
13. =-A13*LN(1+(1-CHANCE)*TANH(1*COST/A13)-CHANCE*TANH(1*NPV/A13))
14. =-A14*LN(1+(1-CHANCE)*TANH(1*COST/A14)-CHANCE*TANH(1*NPV/A14))
15. =-A15*LN(1+(1-CHANCE)*TANH(1*COST/A15)-CHANCE*TANH(1*NPV/A15))
16. =-A16*LN(1+(1-CHANCE)*TANH(1*COST/A16)-CHANCE*TANH(1*NPV/A16))
17. =-A17*LN(1+(1-CHANCE)*TANH(1*COST/A17)-CHANCE*TANH(1*NPV/A17))

Named Cells:
B1 = CHANCE
B2 = NPV
B3 = COST
B5 = RT

CHAPTER 3

Chapter 3 refers to a Monte Carlo simulation of the optimum working interest. The variable working interest spreadsheet (Figure A.1) was used for the calculations.

CHAPTER 4: PORTFOLIO

The spreadsheet of Figure A.3 is used to calculate the optimum working interest in three projects under a constrained budget using the parabolic formula for risk-adjusted value. Although it works well for this three project case, the formulas become very complicated when more than three projects or other formulas such as exponential or hypertangent are used. It would be better to use the linear optimizer provided with Excel in these more complicated circumstances.

	B	C	D	E	F	G	H	I	J	K	L	M	N
1				input = red								limited	
2												0 to 1	
3				PROJECT	V	Ps	Pf	C	RT	E	v^2	OWI	E/C
4				A	110	0.5	0.5	10	30	50	1.44	0.417	5
5				B	200	0.5	0.5	100	30	50	9	0.067	0.5
6				C	300	0.4	0.6	120	30	48	18.375	0.034	0.4
7				totals	610.000			230.000		148.000			
8													
9													
10													
11													
12		x		y								SECOND	SECOND
13		C^2/v^2*E^2	c/v^2*E	bracket	eck cond	cond rt	cond x	cond y	select wi	cost	rav	COND X	COND Y
14		0.028	0.139	0.967872	1	30	0.027778	0.1388889	0.40328	4.032798	10.40591	0.027778	0.138888889
15		0.444	0.222	0.678716	1	30	0.444444	0.2222222	0.045248	4.524773	1.494628	0.444444	0.222222222
16		0.340	0.136	0.598395	1	30	0.340136	0.1360544	0.020354	2.442428	0.684664	0.340136	0.136054422
17		0.812	0.497				0.812	0.497		11	12.58521	0.812	0.497
18													
19													
20							budget=						
21							11						
22													
23													
24													
25												PARABOLIC	
26											WI	COST	RAV
27											40.33%	$4.03	$10.406
28											4.52%	$4.52	$1.495
29											2.04%	$2.44	$0.685
30												$11.00	$12.585

FIGURE A.3 Chapter 4 portfolio spreadsheet.

PORTFOLIO SPREADSHEET

Column C:
Rows 12 – 17:

12. x
13. C^2/v^2*E^2
14. =IF(D14=0,0,(MAX(0,(I4^2/(L4*K4^2)))))
15. =IF(D15=0,0,(MAX(0,(I5^2/(L5*K5^2)))))
16. =IF(D16=0,0,(MAX(0,(I6^2/(L6*K6^2)))))
17. =C16+C15+C14

Column D:
Rows 12 – 17:

12. y
13. c/v^2*E
14. =MAX(0,(I4/(L4*K4)))
15. =MAX(0,(I5/(L5*K5)))
16. =MAX(0,(I6/(L6*K6)))
17. =D16+D15+D14

Column E:
Rows 1 – 16:

1. input = red
2.
3. PROJECT
4. A
5. B
6. C
7. totals
8.
9.
10.
11.
12.
13. bracket
14. =(1-(I4/K4)*((D$17-(H$28/J4))/C$17))
15. =(1-(I5/K5)*((D$17-(H$28/J5))/C$17))
16. =(1-(I6/K6)*((D$17-(H$28/J6))/C$17))

Column F:
Rows 1 – 16:

1.
2.
3. V
4. 110
5. 200
6. 300
7. =F6+F5+F4
8.
9.
10.
11.

INDIVIDUAL CODES 363

12.
13. check cond
14. =IF(M4=0,0,(IF(E14<0,0,1)))
15. =IF(M5=0,0,(IF(E15<0,0,1)))
16. =IF(M6=0,0,(IF(E16<0,0,1)))

Column G:
Rows 1 – 16:

1.
2.
3. Ps
4. 0.5
5. 0.5
6. 0.4
7.
8.
9.
10.
11.
12.
13. cond rt
14. =J4*F14
15. =J5*F15
16. =J6*F16

Column H:
Rows 1 – 28:

1.
2.
3. Pf
4. =1-G4
5. =1-G5
6. =1-G6
7.
8.
9.
10.
11.
12.
13. cond x
14. =C14*F14
15. =C15*F15
16. =C16*F16
17. =H16+H15+H14
18.
19.
20.
21.
22.
23.
24.
25.
26.
27. budget=

28. 11

Column I:
Rows 1 – 17:

1.
2.
3. C
4. 10
5. 100
6. 120
7. =I6+I5+I4
8.
9.
10.
11.
12.
13. cond y
14. =D14*F14
15. =D15*F15
16. =D16*F16
17. =I16+I15+I14

Column J:
Rows 1 – 16:

1.
2.
3. RT
4. 30
5. 30
6. 30
7.
8.
9.
10.
11.
12.
13. select wi
14. =MAX(0,(IF(((G14/(K4*L4))*(1-(I4/K4)*((I$17-(H$28/J4))/H$17))>M4,M4,(G14/(K4*L4))*(1-(I4/K4)*((I$17-(H$28/J4))/H$17)))))
15. =MAX(0,(IF(((G15/(K5*L5))*(1-(I5/K5)*((I$17-(H$28/J5))/H$17))>M5,M5,(G15/(K5*L5))*(1-(I5/K5)*((I$17-(H$28/J5))/H$17)))))
16. =MAX(0,(IF(((G16/(K6*L6))*(1-(I6/K6)*((I$17-(H$28/J6))/H$17))>M6,M6,(G16/(K6*L6))*(1-(I6/K6)*((I$17-(H$28/J6))/H$17)))))

Column K:
Rows 1 – 17:

1.
2.
3. E
4. =G4*F4-I4*H4
5. =G5*F5-I5*H5
6. =G6*F6-I6*H6
7. =K6+K5+K4

INDIVIDUAL CODES 365

8.
9.
10.
11.
12.
13. cost
14. =J14*I4
15. =J15*I5
16. =J16*I6
17. =K16+K15+K14

Column L:
Rows 1 – 29:

1.
2.
3. v^2
4. =(((G4*F4^2+H4*I4^2)/K4^2)-1)
5. =(((G5*F5^2+H5*I5^2)/K5^2)-1)
6. =(((G6*F6^2+H6*I6^2)/K6^2)-1)
7.
8.
9.
10.
11.
12.
13. rav
14. =J14*K4*(1-0.5*J14*L4*K4/J4)
15. =J15*K5*(1-0.5*J15*L5*K5/J5)
16. =J16*K6*(1-0.5*J16*L6*K6/J6)
17. =L16+L15+L14
18.
19.
20.
21.
22.
23.
24.
25. PARABOLIC
26. WI
27. =MAX(0,(IF((G14/(K4*L4))*(1-(I4/K4)*((N$17-(H$28/J4))/M$17))>M4,M4,(G14/(K4*L4))*(1-(I4/K4)*((N$17-(H$28/J4))/M$17)))))
28. =MAX(0,(IF((G15/(K5*L5))*(1-(I5/K5)*((N$17-(H$28/J5))/M$17))>M5,M5,(G15/(K5*L5))*(1-(I5/K5)*((N$17-(H$28/J5))/M$17)))))
29. =MAX(0,(IF((G16/(K6*L6))*(1-(I6/K6)*((N$17-(H$28/J6))/M$17))>M6,M6,(G16/(K6*L6))*(1-(I6/K6)*((N$17-(H$28/J6))/M$17)))))

Column M:
Rows 1 – 30:

1. limited
2. 0 to 1
3. OWI
4. =IF((IF((L4^0.5)^-2*(J4/K4)>1,1,(L4^0.5)^-2*(J4/K4)))<0,0,(IF((L4^0.5)^-2*(J4/K4)>1,1,(L4^0.5)^-2*(J4/K4))))

5. =IF(((IF((L5^0.5)^-2*(J5/K5)>1,1,(L5^0.5)^-2*(J5/K5)))<0,0,(IF((L5^0.5)^-2*(J5/K5)>1,1,(L5^0.5)^-2*(J5/K5))))
6. =IF(((IF((L6^0.5)^-2*(J6/K6)>1,1,(L6^0.5)^-2*(J6/K6)))<0,0,(IF((L6^0.5)^-2*(J6/K6)>1,1,(L6^0.5)^-2*(J6/K6))))
7.
8.
9.
10.
11.
12. SECOND
13. COND X
14. =IF(J14=0,0,H14)
15. =IF(J15=0,0,H15)
16. =IF(J16=0,0,H16)
17. =M16+M15+M14
18.
19.
20.
21.
22.
23.
24.
25.
26. COST
27. =L27*I4
28. =L28*I5
29. =L29*I6
30. =M29+M28+M27

<u>Column N:</u>
<u>Rows 1 – 30:</u>

1.
2.
3. E/C
4. =K4/I4
5. =K5/I5
6. =K6/I6
7.
8.
9.
10.
11.
12. SECOND
13. COND Y
14. =IF(J14=0,0,I14)
15. =IF(J15=0,0,I15)
16. =IF(J16=0,0,I16)
17. =N16+N15+N14
18.
19.
20.
21.
22.
23.
24.

25.
26. RAV
27. =L27*K4*(1-0.5*J14*L4*K4/J4)
28. =L28*K5*(1-0.5*J15*L5*K5/J5)
29. =L29*K6*(1-0.5*J16*L6*K6/J6)
30. =N29+N28+N27

CHAPTER 6: MODIFIED RAV

Two Excel sheets are used to display the modifications to the spreadsheet developed for Chapter 2. These corrections are designed to manage very high gain situations. The first spreadsheet, titled "RAV Comparison," is a modification of the sheet used in Chapter 2. The second spread-sheet (Figure A.4), titled "Modified RAV" does the calculation described in Chapter 6. It first checks to see if the gain exceeds the gain at the optimum working interest. When the gain exceeds the gain at the optimum working interest, it adjusts the calculated working interest to equal the optimum working interest.

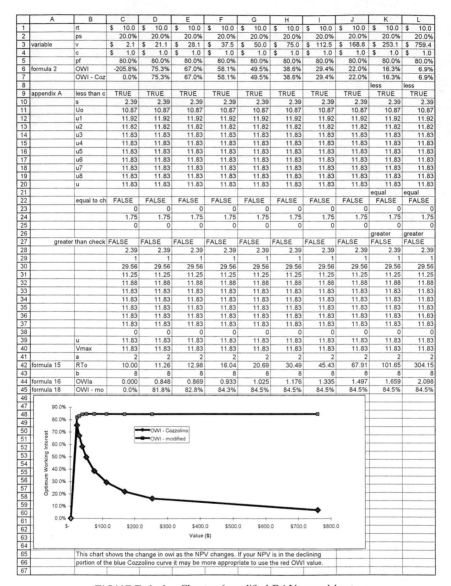

FIGURE A.4 Chapter 6 modified RAV spreadsheet.

INDIVIDUAL CODES 369

MODIFIED RAV:

SHEET NAME:

modified high gain calculation

Column B:
Rows 1 – 45:

1. rt
2. ps
3. v
4. c
5. pf
6. OWI
7. OWI - Cozzolino
8.
9. less than check
10. s
11. Uo
12. u1
13. u2
14. u3
15. u4
16. u5
17. u6
18. u7
19. u8
20. u
21.
22. equal to check
23.
24.
25.
26.
27. greater than check
28.
29.
30.
31.
32.
33.
34.
35.
36.
37.
38.
39. u
40. Vmax
41. a
42. RTo
43. b
44. OWIa
45. OWI - modified

Column C:

Rows 1 – 45:

1. ='rav comparison'!$B$5
2. ='rav comparison'!$B$1
3. =0.1*D3
4. ='rav comparison'!B3
5. =1-C2
6. =C1*LN((C2*C3)/(C5*C4))/(C3+C4)
7. =MIN((MAX(C6,0)),1)
8.
9. =(EXP(1)*C5)>C2
10. =MAX((1-LN(C2/C5)),0)
11. =MAX((EXP(C10)),1)
12. =EXP(C$10+C11^-1)
13. =EXP(C$10+C12^-1)
14. =EXP(C$10+C13^-1)
15. =EXP(C$10+C14^-1)
16. =EXP(C$10+C15^-1)
17. =EXP(C$10+C16^-1)
18. =EXP(C$10+C17^-1)
19. =EXP(C$10+C18^-1)
20. =C19*C9
21. /
22. =C2=EXP(1)*C5
23. 0
24. 1.75
25. =C24*C22
26.
27. =(EXP(1)*C5)<C2
28. =MAX((1-LN(C2/C5)),0)
29. 1
30. =EXP(C$28+C29^-1)
31. =EXP(C$28+C30^-1)
32. =EXP(C$28+C31^-1)
33. =EXP(C$28+C32^-1)
34. =EXP(C$28+C33^-1)
35. =EXP(C$28+C34^-1)
36. =EXP(C$28+C35^-1)
37. =EXP(C$28+C36^-1)
38. =C37*C27
39. =MAX(C20,C25,C38)
40. =C39*C4
41. 2
42. =(((C1^(2*C41))+(((C2*C5)^0.5)*(C3+C4))^(2*C41))^(1/(2*C41))
43. 8
44. =MAX(0,((C3+C4)^-1*LN(C2*C3/(C5*C4))*C42))
45. =MIN(1,((C1/C40)*(TANH((C44/(C1/C40))^C43))^(1/C43)))

Column D:
Rows 1 – 45:

1. =C1
2. =C2
3. =0.75*E3
4. =C4
5. =1-D2

INDIVIDUAL CODES

6. =D1*LN((D2*D3)/(D5*D4))/(D3+D4)
7. =MIN((MAX(D6,0)),1)
8.
9. =(EXP(1)*D5)>D2
10. =MAX((1-LN(D2/D5)),0)
11. =MAX((EXP(D10)),1)
12. =EXP(D$10+D11^-1)
13. =EXP(D$10+D12^-1)
14. =EXP(D$10+D13^-1)
15. =EXP(D$10+D14^-1)
16. =EXP(D$10+D15^-1)
17. =EXP(D$10+D16^-1)
18. =EXP(D$10+D17^-1)
19. =EXP(D$10+D18^-1)
20. =D19*D9
21.
22. =D2=EXP(1)*D5
23. 0
24. 1.75
25. =D24*D22
26.
27. =(EXP(1)*D5)<D2
28. =MAX((1-LN(D2/D5)),0)
29. 1
30. =EXP(D$28+D29^-1)
31. =EXP(D$28+D30^-1)
32. =EXP(D$28+D31^-1)
33. =EXP(D$28+D32^-1)
34. =EXP(D$28+D33^-1)
35. =EXP(D$28+D34^-1)
36. =EXP(D$28+D35^-1)
37. =EXP(D$28+D36^-1)
38. =D37*D27
39. =MAX(D20,D25,D38)
40. =D39*D4
41. 2
42. =((D1^(2*D41))+(((D2*D5)^0.5)*(D3+D4))^(2*D41))^(1/(2*D41))
43. 8
44. =MAX(0,((D3+D4)^-1*LN(D2*D3/(D5*D4))*D42))
45. =MIN(1,((D1/D40)*(TANH((D44/(D1/D40))^D43))^(1/D43)))

Column E:
Rows 1 – 45:

1. =D1
2. =D2
3. =0.75*F3
4. =D4
5. =1-E2
6. =E1*LN((E2*E3)/(E5*E4))/(E3+E4)
7. =MIN((MAX(E6,0)),1)
8.
9. =(EXP(1)*E5)>E2
10. =MAX((1-LN(E2/E5)),0)
11. =MAX((EXP(E10)),1)
12. =EXP(E$10+E11^-1)

13. =EXP(E$10+E12^-1)
14. =EXP(E$10+E13^-1)
15. =EXP(E$10+E14^-1)
16. =EXP(E$10+E15^-1)
17. =EXP(E$10+E16^-1)
18. =EXP(E$10+E17^-1)
19. =EXP(E$10+E18^-1)
20. =E19*E9
21.
22. =E2=EXP(1)*E5
23. 0
24. 1.75
25. =E24*E22
26.
27. =(EXP(1)*E5)<E2
28. =MAX((1-LN(E2/E5)),0)
29. 1
30. =EXP(E$28+E29^-1)
31. =EXP(E$28+E30^-1)
32. =EXP(E$28+E31^-1)
33. =EXP(E$28+E32^-1)
34. =EXP(E$28+E33^-1)
35. =EXP(E$28+E34^-1)
36. =EXP(E$28+E35^-1)
37. =EXP(E$28+E36^-1)
38. =E37*E27
39. =MAX(E20,E25,E38)
40. =E39*E4
41. 2
42. =((E1^(2*E41))+(((E2*E5)^0.5)*(E3+E4))^(2*E41))^(1/(2*E41))
43. 8
44. =MAX(0,((E3+E4)^-1*LN(E2*E3/(E5*E4))*E42))
45. =MIN(1,((E1/E40)*(TANH((E44/(E1/E40))^E43))^(1/E43)))

Column F:
Rows 1 – 45:

1. =E1
2. =E2
3. =0.75*G3
4. =E4
5. =1-F2
6. =F1*LN((F2*F3)/(F5*F4))/(F3+F4)
7. =MIN((MAX(F6,0)),1)
8.
9. =(EXP(1)*F5)>F2
10. =MAX((1-LN(F2/F5)),0)
11. =MAX((EXP(F10)),1)
12. =EXP(F$10+F11^-1)
13. =EXP(F$10+F12^-1)
14. =EXP(F$10+F13^-1)
15. =EXP(F$10+F14^-1)
16. =EXP(F$10+F15^-1)
17. =EXP(F$10+F16^-1)
18. =EXP(F$10+F17^-1)
19. =EXP(F$10+F18^-1)

Individual Codes

20. =F19*F9
21.
22. =F2=EXP(1)*F5
23. 0
24. 1.75
25. =F24*F22
26.
27. =(EXP(1)*F5)<F2
28. =MAX((1-LN(F2/F5)),0)
29. 1
30. =EXP(F$28+F29^-1)
31. =EXP(F$28+F30^-1)
32. =EXP(F$28+F31^-1)
33. =EXP(F$28+F32^-1)
34. =EXP(F$28+F33^-1)
35. =EXP(F$28+F34^-1)
36. =EXP(F$28+F35^-1)
37. =EXP(F$28+F36^-1)
38. =F37*F27
39. =MAX(F20,F25,F38)
40. =F39*F4
41. 2
42. =((F1^(2*F41))+(((F2*F5)^0.5)*(F3+F4))^(2*F41))^(1/(2*F41))
43. 8
44. =MAX(0,((F3+F4)^-1*LN(F2*F3/(F5*F4))*F42))
45. =MIN(1,((F1/F40)*(TANH((F44/(F1/F40))^F43))^(1/F43)))

Column G:
Rows 1 – 45:

1. =F1
2. =F2
3. ='rav comparison'!$B$2
4. =F4
5. =1-G2
6. =G1*LN((G2*G3)/(G5*G4))/(G3+G4)
7. =MIN((MAX(G6,0)),1)
8.
9. =(EXP(1)*G5)>G2
10. =MAX((1-LN(G2/G5)),0)
11. =MAX((EXP(G10)),1)
12. =EXP(G$10+G11^-1)
13. =EXP(G$10+G12^-1)
14. =EXP(G$10+G13^-1)
15. =EXP(G$10+G14^-1)
16. =EXP(G$10+G15^-1)
17. =EXP(G$10+G16^-1)
18. =EXP(G$10+G17^-1)
19. =EXP(G$10+G18^-1)
20. =G19*G9
21.
22. =G2=EXP(1)*G5
23. 0
24. 1.75
25. =G24*G22
26.

27. =(EXP(1)*G5)<G2
28. =MAX((1-LN(G2/G5)),0)
29. 1
30. =EXP(G$28+G29^-1)
31. =EXP(G$28+G30^-1)
32. =EXP(G$28+G31^-1)
33. =EXP(G$28+G32^-1)
34. =EXP(G$28+G33^-1)
35. =EXP(G$28+G34^-1)
36. =EXP(G$28+G35^-1)
37. =EXP(G$28+G36^-1)
38. =G37*G27
39. =MAX(G20,G25,G38)
40. =G39*G4
41. 2
42. =((G1^(2*G41))+(((G2*G5)^0.5)*(G3+G4))^(2*G41))^(1/(2*G41))
43. 8
44. =MAX(0,((G3+G4)^-1*LN(G2*G3/(G5*G4))*G42))
45. =MIN(1,((G1/G40)*(TANH((G44/(G1/G40))^G43))^(1/G43)))

Column H:
Rows 1 – 45:

1. =G1
2. =G2
3. =1.5*G3
4. =G4
5. =1-H2
6. =H1*LN((H2*H3)/(H5*H4))/(H3+H4)
7. =MIN((MAX(H6,0)),1)
8.
9. =(EXP(1)*H5)>H2
10. =MAX((1-LN(H2/H5)),0)
11. =MAX((EXP(H10)),1)
12. =EXP(H$10+H11^-1)
13. =EXP(H$10+H12^-1)
14. =EXP(H$10+H13^-1)
15. =EXP(H$10+H14^-1)
16. =EXP(H$10+H15^-1)
17. =EXP(H$10+H16^-1)
18. =EXP(H$10+H17^-1)
19. =EXP(H$10+H18^-1)
20. =H19*H9
21.
22. =H2=EXP(1)*H5
23. 0
24. 1.75
25. =H24*H22
26.
27. =(EXP(1)*H5)<H2
28. =MAX((1-LN(H2/H5)),0)
29. 1
30. =EXP(H$28+H29^-1)
31. =EXP(H$28+H30^-1)
32. =EXP(H$28+H31^-1)
33. =EXP(H$28+H32^-1)

Individual Codes

34. =EXP(H$28+H33^-1)
35. =EXP(H$28+H34^-1)
36. =EXP(H$28+H35^-1)
37. =EXP(H$28+H36^-1)
38. =H37*H27
39. =MAX(H20,H25,H38)
40. =H39*H4
41. 2
42. =((H1^(2*H41))+(((H2*H5)^0.5)*(H3+H4))^(2*H41))^(1/(2*H41))
43. 8
44. =MAX(0,((H3+H4)^-1*LN(H2*H3/(H5*H4))*H42))
45. =MIN(1,((H1/H40)*(TANH((H44/(H1/H40))^H43))^(1/H43)))

Column I:
Rows 1 – 45:

1. =H1
2. =H2
3. =1.5*H3
4. =H4
5. =1-I2
6. =I1*LN((I2*I3)/(I5*I4))/(I3+I4)
7. =MIN((MAX(I6,0)),1)
8.
9. =(EXP(1)*I5)>I2
10. =MAX((1-LN(I2/I5)),0)
11. =MAX((EXP(I10)),1)
12. =EXP(I$10+I11^-1)
13. =EXP(I$10+I12^-1)
14. =EXP(I$10+I13^-1)
15. =EXP(I$10+I14^-1)
16. =EXP(I$10+I15^-1)
17. =EXP(I$10+I16^-1)
18. =EXP(I$10+I17^-1)
19. =EXP(I$10+I18^-1)
20. =I19*I9
21.
22. =I2=EXP(1)*I5
23. 0
24. 1.75
25. =I24*I22
26.
27. =(EXP(1)*I5)<I2
28. =MAX((1-LN(I2/I5)),0)
29. 1
30. =EXP(I$28+I29^-1)
31. =EXP(I$28+I30^-1)
32. =EXP(I$28+I31^-1)
33. =EXP(I$28+I32^-1)
34. =EXP(I$28+I33^-1)
35. =EXP(I$28+I34^-1)
36. =EXP(I$28+I35^-1)
37. =EXP(I$28+I36^-1)
38. =I37*I27
39. =MAX(I20,I25,I38)
40. =I39*I4

41. 2
42. =((I1^(2*I41))+(((I2*I5)^0.5)*(I3+I4))^(2*I41))^(1/(2*I41))
43. 8
44. =MAX(0,((I3+I4)^-1*LN(I2*I3/(I5*I4))*I42))
45. =MIN(1,((I1/I40)*(TANH((I44/(I1/I40))^I43))^(1/I43)))

Column J:
Rows 1 – 45:

1. =I1
2. =I2
3. =1.5*I3
4. =I4
5. =1-J2
6. =J1*LN((J2*J3)/(J5*J4))/(J3+J4)
7. =MIN((MAX(J6,0)),1)
8.
9. =(EXP(1)*J5)>J2
10. =MAX((1-LN(J2/J5)),0)
11. =MAX((EXP(J10)),1)
12. =EXP(J$10+J11^-1)
13. =EXP(J$10+J12^-1)
14. =EXP(J$10+J13^-1)
15. =EXP(J$10+J14^-1)
16. =EXP(J$10+J15^-1)
17. =EXP(J$10+J16^-1)
18. =EXP(J$10+J17^-1)
19. =EXP(J$10+J18^-1)
20. =J19*J9
21.
22. =J2=EXP(1)*J5
23. 0
24. 1.75
25. =J24*J22
26.
27. =(EXP(1)*J5)<J2
28. =MAX((1-LN(J2/J5)),0)
29. 1
30. =EXP(J$28+J29^-1)
31. =EXP(J$28+J30^-1)
32. =EXP(J$28+J31^-1)
33. =EXP(J$28+J32^-1)
34. =EXP(J$28+J33^-1)
35. =EXP(J$28+J34^-1)
36. =EXP(J$28+J35^-1)
37. =EXP(J$28+J36^-1)
38. =J37*J27
39. =MAX(J20,J25,J38)
40. =J39*J4
41. 2
42. =((J1^(2*J41))+(((J2*J5)^0.5)*(J3+J4))^(2*J41))^(1/(2*J41))
43. 8
44. =MAX(0,((J3+J4)^-1*LN(J2*J3/(J5*J4))*J42))
45. =MIN(1,((J1/J40)*(TANH((J44/(J1/J40))^J43))^(1/J43)))

Column K:

INDIVIDUAL CODES

Rows 1 – 45:

1. =J1
2. =J2
3. =1.5*J3
4. =J4
5. =1-K2
6. =K1*LN((K2*K3)/(K5*K4))/(K3+K4)
7. =MIN((MAX(K6,0)),1)
8. less
9. =(EXP(1)*K5)>K2
10. =MAX((1-LN(K2/K5)),0)
11. =MAX((EXP(K10)),1)
12. =EXP(K$10+K11^-1)
13. =EXP(K$10+K12^-1)
14. =EXP(K$10+K13^-1)
15. =EXP(K$10+K14^-1)
16. =EXP(K$10+K15^-1)
17. =EXP(K$10+K16^-1)
18. =EXP(K$10+K17^-1)
19. =EXP(K$10+K18^-1)
20. =K19*K9
21. equal
22. =K2=EXP(1)*K5
23. 0
24. 1.75
25. =K24*K22
26. greater
27. =(EXP(1)*K5)<K2
28. =MAX((1-LN(K2/K5)),0)
29. 1
30. =EXP(K$28+K29^-1)
31. =EXP(K$28+K30^-1)
32. =EXP(K$28+K31^-1)
33. =EXP(K$28+K32^-1)
34. =EXP(K$28+K33^-1)
35. =EXP(K$28+K34^-1)
36. =EXP(K$28+K35^-1)
37. =EXP(K$28+K36^-1)
38. =K37*K27
39. =MAX(K20,K25,K38)
40. =K39*K4
41. 2
42. =((K1^(2*K41))+(((K2*K5)^0.5)*(K3+K4))^(2*K41))^(1/(2*K41))
43. 8
44. =MAX(0,((K3+K4)^-1*LN(K2*K3/(K5*K4))*K42))
45. =MIN(1,((K1/K40)*(TANH((K44/(K1/K40))^K43))^(1/K43)))

Column L:
Rows 1 – 45:

1. =K1
2. =K2
3. =3*K3
4. =K4
5. =1-L2

6. =L1*LN((L2*L3)/(L5*L4))/(L3+L4)
7. =MIN((MAX(L6,0)),1)
8. less
9. =(EXP(1)*L5)>L2
10. =MAX((1-LN(L2/L5)),0)
11. =MAX((EXP(L10)),1)
12. =EXP(L$10+L11^-1)
13. =EXP(L$10+L12^-1)
14. =EXP(L$10+L13^-1)
15. =EXP(L$10+L14^-1)
16. =EXP(L$10+L15^-1)
17. =EXP(L$10+L16^-1)
18. =EXP(L$10+L17^-1)
19. =EXP(L$10+L18^-1)
20. =L19*L9
21. equal
22. =L2=EXP(1)*L5
23. 0
24. 1.75
25. =L24*L22
26. greater
27. =(EXP(1)*L5)<L2
28. =MAX((1-LN(L2/L5)),0)
29. 1
30. =EXP(L$28+L29^-1)
31. =EXP(L$28+L30^-1)
32. =EXP(L$28+L31^-1)
33. =EXP(L$28+L32^-1)
34. =EXP(L$28+L33^-1)
35. =EXP(L$28+L34^-1)
36. =EXP(L$28+L35^-1)
37. =EXP(L$28+L36^-1)
38. =L37*L27
39. =MAX(L20,L25,L38)
40. =L39*L4
41. 2
42. =((L1^(2*L41))+(((L2*L5)^0.5)*(L3+L4))^(2*L41))^(1/(2*L41))
43. 8
44. =MAX(0,((L3+L4)^-1*LN(L2*L3/(L5*L4))*L42))
45. =MIN(1,((L1/L40)*(TANH((L44/(L1/L40))^L43))^(1/L43)))

CHAPTER 8: NIND'S FORMULA

The simple spreadsheet of Figure A.5 calculates the optimum number of drilling locations as discussed in Chapter 8. There are only three lines of code and seven input cells. The rest of the cells are text. Note that the input and intermediate calculations are referenced in Excel using names.

INDIVIDUAL CODES 379

	A	B	C	D	E	F
1	ninds formula					
2						
3	formula 5	pdv=(n*u*Q/(j*R+n*Q))-n*(L-C)-D				
4	formula 6	onw=R*Q^-1*((j*Q*u/(C+L))^.5-j)				
5	formula 7a	PDVmax=R*((u^.5)-(j*(C+L)/Q)^.5)^2-D				
6	formula 7b	PDVmax=Q*((R*j)^-1)*(C+L)*(ONW)^2-D				
7						
8						
9						
10						
11	EXAMPLE:		variables	cell names		
12	average initial production per well	Q/365	INITP	77	bbls/day	
13	Development costs per well	C	COST	$0.72	MM	
14	Lifting costs per well	L	LIFT	$0.09	MM	
15	Oil value	u	OVAL	$10.50	MM	
16	Discount rate	j	DRATE	0.1	per year	
17	Specific volume oil recoverability	V	RECOV	400	bbls/acrefoot	
18	Reservoir A1 sand thickness	T	THICK	20	feet	
19	recoverable oil	R=V*T	OIL	8000	bbls /acre	
20	Optimum number of wells		ONW	2.82		
21	Present Day Value (less other dev costs)		PDV	$225.55	MM	
22						

FIGURE A.5   Chapter 8 spreadsheet for Nind's formula.

Nind's Formula

Column D
Rows 12-21

12. 77
13. 0.72
14. 0.09
15. 10.5
16. 0.1
17. 400
18. 20
19. =RECOV*THICK
20. =OIL*(INITP*365)^-1*((DRATE*INITP*OVAL/(COST+LIFT))^0.5-DRATE)
21. =(INITP*365)*((OIL*DRATE)^-1)*(COST+LIFT)*ONW^2

Names cells:

D12 INITP
D13 COST
D14 LIFT
D15 OVAL
D16 DRATE
D17 RECOV
D18 THICK
D19 OIL
D20 ONW
D21 PDV

CHAPTER 12: BAYESIAN UPDATING

The simple plot of Figure A.6 is used to invert the probabilities as discussed in Chapter 12. It is easy to program using simple instructions of

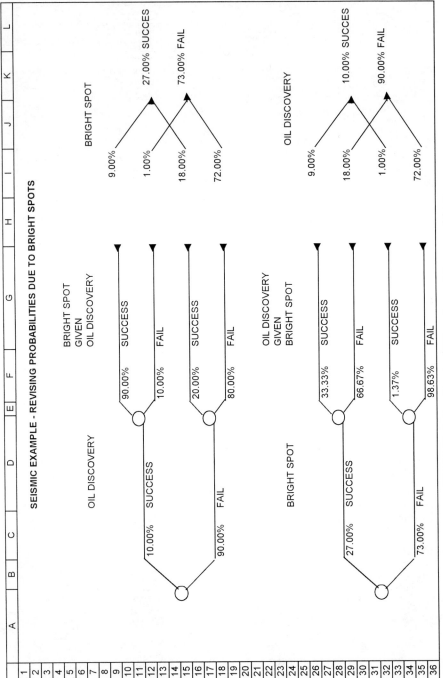

FIGURE A.6 Chapter 12 Bayesian updating spreadsheet.

Bayes Worksheet

Column C
Rows

12. 0.1
18. =1-C12
29. =K12
35. =1-C29

Column F
Rows

10. 0.9
13. =1-F10
16. 0.2
19. =1-F16
27. =I26/C29
30. =1-F27
33. =I32/C35
36. =1-F33

Column I
Rows

9. =C12*F10
12. =F13*C12
15. =F16*C18
18. =F19*C18
26. =I9
29. =I15
32. =I12
35. =I18

Column K
Rows

12. =I9+I15
15. =I12+I18
29. =I26+I32
32. =I29+I35

addition, multiplication, or division. Only input and calculation cells are included in the description; the rest of the cells are text.

CHAPTER 13: OPTIONS

The spreadsheet of Figure A.7 was used to develop the plots included with Chapter 13. It requires several pages of code. An "x" in the code indicates the cell is blank or empty.

	B	C	D	E	F	G	H	I	J	K
1		case 1	case 2	case 3	case 4					
2	gain	100	200	100	100					
3	cost	10	10	5	10					
4	chance	0.5	0.25	0.25	1					
5	interest	0.5	0.5	0.5	0.5					
6	Expected V	20	20	10	45	=EValue at interest				
7	ovprelim	5	15	7.5	0	=(Chance*Interest*(gain-cost)-ev)/Chance				
8	check1	TRUE	TRUE	TRUE	TRUE	=ovprelim<interest*(gain-cost)				
9	check2	TRUE	TRUE	TRUE	TRUE	=Interest*EValue>ovprelim				
10	**Option V**	5	15	7.5	0	**=check1*check2*ovprelim**				
11										
12		case 1 variable gain								
13	gain	15	20	35	40	55	60	65	70	75
14	cost	10	10	10	10	10	10	10	10	10
15	chance	0.5	0.5	0.5	0.5	0.5	0.5	0.5	0.5	0.5
16	interest	1	1	1	1	1	1	1	1	1
17	Expected V	-2.5	0	7.5	10	17.5	20	22.5	25	27.5
18	ovprelim	10	10	10	10	10	10	10	10	10
19	check1	TRUE	TRUE	TRUE	TRUE	TRUE	TRUE	TRUE	TRUE	TRUE
20	check2	FALSE	FALSE	FALSE	FALSE	TRUE	TRUE	TRUE	TRUE	TRUE
21	ov	0	0	0	0	10	10	10	10	10
22	**Option V**	0	0	7.5	10	10	10	10	10	10
23		case 2 variable cost								
24	gain	40	40	40	40	40	40	40	40	40
25	cost	0	2	4	6	8	10	15	20	25
26	chance	0.5	0.5	0.5	0.5	0.5	0.5	0.5	0.5	0.5
27	interest	1	1	1	1	1	1	1	1	1
28	Expected V	20	18	16	14	12	10	5	0	-5
29	ovprelim	0	2	4	6	8	10	15	20	25
30	check1	TRUE	TRUE	TRUE	TRUE	TRUE	TRUE	TRUE	TRUE	TRUE
31	check2	TRUE	TRUE	TRUE	TRUE	TRUE	FALSE	FALSE	FALSE	FALSE
32	ov	0	2	4	6	8	0	0	0	0
33	**Option V**	0	2	4	6	8	10	5	0	0
34		case 2 variable chance								
35	gain	40	40	40	40	40	40	40	40	40
36	cost	10	10	10	10	10	10	10	10	10
37	chance	0.2	0.25	0.4	0.5	0.6	0.7	0.8	0.9	1
38	interest	1	1	1	1	1	1	1	1	1
39	Expected V	-2	0	6	10	14	18	22	26	30
40	ovprelim	40	30	15	10	6.6666667	4.285714	2.5	1.111111	0
41	check1	TRUE	TRUE	TRUE	TRUE	TRUE	TRUE	TRUE	TRUE	TRUE
42	check2	FALSE	FALSE	FALSE	FALSE	TRUE	TRUE	TRUE	TRUE	TRUE
43	ov	0	0	0	0	6.6666667	4.285714	2.5	1.111111	0
44	**Option V**	0	0	6	10	6.6666667	4.285714	2.5	1.111111	0

FIGURE A.7 Chapter 13 options spreadsheet.

Options Formula

Column B
Rows 1-44
1. x
2. gain
3. cost
4. chance
5. interest
6. Expected V
7. ovprelim
8. check1
9. check2
10. Option V
11. x
12. x
13. gain
14. cost
15. chance
16. interest
17. Expected V
18. ovprelim
19. check1
20. check2
21. ov
22. Option V
23. x
24. gain
25. cost
26. chance
27. interest
28. Expected V
29. ovprelim
30. check1
31. check2
32. ov
33. Option V
34. x
35. gain
36. cost
37. chance
38. interest
39. Expected V
40. ovprelim
41. check1
42. check2
43. ov
44. Option V

Column C
Rows 1-44
1. case 1
2. 100
3. 10
4. 0.5
5. 0.5

6. =C2*C5*C4-C3*C5
7. =(C4*C5*(C2-C3)-C6)/C4
8. =C7<C2*(C2-C3)
9. =C6>C7
10. =C9*C8*C7
11.
12. case 1
13. 15
14. 10
15. 0.5
16. 1
17. =C13*C16*C15-C14*C16
18. =(C15*C16*(C13-C14)-C17)/C15
19. =C18<C13*(C13-C14)
20. =C17>C18
21. =C20*C19*C18
22. =IF(C17<C18,(MAX(C17,C20*C19*C18)),C18)
23. case 2
24. 40
25. 0
26. 0.5
27. 1
28. =C24*C27*C26-C25*C27
29. =(C26*C27*(C24-C25)-C28)/C26
30. =C29<C24*(C24-C25)
31. =C28>C29
32. =C31*C30*C29
33. =IF(C28<C29,(MAX(C28,C31*C30*C29)),C29)
34. case 2
35. 40
36. 10
37. 0.2
38. 1
39. =C35*C38*C37-C36*C38
40. =(C37*C38*(C35-C36)-C39)/C37
41. =C40<C35*(C35-C36)
42. =C39>C40
43. =C42*C41*C40
44. =IF(C39<C40,(MAX(C39,C42*C41*C40)),C40)

Column D
Rows 1-44
1. case 2
2. 200
3. 10
4. 0.25
5. 0.5
6. =D2*D5*D4-D3*D5
7. =(D4*D5*(D2-D3)-D6)/D4
8. =D7<D2*(D2-D3)
9. =D6>D7
10. =D9*D8*D7
11. x
12. variable gain
13. 20
14. 10

INDIVIDUAL CODES 385

15. 0.5
16. 1
17. =D13*D16*D15-D14*D16
18. =(D15*D16*(D13-D14)-D17)/D15
19. =D18<D13*(D13-D14)
20. =D17>D18
21. =D20*D19*D18
22. =IF(D17<D18,(MAX(D17,D20*D19*D18)),D18)
23. variable cost
24. 40
25. 2
26. 0.5
27. 1
28. =D24*D27*D26-D25*D27
29. =(D26*D27*(D24-D25)-D28)/D26
30. =D29<D24*(D24-D25)
31. =D28>D29
32. =D31*D30*D29
33. =IF(D28<D29,(MAX(D28,D31*D30*D29)),D29)
34. variable chance
35. 40
36. 10
37. 0.25
38. 1
39. =D35*D38*D37-D36*D38
40. =(D37*D38*(D35-D36)-D39)/D37
41. =D40<D35*(D35-D36)
42. =D39>D40
43. =D42*D41*D40
44. =IF(D39<D40,(MAX(D39,D42*D41*D40)),D40)

Column E
Rows 1-44
1. case 3
2. 100
3. 5
4. 0.25
5. 0.5
6. =E2*E5*E4-E3*E5
7. =(E4*E5*(E2-E3)-E6)/E4
8. =E7<E2*(E2-E3)
9. =E6>E7
10. =E9*E8*E7
11. x
12. x
13. 35
14. 10
15. 0.5
16. 1
17. =E13*E16*E15-E14*E16
18. =(E15*E16*(E13-E14)-E17)/E15
19. =E18<E13*(E13-E14)
20. =E17>E18
21. =E20*E19*E18
22. =IF(E17<E18,(MAX(E17,E20*E19*E18)),E18)
23. x

24. 40
25. 4
26. 0.5
27. 1
28. =E24*E27*E26-E25*E27
29. =(E26*E27*(E24-E25)-E28)/E26
30. =E29<E24*(E24-E25)
31. =E28>E29
32. =E31*E30*E29
33. =IF(E28<E29,(MAX(E28,E31*E30*E29)),E29)
34. x
35. 40
36. 10
37. 0.4
38. 1
39. =E35*E38*E37-E36*E38
40. =(E37*E38*(E35-E36)-E39)/E37
41. =E40<E35*(E35-E36)
42. =E39>E40
43. =E42*E41*E40
44. =IF(E39<E40,(MAX(E39,E42*E41*E40)),E40)

Column F
Rows 1-44
1. case 4
2. 100
3. 10
4. 1
5. 0.5
6. =F2*F5*F4-F3*F5
7. =(F4*F5*(F2-F3)-F6)/F4
8. =F7<F2*(F2-F3)
9. =F6>F7
10. =F9*F8*F7
11. x
12. x
13. 40
14. 10
15. 0.5
16. 1
17. =F13*F16*F15-F14*F16
18. =(F15*F16*(F13-F14)-F17)/F15
19. =F18<F13*(F13-F14)
20. =F17>F18
21. =F20*F19*F18
22. =IF(F17<F18,(MAX(F17,F20*F19*F18)),F18)
23. x
24. 40
25. 6
26. 0.5
27. 1
28. =F24*F27*F26-F25*F27
29. =(F26*F27*(F24-F25)-F28)/F26
30. =F29<F24*(F24-F25)
31. =F28>F29
32. =F31*F30*F29

33. =IF(F28<F29,(MAX(F28,F31*F30*F29)),F29)
34. x
35. 40
36. 10
37. 0.5
38. 1
39. =F35*F38*F37-F36*F38
40. =(F37*F38*(F35-F36)-F39)/F37
41. =F40<F35*(F35-F36)
42. =F39>F40
43. =F42*F41*F40
44. =IF(F39<F40,(MAX(F39,F42*F41*F40)),F40)

Column G
Rows 1-44
1. x
2. x
3. x
4. x
5. x
6. x
7. x
8. x
9. x
10. x
11. x
12. x
13. 55
14. 10
15. 0.5
16. 1
17. =G13*G16*G15-G14*G16
18. =(G15*G16*(G13-G14)-G17)/G15
19. =G18<G13*(G13-G14)
20. =G17>G18
21. =G20*G19*G18
22. =IF(G17<G18,(MAX(G17,G20*G19*G18)),G18)
23. x
24. 40
25. 8
26. 0.5
27. 1
28. =G24*G27*G26-G25*G27
29. =(G26*G27*(G24-G25)-G28)/G26
30. =G29<G24*(G24-G25)
31. =G28>G29
32. =G31*G30*G29
33. =IF(G28<G29,(MAX(G28,G31*G30*G29)),G29)
34. x
35. 40
36. 10
37. 0.6
38. 1
39. =G35*G38*G37-G36*G38
40. =(G37*G38*(G35-G36)-G39)/G37
41. =G40<G35*(G35-G36)

42. =G39>G40
43. =G42*G41*G40
44. =IF(G39<G40,(MAX(G39,G42*G41*G40)),G40)

Column H
Rows 1-44
1. x
2. x
3. x
4. x
5. x
6. x
7. x
8. x
9. x
10. x
11. x
12. x
13. 60
14. 10
15. 0.5
16. 1
17. =H13*H16*H15-H14*H16
18. =(H15*H16*(H13-H14)-H17)/H15
19. =H18<H13*(H13-H14)
20. =H17>H18
21. =H20*H19*H18
22. =IF(H17<H18,(MAX(H17,H20*H19*H18)),H18)
23. x
24. 40
25. 10
26. 0.5
27. 1
28. =H24*H27*H26-H25*H27
29. =(H26*H27*(H24-H25)-H28)/H26
30. =H29<H24*(H24-H25)
31. =H28>H29
32. =H31*H30*H29
33. =IF(H28<H29,(MAX(H28,H31*H30*H29)),H29)
34. x
35. 40
36. 10
37. 0.7
38. 1
39. =H35*H38*H37-H36*H38
40. =(H37*H38*(H35-H36)-H39)/H37
41. =H40<H35*(H35-H36)
42. =H39>H40
43. =H42*H41*H40
44. =IF(H39<H40,(MAX(H39,H42*H41*H40)),H40)

Column I
Rows 1-44
1. x
2. x
3. x

Individual Codes

4. x
5. x
6. x
7. x
8. x
9. x
10. x
11. x
12. x
13. 65
14. 10
15. 0.5
16. 1
17. =I13*I16*I15-I14*I16
18. =(I15*I16*(I13-I14)-I17)/I15
19. =I18<I13*(I13-I14)
20. =I17>I18
21. =I20*I19*I18
22. =IF(I17<I18,(MAX(I17,I20*I19*I18)),I18)
23. x
24. 40
25. 15
26. 0.5
27. 1
28. =I24*I27*I26-I25*I27
29. =(I26*I27*(I24-I25)-I28)/I26
30. =I29<I24*(I24-I25)
31. =I28>I29
32. =I31*I30*I29
33. =IF(I28<I29,(MAX(I28,I31*I30*I29)),I29)
34. x
35. 40
36. 10
37. 0.8
38. 1
39. =I35*I38*I37-I36*I38
40. =(I37*I38*(I35-I36)-I39)/I37
41. =I40<I35*(I35-I36)
42. =I39>I40
43. =I42*I41*I40
44. =IF(I39<I40,(MAX(I39,I42*I41*I40)),I40)

Column J
Rows 1-44

1. x
2. x
3. x
4. x
5. x
6. x
7. x
8. x
9. x
10. x
11. x
12. x

13. 70
14. 10
15. 0.5
16. 1
17. =J13*J16*J15-J14*J16
18. =(J15*J16*(J13-J14)-J17)/J15
19. =J18<J13*(J13-J14)
20. =J17>J18
21. =J20*J19*J18
22. =IF(J17<J18,(MAX(J17,J20*J19*J18)),J18)
23. x
24. 40
25. 20
26. 0.5
27. 1
28. =J24*J27*J26-J25*J27
29. =(J26*J27*(J24-J25)-J28)/J26
30. =J29<J24*(J24-J25)
31. =J28>J29
32. =J31*J30*J29
33. =IF(J28<J29,(MAX(J28,J31*J30*J29)),J29)
34. x
35. 40
36. 10
37. 0.9
38. 1
39. =J35*J38*J37-J36*J38
40. =(J37*J38*(J35-J36)-J39)/J37
41. =J40<J35*(J35-J36)
42. =J39>J40
43. =J42*J41*J40
44. =IF(J39<J40,(MAX(J39,J42*J41*J40)),J40)

Column K
Rows 1-44
1. x
2. x
3. x
4. x
5. x
6. x
7. x
8. x
9. x
10. x
11. x
12. x
13. =J13+5
14. 10
15. 0.5
16. 1
17. =K13*K16*K15-K14*K16
18. =(K15*K16*(K13-K14)-K17)/K15
19. =K18<K13*(K13-K14)
20. =K17>K18
21. =K20*K19*K18

22. =IF(K17<K18,(MAX(K17,K20*K19*K18)),K18)
23. x
24. 40
25. 25
26. 0.5
27. 1
28. =K24*K27*K26-K25*K27
29. =(K26*K27*(K24-K25)-K28)/K26
30. =K29<K24*(K24-K25)
31. =K28>K29
32. =K31*K30*K29
33. =IF(K28<K29,(MAX(K28,K31*K30*K29)),K29)
34. x
35. 40
36. 10
37. 1
38. 1
39. =K35*K38*K37-K36*K38
40. =(K37*K38*(K35-K36)-K39)/K37
41. =K40<K35*(K35-K36)
42. =K39>K40
43. =K42*K41*K40
44. =IF(K39<K40,(MAX(K39,K42*K41*K40)),K40)

BIBLIOGRAPHY

AAPG Explorer, January, 1994.

Airy, G. B. (1855). On the computation of the effect of the attraction of mountain-masses as distributing the apparent astronomical latitude of station of geodetic surveys. *Philos. Trans. R. Soc. London* **145,** 101–104.

Anderson, C. M., Johnson, W. R., Marshall, C. F., and Lear, E. M. (1995). "Oil Spill Risk Analysis: Outer Continental Shelf Lease Sale 144, Beaufort Sea," Rep. OCS/MMS 95-0046. U.S. Department of the Interior, Washington, DC.

Anderson, R. N., Langseth, M., and Sclater, J. G. (1977). The mechanisms of heat transfer through the floor of the Indian Ocean. *J. Geophys. Res.* **82,** 3391–3409.

Arps, J. J. (1962). Estimation of primary oil & gas reserves. *In* "Petroleum Production Handbook" (T. C. Frick, ed.), Vol. 2, Chapter 37. McGraw-Hill, New York.

Arps, J. J., and Roberts, T. G. (1958). Economics of drilling for Cretaceous oil on east flank of Denver-Julesburg Basin. *Am. Assoc. Pet. Geol. Bull.* **42,** 2549–2566.

Arps, J. J., Smith, M. B., and Mortada, J. (1971). Relationship between proved reserves and exploratory effort. *J. Pet. Technol.* **23,** 671–675.

Atchison, J., and Brown, J. A. C. (1957). "The Lognormal Distribution." Cambridge University Press, London.

Aurell, E., and Zyckzkowski, K. (1996). Option pricing and partial hedging: Theory of Polish options. *J. Political Econ.* pp 1–30.

Aylor, W. K. (1995). Business performance and value of exploitation 3-D seismic. *The Leading Edge,* July, pp. 797–801.

Baker, R. A. (1988). When is a prospector play played out? *Oil Gas J.* January 11, pp. 77–80.

Bayliss, G. S., Martin, S. J., Cernock, P. J., Carlisle, C. T., and Odiorne, J. D. (1977). "Offshore Basin Assessment for Oil and Gas Using Geochemical Data," OTC Pap. No. 2936. presented at 9th Annual OTC, Houston, TX.

Black, F., and Scholes, M. S. (1973). *J. Political Econ.*

Bouchard, J.-P., and Sornette, D. (1994). *J. Phys. 1.* **4,** 863–881.

Bouchard, J.-P., and Sornette, D. (1995). *J. Phys. 1* **5,** 219–220.

Bouchard, J.-P., and Sornette, D., (1996). Derivatives trading: Physicists favor less complex and risky theory. *Phys. Today,* March, pp. 15 and 128. (see also B. Riley, *Financial Times,* 17 April 1996).

Brennan, M. J., and Schwartz, E. S. (1978). Finite difference methods and jump processes arising in the pricing of contingent claims: A synthesis. *J. Financial Quant. Anal.* **13**, 461–473.
Brennan, M. J., and Schwartz, E. S. (1985). Evaluating natural resource investments. *J. Bus.* **58**, 135–157.
Brons, F., and McGarry M. W. (1957). Methods for calculating profitabilities. *32nd Annu. Fall Meet., Soc. Pet. Eng. AIME,* Dallas, TX, Pap. 870-G.
Brooks, J., and Welte, D. (1984). "Advances in Petroleum Geochemistry," Vol. 1. Academic Press, New York.
Bullion, J., and Waddy, M. (1956). Tax considerations in oil transactions. *J. Pet. Technol.* **8**, 12.
Campbell, J. M. (1959). "Oil Property Evaluation." Prentice-Hall, Englewood Cliffs, NJ.
Campbell, J. M. (1973). "Petroleum Reservoir Property Evaluation." Campbell Petroleum Series, Norman, OK.
Cao, S. (1988). Sensitivity analysis of 1-D dynamical model for basin analysis. Ph.D. dissertation, University of South Carolina, Columbia.
Capen, E. C., Clapp, R. B., and Campbell, W. M. (1971). Competitive bidding in high-risk situations. *J. Pet. Technol.* **23**, 641–653.
Capen, E. C., Clapp, R. B., and Phelps, W. W. (1976). Growth rate—A rate of return measure of investment efficiency. *J. Pet. Technol.* **28**, 531–543.
Central Intelligence Agency (1977). "Major Oil and Gas Fields of the Free World," ER 77-10313. Directorate of Intelligence, Washington, DC.
Chatellier, J.-Y. and Slevin, A. (1988). Review of African petroleum and gas deposits. *J. Afr. Earth Sci.* **7**, 561–578.
Cleland, D. I., and King, W. R. (1975). "Systems Analysis and Project Management." McGraw-Hill, New York.
Cozzolino, J. M. (1977a). A simplified utility framework for the analysis of financial risk. *Econ. Eval. Symp. Soc. Pet. Eng.,* Dallas, TX *1977,* SPE No. 6359.
Cozzolino, J. M. (1977b). "Management of Oil and Gas Exploration Risk." Cozzolino Associates, West Berlin, NJ.
Cozzolino, J. M. (1978). A new method for measurement and control of exploration risk. *Soc. Pet. Eng. AIME,* SPE No. 6632.
Craig, F. F., Jr., Willcox, P. J., Ballard, J. R., and Nation, W. R. (1977). Optimized recovery through continuing interdisciplinary cooperation. *J. Pet Technol.* **29**, 755–760.
Cronquist, C. (1991). Reserves and probabilities—Synergism or anachronism? *J. Pet. Technol.,* October, pp. 1258–1264, Pap. SPE 23586.
Crovelli, R. A., and Balay, R. H. (1991). A microcomputer program for energy assessment and aggregation using the triangular probability distribution. *Comput. Geosci.* **17**, 197–225.
Dahlberg, E. (1982). "Applied Hydrodynamics in Petroleum Exploration." New York, Springer-Verlag.
Dahlstrom, C.D.A. (1969). Balanced cross-sections. *Can. J. Earth Sci.* **6**, 743–757.
Davidson, L. B., and Cooper, D. D. (1976). A simple way of developing a probability distribution of present value. *SPE Trans.* **261**, 1069–1078.
Dean, J. (1954). Measuring the productivity of capital. *Harv. Bus. Rev.,* p. 120.
Dixit, A. K., and Pindyck, R. S. (1994). "Investment Under Uncertainty." Princeton University Press, Princeton, NJ.
Dixit, A. K., and Pindyck, R. S. (1995). The options approach to capital investment. *Harv. Bus. Rev.* May–June, pp. 105–115.
Domenico, S. N. (1974). Effect of water saturation on seismic reflectivity of sand reservoirs encased in shale. *Geophysics* **39**, 759–769.
Dougherty, E. L., and Lohrenz, J. (1976). Statistical analysis of bids for federal offshore leases. *J. Pet. Technol.* **28**, 1377–1390.
Dougherty, E. L., and Nozaki, M. (1973). Analysis of bidding strategies. Pap. SPE-4566, *SPE 48th Annu. Fall. Meet.,* Las Vegas, NV.

Ekern, S. (1988). An option pricing approach to evaluating petroleum projects. *Energy Econ.* **10,** 91–99.

Exxon, "Energy Outlook 1975–1990," Public Affairs Department, P.O. Box 2180, Houston, TX 77001.

Falvey, D. A., and Middleton, M. F. (1981). Passive continental margins: Evidence for a prebreakup deep crustal metamorphic subsidence mechanism. *Int. Geol. Cong. 26th, 1980:* Colloq. C3.3, Geol. Continental Margins, Suppl. to Vol. 4, pp. 103–114.

Feller, J. (1968). "Elements of Probability Theory," Vols. 1 and 2. McGraw-Hill, Englewood Cliffs, NJ.

Frick, T. C., ed. (1962). Estimation of primary oil and gas reserves. "Petroleum Production Handbook," Vol. II, pp. 37-1 to 37-56. McGraw-Hill, New York.

Gebelein, C. A., Pearson, C. E., and Silbergh, M. (1978). Assessing political risk of oil investment ventures. *J. Pet. Technol.* **30,** 725–730.

Gemmil, G. (1993). "Options Pricing—An International Perspective." McGraw-Hill, Englewood Cliffs, NJ.

Geske, R., and Shastri, K. (1985). Valuation by approximation: A comparison of alternative option valuation techniques. *J. Financial Quant. Anal.* **20,** 45–71.

Ghauri, W. K. (1960). Results of well simulation by hydraulic fracturing and high rate oil backflush. *J. Pet. Technol.* **12,** 19–26.

Gibbs, A. D. (1983). Balanced cross-sections construction from the seismic sections in areas of extensional tectonics. *J. Struct. Geol.* **5,** 153–160.

Glanville, J. W. (1957). Rate of return calculations as a measure of investment opportunities. *J. Pet. Technol.* **9,** 12–19.

Gradshteyn, I. S., and Ryzhik, I. M. (1965). "Table of Integrals, Series and Products." Academic Press, New York.

Grayson, C. J. (1960). "Decisions Under Uncertainty: Drilling Decisions by Oil and Gas Operators." Harvard University Press, Boston.

Guidish, T. M., Kendall, C. G. St. C., Lerche, I., Toth, D. J., and Yarzab, R. F. (1984). Basin evaluation using burial history calculations: An overview. *AAPG Bull.* **69,** 92–105.

Harbaugh, J. W., Doveton, J. H., and Davis, J. C. (1977). "Probability Methods in Oil Exploration." Wiley, New York.

Harbaugh, J. W., Davis, J. C., and Wendebourg, J. (1996). "Computing Risk for Oil Prospects: Principles and Programs." Elsevier, Amsterdam.

Hardin, G. C., and Mygdal, K. (1968). Geologic success and economic failure *Am. Assoc. Pet. Geol. Bull.* **52,** 2079–2091.

Harding, T. P. (1974). Petroleum traps associated with wrench faults. *AAPG Bull.* **58,** 1290–1304.

Harding, T. P. (1985). Seismic characteristics and identification of negative flower structures, positive flower structures, and positive structural inversion. *AAPG Bull.* **69,** 582–600.

Harris, D. G., and Hewitt, C. H. (1977). Synergism in reservoir management—The geologic perspective. *J. Pet. Technol.* **29,** 761–770.

Haun, J. D., ed. (1975). "Methods of Estimating the Volume of Undiscovered Oil and Gas Resources," Stud. Geol., No. 1. Am. Assoc. Pet. Geol., Tulsa, OK.

Horner, W. L., and Roebuck, I. F. (1957). Economics and prediction of oil recovery by fluid injection operations. *In* "Improving Oil Recovery." Department of Petroleum Engineering, University of Texas, Austin.

Hubbard, R. G. (1994). Investment under uncertainty: Keeping one's options open *J. Econ. Literature* **32,** 1816–1831.

Hubbert, M. K. (1940). The theory of ground-water motion. *J. Geol.* **48,** 785–944.

Hubbert, M. K. (1962). "Energy Resources," Nat. Res. Publ. 1000-D. Washington, DC.

Hull, J., and White, A. (1990). Valuing derivative securities using the explicit finite difference method. *J. Financial Quant. Anal.* **25,** 87–100.

Hunt, J. M. (1979). "Petroleum Geochemistry and Geology" Freeman, San Francisco.

Ion, D. C. (1975). "Availability of World Energy Sources." Graham & Trotman, London.

Ivanoe, L. F. (1976). Oil/gas potential of basins estimated. *Oil and Gas J.,* pp. 154–156.

Jacoby, H. D., and Laughton, D. G. (1988). "Project Evaluation using a Probabilistic-Process Representation of Uncertainty," MIT Center for Energy Policy Research, Working Manuscript MIT-EL 88-001 WP. MIT University Press, Cambridge MA.

Jacoby, H. D., and Laughton, D. G. (1991). "Project Evaluation: A Practical Modern Assest Pricing Approach," Workng Paper, pp. 1–91 University of Alberta, Institute for Financial Research, Edmonton, Alberta, Canada.

Jaynes, E. T. (1978). Where do we stand on Maximum Entropy? *In* "The Maximum Entropy Formalism" (R. D. Levine and M. Tribus, eds.), pp. 15–118. MIT Press, Cambridge, MA.

Johnson, B. E. (1994). Modeling energy technology choices—Which investment analysis tools are appropriate? Energy Policy **22,** 877–883.

Jones, R. W. (1975). A quantitative geological approach to prediction of petroleum resources. *Stud. Geol. (Tulsa, Okla.)* **1,** 186–195.

Kaufman, G. M., Balcer, Y., and Kruyt, D. (1975). A probabilistic model of oil and gas discovery. *Stud. Geol. (Tulsa, Okla.)* **1,** 113–142.

Kensinger, J. W. (1987). Adding the value of active management into the capital budgeting equation. *Interfaces,* pp. 31–42.

Klemme, H. D. (1975). Giant oil fields related to their geologic setting. *Bull. Can. Pet. Geol.* **23,** 30–66.

Klemme, H. D. (1978). Worldwide petroleum exploration and prospects. *Explor. Econ. Petr. Ind.* **16,** 39–101.

Krumbein, W. C., and Grayhill, F. A. (1965)." *An Introduction to Statistical Models in Geology.*" McGraw-Hill, Englewood Cliffs, NJ.

Lerche, I. (1990a). "Basin Analysis: Quantitative Methods," Vol. 1. Academic Press, San Diego, CA.

Lerche, I. (1990b). "Basin Analysis: Quantitative Methods," Vol. 2. Academic Press, San Diego, CA.

Lerche, I. (1991). "Oil Exploration: Basin Analysis and Economics." Academic Press, San Diego, CA.

Lerche, I. (1994). Anatomy of a lease sale: Sub-salt bids and strategies at Gulf of Mexico lease sale 147. *Oil Gas J.,* October 3, pp. 68–71.

Lerche, I. (1997). "Geological Risk and Uncertainty in Oil Exploration." Academic Press, San Diego, CA.

Lerche, I., and O'Brien, J. J. (1987). "Dynamical Geology of Salt and Related Structures." Academic Press, Orlando, FL.

Lerche, I., and O'Brien, J. J. (1994). Understanding subsalt overpressure may reduce drilling risk. *Oil Gas. J.,* January 24, pp. 28–34.

Lerche, I., Yarzab, R. F., and Kendall, C. G. St. C. (1984). Determination of paleoheatflux from vitrinite reflectance data. *AAPG Bull.* **68,** 1704–1717.

Lewis, C. R., and Rose, S. C. (1970). A theory relating high temperatures and overpressures. *J. Pet. Technol.* **22,** 11–16.

Lohrenz, J., and Bailey, A. J. (1995). Evidence and results of present value maximization for oil and gas development projects. *Soc. Pet. Eng. Hydrocarbon Econ. Eval. Symp.,* Dallas, TX, *1995,* Pap. SPE 30050, pp. 163–177.

Lohrenz, J., and Dickens, R. N. (1993). Option theory for evaluation of oil and gas assets: The upsides and downsides. *Soc. Pet. Eng.* Pap. SPE-25837, 179–188.

Lopatin, N. V. (1971). Temperature and geologic time as factors in coalification. *Izv. Akad. Nauk SSSR, Ser. Geol.* **3,** 95–106.

Lumley, J. L. (1970). "Stochastic Tools in Turbulence." Academic Press, New York.

MacKay, J. A., and Lerche, I. (1996a). On the influence of uncertainties in estimating risk aversion and working interest. *Energy Explor. Exploit.* **14,** 13–46.
MacKay, J. A., and Lerche, I. (1996b). On the value of options in exploration economics. *Energy Explor. Exploit.* **14,** 47–61.
MacKay, J. A., and Lerche, I. (1996c). Model dependencies of risk aversion and working interest estimates. *Energy Explor. Exploit.* **14,** 183–196.
MacKay, J. A., and Lerche, I. (1996d). Portfolio balancing and risk adjusted values under constrained budget conditions. *Energy Explor. Exploit.* **14,** 197–225.
Markland, J. T. (1992). Options theory: A new way forward for exploration and engineering economics. *Soc. Pet. Eng.* Pap. SPE-24232, 51–67.
Martin, C. C. (1976). "Project Management." American Management Association, New York.
Matheson, J. E. (1968). The economic value of analysis and computation. *IEEE Trans. Syst. Sci. Cybernet.* **SSC-4,** No. 3.
McCray, A. W. (1969). Evaluation of exploratory drilling ventures by statistical decision methods. *J. Pet. Technol.* **21,** 1199–1209.
McCray, A. W. (1975). "Petroleum Evaluations and Economic Decisions," Prentice-Hall, New York.
McDonald, R., and Siegal, D. (1986). The value of waiting to invest. *Q. J. Econ.* **101,** 707–728.
McDowell, A. N. (1975). What are the problems in estimating the oil potential of a basin? *Oil Gas J.* (Jan.), pp. 85–90.
McKenzie, D. (1978). Some remarks on the development of sedimentary basins. *Earth Planet. Sci. Lett.* **40,** 25–32.
Megill, R. E. (1971). "Exploration Economics." Petroleum Publishing Co., Tulsa, OK.
Megill, R. E. (1977). "An Introduction to Risk Analysis." Petroleum Publishing Co., Tulsa, OK.
Milliman, T. H. (1974). "Marine Geology," Part 1. Springer-Verlag, New York.
Muskat, M. (1949). "Physical Principles of Oil Production." McGraw-Hill, New York.
Nakayama, K. (1987). A dynamic, two dimensional, fluid flow model for basin analysis. Ph.D. dissertation, University of South Carolina, Columbia.
Newendorp, P. D. (1975). "Decision Analysis for Petroleum Exploration." Petroleum Publishing Co., Tulsa, OK.
Newendorp, P. D., and Root, P. J. (1968). Risk analysis in drilling investment decisions, *J. Pet. Technol.* **20,** 579–585.
Nind, T. E. W. (1959). Profitability of oilfield projects. *Southam-MacLean's Oil/Gas World,* p. 14.
Nind, T. E. W. (1981). "Principles of Oil Well Production." McGraw-Hill, New York.
Northern, I. G. (1967). Risk, probability and decision-making in oil and gas operations. *J. Can. Pet. Technol.* **6,** 150–154.
O'Brien, J. J., and Lerche, I. (1994). Understanding subsalt overpressure may reduce drilling risk. *Oil Gas J.* January 24, pp. 28–34.
Oil and Gas Journal (1989). December 25, pp. 95–98.
Otis, R. M., and Schneidermann, N. (1997). A process for evaluating exploration prospects. *AAPG Bull.* **81,** 1087–1109.
Paddock, J. L., Siegal, D. R., and Smith, J. L. (1988). Option valuation of claims on real assets: The case of offshore petroleum leases. *Q. J. Econ.* **103,** 479–508.
Phillips, C. E. (1958). The relationship between rate of return, payout, and ultimate return in oil and gas properties. *J. Pet. Technol.* **10,** 26–32.
Porter, E. D. (1992). U.S. petroleum supply: History, prospects and policy implications. *Am. Pet. Inst. Res. Study,* No. 064, pp. 1–57.
Riley, B. (1996). *Financial Times,* April 17.
Rose, P. R. (1987). Dealing with risk and uncertainty in exploration: How can we improve? *AAPG Bull.* **71,** 1–16.

Rose, P. R. (1992). Chances of success and its use in petroleum exploration. In "The Business of Petroleum Exploration" (R. Steinmetz, ed.), AAPG Treatise Pet. Geol.: Handb. Pet. Geol., pp. 71–86. AAPG, Tulsa, OK.

Rose, P. R. (1994). Implementing risk analysis in exploration organizations: What works? What doesn't? Am. Assoc. Pet. Geol. Meet., 1994, Notes.

Rose, P. R. (1996). Private communication.

Royden, L., Sclater, J. G., and von Herzen, R. P. (1980). Continental Margin subsidence and heat flow: Important parameters in formation of petroleum hydrocarbons. *AAPG Bull.* **64,** 173–187.

Sampson, A. (1975). "The Seven Sisters." Viking Press, New York.

Sclater, J. G., and Christie, P. A. F. (1980). Continental stretching: An explanation of the post-mid-Cretaceous subsidence of the central North Sea basin. *J. Geophys. Res.* **85,** 3711–3739.

Silverman, M. (1976). "Project Management." Wiley, New York.

Slider, H. C. (1976). "Practical Petroleum Reservoir Engineering Methods." Petroleum Publishing Co., Tulsa, OK.

Smith, M. B. (1968). Estimate reserves by using computer simulation methods. *Oil Gas J.* (Jan.), pp. 81–84.

Smith, M. B. (1974). Probability estimates for petroleum drilling decisions. *J. Pet. Technol.* **26,** 687–695.

Sprague, A., ed. (1990). "British Petroleum Statistical Review of World Energy," British Petroleum Company, Britannic House, London.

Stauffer, T. R. (1995). Estimating the full costs of oil and gas production. *Soc. Pet. Eng. Hydrocarbon Econ. Eval. Symp.,* Dallas, TX, 1995, Pap. SPE 30064, pp. 295–305.

Thomsen, R. O., and Lerche, I. (1997). Relative contributions to uncertainties in reserve estimates. *Mar. Pet. Geol.* **14,** 65–74.

Tissot, B. (1969). Premières données sur les mécanismes et la cinétique de la formation du petrole dans les sediments: Simulation d'un schema réactionnel sur ordinateur. *Rev. Inst. Fr. Pet.* **24,** 470–501.

Tissot, B., and Espitalie, J. (1975). L'évolution thermique de la matière organique des sediments: Applications d'une simulation mathématique. *Rev. Inst. Fr. Pet.* **30,** 743–777.

Tissot, B. P., and Welte, D. F. (1978). "Petroleum Formation and Occurrence." Springer-Verlag, Berlin.

Veevers, J. J. (1986). Breakup of Australia and Antarctica estimated as mid-Cretaceous (95 ± 5 Ma) from magnetic and seismic data at the continental margin. *Earth Planet. Sci. Lett.* **77,** 91–99.

Wagner, C. W. (1982). Simpson's paradox (UMAP unit 587). "Modules in Undergraduate Mathematics and its Application." Consortium for Mathematics and Its Applications (COMAP), Lexington, MA.

Waples, D. W. (1980). Time and Temperature in petroleum formation—application of Lopatin's method to petroleum exploration. *AAPG Bull.* **64,** 916–926.

Waples, D. W. (1988). Maturity modeling of sedimentary basins: Approaches, limitations, and assessment of future developments, *Abstr., Workshop Quant. Dyn. Stratigr.* Denver, CO, p. 19.

Warren, J. E. (1975). "Petroleum Exploration and Development: Management, Money, Reserves and Risk." International Human Resources Development Corp., Boston.

Warren, J. E. (1978). The development decision for frontier areas: The North Sea. *Eur. Offshore Pet. Conf. Exhib.,* London.

Watkins, P. B. (1959). Economic evaluations. *J. Pet. Technol.* **11,** 20–23.

White, D. Q. (1993). Geological risking guide for prospects and plays. *AAPG Bull.* **77,** 2048–2061.

White, D. Q., and Gehman, H. M. (1979). Methods of estimating oil and gas resources. *AAPG Bull.* **63,** 2183–2192.

White, D. Q., Garrett, R. W., Marsh, G. R., Baker, R. A., and Gehman, H. M. (1975). Assessing regional oil and gas potential. *Stud. Geol. (Tulsa, Okla.)* **1,** 143–159.

Withers, R. (1992). The value of reservoir geophysics. *The Leading Edge,* March, pp. 35–39.

Woody, L. D., Jr., and Capshaw, T. D. (1960). Investment evaluation by present-value profile. *J. Pet. Technol.* **12,** 15–22.

Zapp, A. D. (1962). Future petroleum producing capacity of the United States. *Geol. Surv. Bull. (U.S.)* **1142-H.**

Zisook, A. (1994). The physics of high finance. *Phys. Today,* June, pp. 55–56, as reported by A. L. Robinson.

INDEX

Abandonment, 285ff
Acreage, 192–195, 229, 248, 259
Activation energy, 19
Added information, 211ff
AIDS, 283
Alabama, 102
Amoco, 233–243
Anadarko, 233–243
Apparent risk tolerance, 20–21
Arrhenius formula, 19
Available information, 306

Bayesian, 281ff, 379–381
Benefits, 225–226
BHP Petroleum, 246
Bias weighting, 220–225
Bid analysis, 232ff, 250ff
Bidding statistics, 227ff
Bid distributions, 233–252, 260–264
Bid ratios, 242–243
Bid worth, 265
Bids with options, 337ff, 345ff
Blocks, 237–238, 243–244
BP Exploration, 246
Break-even probability, 25, 38
Bright spots, 285ff
Budget, 55–135
    constraints, 58–59
    corporate, 163ff
    cycle time, 180
    exhaustion, 180
    fixed, 172ff
    high, 71–73, 75–76
    low, 73–75
Buy-out, 48–51

Cash flow, 189, 269, 272
Chance minimum acceptable, 291
Chance of success, 291
Column of oil, 290
Corporate budget, 163ff
Corporations, 238–242
Correlation, 87ff
    fiscal, 113ff
    geological, 99ff
    multiple opportunities, 102, 105
    rank, 150
    single opportunities, 101, 104
    value, 93ff
Cost, 18–30, 57–58, 165ff
    balancing, 163ff
    development, 271–273
    expenditure, 56
    exploration, 271
    exposure, 56
    production, 270–271
    total, 340ff
    variable, 318ff
Cozzolino formula, 18–30, 59, 63, 137ff
    modifications, 139ff

Cross-over, 143–144
Cumulative depth, 135–136
Current option value, 303–306

Data acquisition, 211–226
Decision analysis, 3
Decisions, 303–306
Decline constant, 269–270
Depreciation recovery, 272
Development costs, 190ff, 271–272
Discount rate, 188ff
Dollars spent, 168–170
Dry hole, 286, 290

East Cameron South, 234ff, 359
Economic analysis, 2
  models, 267ff, 274
Elf Exploration, 246
Engineering design, 2
Estimates,
  economic, 9
  geotechnical, 8
  risk, 10–12
Eugene Island, 65
Eugene Island South, 234, 345, 346
Exchange model, 342–344
Expected value, 18–30, 40, 95, 211ff, 252ff, 302ff
  constant, 220–225
  decreased, 220–225
  increased, 213–220
  negative, 28–29, 48–51
Expected worth, 260–261, 266
  maximum, 262
  maximum uncertainty, 263–264
  uncertainty, 261–264, 266
Exploration costs, 271

Field development costs, 191
Field lifetime, 198–200
First year updating, 296
Fixed budget, 163ff
Flex trend, 113
Fourth year updating, 298
Funding requests, 164ff
Funding strategies, 172ff
Future models, 342ff

Gain, 137ff, 164ff
  high, 149, 308
  low, 149
  maximal option, 308
  variable, 318ff
Gas, 116–133
Geological concept, 1
Geological correlation, 99ff
Gulf of Mexico, 232–259

HIV, 283–285

Information, 211ff, 304
  perfect, 211
Investment, 310

Lease sale, 147, 157, 232–258
Lifetime, 198–200
  sensitivity, 205–206
Lifting costs, 188ff
Lognormal distribution, 7–8
Loss, 311
Louisiana, 113

Mahogany prospect, 232, 238
Main Pass, 245, 280
Majil Oil Company, 228–232
Massive changes, 325–327
Maximum well number, 200–201
Mexico, Gulf of, 232–258
Minerals Management Service, 232, 238
Minimal loss option, 310–311
Minimum acceptable chance, 292–293
Minimum well number, 200–201
Mississippi Canyon, 245, 728, 772
Modified RAV, 367–378
Multiple requests, 172ff

Nind's formula, 187ff, 378–379
Norphlet formation, 103
Number of wells, 188ff
  maximum, 200–201
  minimum, 200–201
  optimum, 197–198
Numerical methods, 351ff

Oil, 113–116, 116–133
Oil recoverability, 190ff
Oilfield, 187ff, 285ff
Opportunities, 64–83
  abandonment, 285ff
  correlation, 101ff
  dissimilar, 92–93
  similar, 90–91
Optimum number wells, 197–198
Optimum working interest, 41–42, 47–48, 95
Options, 301ff, 381–391
  current value, 303–306
  mean value, 323
  minimal loss, 310–311
  partial, 309–310

Parabolic RAV, 59ff, 161–162
Partial options, 309–310
Phillips Petroleum, 232ff
Pipeline spills, 294ff
Portfolio balancing, 52–86, 327–334, 360–367
Practical concerns, 398–400
  positive, 12–13
  negative, 13–14
Preference indices, 167–168
Present-day worth, 187ff
  maximum, 198
Probability distributions, 5–8, 182–185, 193–208
  portfolio, 66–83
  ranges, 194–208, 314ff
  requests, 172ff
  spill, 294ff, 298–300
  success, 18–30, 265, 282ff
  unequal, 91
  values, 98
Probable profits, 276–279
Producible reserves, 193
Production, 188ff
  costs, 270–271
  model, 268ff
Profitability, 57, 67
Project requests, 167ff
Projects, 170–173

Rank correlation, 100
Recoverable oil, 190ff, 195
  sensitivity, 201–208

Relative importance, 37–51, 57, 66, 77, 83–84, 156–157, 193–208, 324–325
Requests multiple, 172–175
Reserves, 101, 187ff, 195–196, 270–279
Reservoir thickness, 190ff
Resolution probability, 337ff, 345ff
Resource assessment, 2
Returns, 310
Revenue, 270
Risk-adjusted value (RAV), 17–30, 39, 44–46, 59–86, 87ff, 95, 137ff
  modified, 367–378
  weighted, 85–86
Risk assessment, 2
Risk aversion
  Cozzolino, 23–26
  hyperbolic, 21–30
Risk tolerance, 18–30, 314ff
  apparent, 20–21
  maximum, 21
  variable, 358–360

Second year updating, 297
Seismic, 163–164, 211–213
Selling price, 188ff, 195, 275–276, 278–279
Sensitivity, 201–208
  field lifetime, 205–206
  maximum well number, 206
  maximum worth, 203
  minimum well number, 206
  optimum well number, 202
  recoverable reserves, 202
Ship Shoal South, 233ff, 337
Similarity, 87ff
Simpson's paradox, 228–232
Skewness, 94, 98, 220–225
South Marsh Island, 245
South Timbalier, 245
Spill, pipeline, 294ff
  probability, 298–300
Spreadsheets, 351ff
Statistical results, 79–83
Statistics, 5–8
  bidding, 227ff
  Gaussian, 141
Success, 228–232, 291

Texaco, 248
Thailand, Gulf of, 103, 116
Third year updating, 297

Tie-downs, 294ff
Timescale, 192
Total costs, 340ff
Two-commodity model, 342–344

Unanticipated benefits, 225–226
Uncertainty,
  costs, 35–37
  decreased, 213–219, 220–225
  expected value, 261–262
  expected worth, 261–262
  geological, 3–5
  improvement priorities, 208
  probability, 35–37
  value, 34, 54ff
  weighted, 54ff
Updating, 296ff
  first year, 296
  fourth year, 299
  second year, 297
  third year, 297

Value, 18–30, 138ff, 168–170, 211ff, 311–314
Variable costs, 69–71, 156ff
  probability, 72–73, 156ff

  risk tolerance, 358–360
  values, 68–69, 156ff
  working interest, 354–357
Variance, 18–30, 37–51, 77, 93, 201–208, 340
Vermilion South, 233, 234ff, 295, 307, 375–376
Viosca Ridge, 179
Volatility, 18–30, 39, 66, 140ff, 195–196, 220–225, 319–323
  weighted, 156ff

Weighted RAV, 85–86
  uncertainties, 154ff
West Cameron, 429
West Texas, 103, 116
Working interest, 17–30, 59–60, 87ff, 112ff, 137ff, 327–334
  break-even, 21, 22
  common, 334–337
  maximum, 20, 22
  optimum, 41–42, 47–48, 112, 137ff
  uncertainties, 43–44
  variable, 314ff, 354–358

Zilkha Energy, 248